# Resilience of Educators in Extraordinary Circumstances:

## War, Disaster, and Emergencies

Enakshi Sengupta
*Borders Without Barbed Wires, India*

Arni Thor Arnthorsson
*American University of Afghanistan, Afghanistan*

A volume in the Advances
in Educational Marketing,
Administration, and Leadership
(AEMAL) Book Series

IGI Global
PUBLISHER of TIMELY KNOWLEDGE

Published in the United States of America by
IGI Global
Information Science Reference (an imprint of IGI Global)
701 E. Chocolate Avenue
Hershey PA, USA 17033
Tel: 717-533-8845
Fax: 717-533-8661
E-mail: cust@igi-global.com
Web site: http://www.igi-global.com



Library of Congress Cataloging-in-Publication Data

Names: Sengupta, Enakshi, editor. | Arnthorsson, Arni, 1963- DATE.
Title: Resilience of educators in extraordinary circumstances : war, disaster, and emergencies / Enakshi Sengupta, Arni Arnthorsson.
Description: Hershey, PA : Information Science Reference, 2024. | Includes bibliographical references and index. | Summary: "This book aims to incorporate the narratives of Afghan students and faculty members as they navigated their way to safety and their take on higher education"-- Provided by publisher.
Identifiers: LCCN 2024007198 (print) | LCCN 2024007199 (ebook) | ISBN 9798369314838 (hardcover) | ISBN 9798369314845 (ebook)
Subjects: LCSH: War and education--Afghanistan. | War victims--Education--Education. | Education--Afghanistan. | Right to education--Afghanistan.
Classification: LCC LC191.8.A34 R47 2024 (print) | LCC LC191.8.A34 (ebook) | DDC 370.9581--dc23/eng/20240318
LC record available at https://lccn.loc.gov/2024007198
LC ebook record available at https://lccn.loc.gov/2024007199

This book is published in the IGI Global book series Advances in Educational Marketing, Administration, and Leadership (AEMAL) (ISSN: 2326-9022; eISSN: 2326-9030)

British Cataloguing in Publication Data
A Cataloguing in Publication record for this book is available from the British Library.
All work contributed to this book is new, previously-unpublished material.
The views expressed in this book are those of the authors, but not necessarily of the publisher.
For electronic access to this publication, please contact: eresources@igi-global.com.

# Advances in Educational Marketing, Administration, and Leadership (AEMAL) Book Series

ISSN:2326-9022
EISSN:2326-9030

Editor-in-Chief: Siran Mukerji, IGNOU, India, Purnendu Tripathi, IGNOU, India

**MISSION**

With more educational institutions entering into public, higher, and professional education, the educational environment has grown increasingly competitive. With this increase in competitiveness has come the need for a greater focus on leadership within the institutions, on administrative handling of educational matters, and on the marketing of the services offered.

The **Advances in Educational Marketing, Administration, & Leadership (AEMAL) Book Series** strives to provide publications that address all these areas and present trending, current research to assist professionals, administrators, and others involved in the education sector in making their decisions.

**COVERAGE**

- Enrollment Management
- Educational Management
- Advertising and Promotion of Academic Programs and Institutions
- Direct marketing of educational programs
- Governance in P-12 and Higher Education
- Consumer Behavior
- Educational Leadership
- Students as Consumers
- Marketing Theories within Education
- Academic Pricing

IGI Global is currently accepting manuscripts for publication within this series. To submit a proposal for a volume in this series, please contact our Acquisition Editors at Acquisitions@igi-global.com or visit: http://www.igi-global.com/publish/.

The Advances in Educational Marketing, Administration, and Leadership (AEMAL) Book Series (ISSN 2326-9022) is published by IGI Global, 701 E. Chocolate Avenue, Hershey, PA 17033-1240, USA, www.igi-global.com. This series is composed of titles available for purchase individually; each title is edited to be contextually exclusive from any other title within the series. For pricing and ordering information please visit http://www.igi-global.com/book-series/advances-educational-marketing-administration-leadership/73677. Postmaster: Send all address changes to above address. 

## Titles in this Series

*For a list of additional titles in this series, please visit:*
*http://www.igi-global.com/book-series/advances-educational-marketing-administration-leadership/73677*

***Improving Literacy Through Home, School, and Community Partnerships***
Al Ryanne Gabonada Gatcho (Hunan Institute of Science and Technology, China) Cecille Marie Titar Improgo (Bukidnon State University, Philippines) and Merry Ruth Morauda Gutierrez (Philippine Normal University, Philippines)
Information Science Reference • © 2024 • 328pp • H/C (ISBN: 9798369317778) • US $235.00

***Discussions of Inclusive Education Within African Contexts***
Jonathan Chitiyo (University of Pittsburgh at Bradford, USA)
Information Science Reference • © 2024 • 209pp • H/C (ISBN: 9781668488416) • US $195.00

***Minority Women in K-12 Education Leadership Challenges, Resilience, and Support***
Annette G. Walters (Independent Researcher, USA)
Information Science Reference • © 2024 • 290pp • H/C (ISBN: 9798369317730) • US $245.00

***History and Educational Philosophy for Social Justice and Human Rights***
Jahid Siraz Chowdhury (Lincoln University College, Malaysia) Kumarashwaran Vadevelu (Universiti Malaya, Malaysia) A.F.M. Zakaria (Shahjalal University of Science and Technology, Bangladesh) Abdullah Al-Mamun (Sunway University, Malaysia) and Sajib Ahmed (Universiti Malaya, Malaysia)
Information Science Reference • © 2024 • 277pp • H/C (ISBN: 9781668499535) • US $235.00

*For an entire list of titles in this series, please visit:*
*http://www.igi-global.com/book-series/advances-educational-marketing-administration-leadership/73677*

IGI Global
PUBLISHER of TIMELY KNOWLEDGE

701 East Chocolate Avenue, Hershey, PA 17033, USA
Tel: 717-533-8845 x100 • Fax: 717-533-8661
E-Mail: cust@igi-global.com • www.igi-global.com

# Table of Contents

# Detailed Table of Contents



This chapter tells the narrative of a young Afghan girl determined to change the world for the better and inspire her generation and the world. However, the political circumstances of her country had a devastating impact on her life. She had to overcome numerous obstacles to reach her life milestones. Resilience, patience, and diligence enabled her to conquer life's challenges. This chapter intends to advise people from war-torn countries, particularly Afghan girls, on how to fight for their aspirations and speak up for their rights.



Education is the most potent weapon that can be used to alter the world; this is a well-known adage in today's society. Individuals are increasingly interested in acquiring diverse information that they may apply in their future undertakings. Afghan professors and students faced risks to their lives and well-being, as well as academic losses. They were and still are repressed between resistance and the possibility of survival. Exiles frequently suffer from mental trauma as a result of being exposed to violent, contentious, and abusive settings. Different international NGOs and donor organisations give such services to Afghan exile professors and students so that they may cope with the mental disturbance much better. Scholarships, language development programs, recognition, accreditation, and financial assistance support such groups to deal with the challenges.

This scholarly inquiry explores the narratives of escape and resilience among Afghan refugees in Nordhausen (Germany), employing a multidisciplinary approach. Utilizing qualitative methods, particularly open-ended structured interviews, the study illuminates the psychosocial dynamics and lived experiences of Afghan refugees. It seeks to unravel the complexities of migration, decode reasons for leaving one's country, and understand asylum-seeking processes. The research identifies adaptive mechanisms and coping strategies, shedding light on socio-cultural, economic, and political factors shaping refugee narratives. Ethically sensitive, culturally responsive engagement underscores respect for refugee voices and agency. Narratives underscore the significance of agency, resourcefulness, and community networks. The study emphasizes the role of host-country policies and social structures in shaping refugee lives. This research bears implications for academia and policy, offering insights into refugee dynamics and factors facilitating successful integration amid the global refugee crisis.



Education is vital for professional development and is a fundamental human necessity. Unfortunately, individuals in many nations lack access to high-quality education. This study explores curriculum adjustments in Afghan education, emphasizing the narratives of academics and students. To meet evolving needs, institutions must prioritize ongoing curriculum modifications for improved educational quality. Afghan public universities can benefit from incorporating outcome-based education and student-centered learning. Local governments must understand displaced people's needs, adapting the curriculum accordingly. Implementing changes may pose psychological and cultural challenges for students and teachers. International collaboration is crucial for successful curriculum adoption. Decision-makers should focus on developing flexible curricula to address displaced individuals' unique needs. The resilience of Afghan educators and students showcases the human spirit's power. Collaborative efforts can break barriers, fostering a globally accessible education landscape.

This chapter explores ALPA's initiative to sustain education for Afghan girls, detailing its history, exile post-Taliban, and the growth of a robust online program for 500+ students. Despite the 70% exile of Afghan academics, ALPA engages educators to teach law and English, highlighting its vision for global collaboration. The concise narrative delves into challenges faced by Afghan girls in online education and ALPA's commitment to overcoming obstacles, concluding with the inspiring story of hope, resilience, and collective efforts in exile.



This chapter recounts a poignant and inspiring narrative of resilience and personal growth in Afghanistan's political instability. The author's journey began at the American University of Afghanistan, where education became a symbol of defiance and optimism during the Taliban's rise. Education and empowerment remain the author's priorities despite the COVID-19 pandemic and political instability. The narrative vividly recounts the author's harrowing experiences of evacuation, the transition to life in Bishkek at the American University of Central Asia, and the author's eventual relocation to the United States to attend Bard College. The journey explores resilience, personal growth, and the relentless search for freedom and empowerment. This chapter celebrates family, self-discovery, and the resilience of optimism and persistence in the face of adversity, demonstrating the tenacity of individuals who pursue education and improvement.



In the wake of the collapse of the Islamic Republic of Afghanistan by the Taliban in mid-August 2021, the lives of Hamidullah Ahmadi and Gul Hassan Mohammadi, two friends and classmates from the American University of Afghanistan (AUAF), took an unexpected turn. As the country fell into the hands of the Taliban, they faced many challenges that challenged their resilience and determination. Among the chaos and uncertainty, Hamidullah and Gul Hassan struggled to maintain their studies, jobs, and personal lives. The security situation deteriorated rapidly, making it increasingly difficult for them to pursue their dreams. However, undiscouraged by the hardships they faced, they embarked on a journey of determination and unwavering hope.



This book chapter explores the critical role of professional development in equipping teachers to navigate the challenges and barriers of emergency contexts. It focuses specifically on the resilience exhibited by teachers in teaching the Rohingya refugees in Bangladesh. The Rohingya refugee crisis, characterised by forced displacement and limited access to education, presents unique challenges and barriers for both teachers and students. Drawing upon secondary sources and theoretical frameworks, this chapter unpacks the various dimensions of teachers' professional development for both the Rohingya and the host community teachers in such adverse circumstances in refugee camps in Bangladesh. It highlights the adaptive strategies employed by the Cox's Bazar Education Sector and the teachers themselves to enhance their pedagogical skills, foster socio-emotional well-being, and address the diverse needs of the Rohingya learners. Furthermore, the chapter analyses the organisational support systems, policies, and interventions aimed at promoting teacher resilience and effectiveness in emergency settings.



This chapter narrates the remarkable journey of students from the American University of Afghanistan (AUAF) following the Taliban's takeover in August 2021. The evacuation, initiated amidst the chaos at Hamid Karzai International Airport, and finally getting completed two years later, encountered numerous challenges. The narrative details the phased evacuation process, where AUAF students were dispersed to different countries in Asia, such as Kyrgyzstan, Iraq, and Qatar. The chapter delves into the mental health concerns that emerged, exploring the anxiety and uncertainty faced by students as they grappled with the abrupt regime change, the efficacy of online learning, and the challenges of continuing education amid geopolitical upheaval. The narrative also highlights the resilience of female students who faced numerous obstacles, including delays, flight cancellations, and the imposition of restrictive travel norms. The chapter captures the inspiring resilience of AUAF students, emphasizing their unwavering commitment to education amidst adversity.



In a war-torn country, your foremost struggle is not the pursuit of wealth or riches; it is the fight for survival and then obtaining an education. In 2016, when the author enthusiastically embarked on the author's higher education journey at the American University of Afghanistan, the author had to confront death to pursue that education. It was just the beginning of the second semester when the Taliban, now in control of the country, launched a terrorist attack on my university on August 24, 2016, resulting in the tragic loss of friends, classmates, and teachers. To survive that night, the author leaped from the second floor, landing on the ground, feigning death beside a classmate who wasn't breathing and was covered in blood. The author could witness the terror and the horror right before the author's eyes, resembling a scene from a Hollywood movie. That night, not only did the author lose the author's friends, classmates, and teachers but the author also lost a piece of the author's own mental well-being. Yet, amidst the ruins, the author did not lose hope. In the author's heart, the author still carried hope and optimism for a better beginning.



After a long military buildup, Russian recognition of the unilaterally declared Donetsk People's Republic and the Lugansk People's Republic, and entry of the Russian Armed Forces into the Donbass region of Eastern Ukraine on February 21, 2022, the Russian forces launched a military campaign against Ukraine on February 24, 2022. Kyiv and other cities around Ukraine were bombed on February 24 after Russian President Vladimir Putin announced a military action to "disarm and de-Nazify Ukraine" in a speech. The purpose of this research is to shed light on why Russia and Ukraine are at odds and how NATO has played a part in fueling this old war. Since Ukraine was a part of the Soviet Union, this study gives some context for understanding both Russia and Ukraine. In order to better comprehend the nature of

# Preface

As editors of this volume, we are humbled to present *Resilience of Educators in Extraordinary Circumstances: War, Disaster, and Emergencies*. In August 2021, the world witnessed the harrowing plight of Afghan academia and its students as they grappled with the resurgence of Taliban rule. The echoes of their struggles reverberate through these pages, encapsulating a narrative of perseverance amidst adversity.

The events of August 15th, 2021, marked a watershed moment for Afghanistan, plunging its people into uncertainty and despair. Yet, amidst the chaos, educators and students alike demonstrated remarkable resilience, seeking refuge and safeguarding their pursuit of knowledge. Their stories serve as a testament to the indomitable spirit of the human intellect in the face of tyranny.

This volume endeavors to amplify the voices of Afghan scholars and students as they recount their journey to safety and reflect on the future of higher education in their homeland. Each chapter, meticulously curated, offers a profound insight into the challenges confronted and the resilience exhibited in extraordinary circumstances.

From the plea for evacuation to the reconfiguration of academic pursuits under Taliban rule, this book delves into the intricacies of educational upheaval amidst conflict. Through empirical analysis and firsthand accounts, it seeks to elucidate the gravity of the crisis and identify avenues for intervention and support.

Moreover, this volume serves as a beacon of hope, highlighting exemplary initiatives and best practices in education amidst turmoil. By documenting these narratives and practices, we aspire to contribute to a collective understanding of effective responses to educational crises worldwide.

We extend our gratitude to the contributors who have shared their experiences and expertise, enriching this discourse with their invaluable insights. It is our sincere hope that this book will serve as a vital resource for academics, policymakers, humanitarian actors, and all those committed to the preservation of education in times of adversity.

## ORGANIZATION OF THE BOOK

### Chapter 1: A Shattered, Lost, and Helpless Afghan Girl Found Her Way to a Promising Future: I Think It Will Impress You

Authored by Somaya Salem

This chapter chronicles the journey of a determined young Afghan girl amid the turmoil of her country's upheaval. Facing insurmountable obstacles following the fall of her government, she exemplifies resilience, patience, and diligence in pursuing her dreams. This narrative serves as an empowering guide for individuals, particularly Afghan girls, navigating the complexities of life's challenges, advocating for their aspirations, and asserting their rights.

### Chapter 2: Academics in Exile: Navigating New Educational Landscapes Amidst Displacement and Uncertainty

Authored by Sara Aldous and Ghaith Ismail

Focusing on the plight of Afghan professors and students, this chapter delves into the risks and losses endured in the pursuit of education amidst displacement. It explores the psychological trauma experienced by exiles and the support provided by international NGOs and donor organizations. Scholarships, language programs, and financial aid are vital in aiding these individuals to cope with mental distress and rebuild their academic lives.

### Chapter 3: Capturing the Narratives of Escape and Resilience: A Study of Afghan Refugees in Nordhausen

Authored by Saad Ullah Khan, Fahim Sadat, and Sadaf Khan

Utilizing qualitative methods, this scholarly inquiry examines the narratives of escape and resilience among Afghan refugees in Nordhausen, Germany. Through open-ended structured interviews, the study sheds light on the psychosocial dynamics and lived experiences of Afghan refugees, unraveling the complexities of migration and asylum-seeking processes. It emphasizes the significance of agency, resourcefulness, and community networks in shaping refugee narratives, with implications for academia and policy.

### Chapter 4: Changing and Adapting Curriculum While Teaching: Narratives From Afghan Academics and Students

Authored by Sara Aldous and Ghaith Ismail

This chapter explores the evolving landscape of Afghan education, focusing on curriculum adjustments and the narratives of academics and students. It advocates for ongoing modifications to improve educational quality, emphasizing the importance of outcome-based education and student-centered learning. The resilience of Afghan educators and students highlights the transformative power of collaborative efforts in fostering a globally accessible education landscape.

## Chapter 5: Disrupted Educational Pathways in Afghanistan: Evaluating the Disparities in Education Under the Taliban Affecting Women's Educational Access and Capabilities

Authored by Nassir Wani, Farha Khan, and Shukriya Hashmi

Examining the impact of education disparities under the Taliban, this chapter explores the challenges faced by Afghan women in accessing education and developing capabilities. It argues for the transformative potential of women's education in advancing Afghan society and securing the country's progress. The chapter concludes with reflections on implications for government, development partners, and educational institutions in strengthening education services and building resilience in fragile contexts.

## Chapter 6: Educational Challenges in Refugee Camps and Conflict Zones

Authored by Mustafa Kayyali

This chapter delves into the multifaceted educational challenges confronting individuals in refugee camps and conflict zones. It examines the critical roles of educators in navigating limited resources, security concerns, and the responsibility of nurturing the next generation. Through a comprehensive lens, it calls for recognition of the significance of addressing these challenges and fostering resilience in pursuit of education.

## Chapter 7: Pioneering Education for Afghan Girls in a Time of Crisis

Authored by Ahmad Vali Behnood

Highlighting ALPA's initiative to sustain education for Afghan girls, this chapter details the organization's history and commitment to global collaboration. It explores challenges faced in online education and ALPA's efforts to overcome obstacles, culminating in a narrative of hope, resilience, and collective efforts in exile.

## Chapter 8: Planning the Big Escape: Perilous Journeys and Lessons Learnt

Authored by Mimla Wardak

This chapter recounts a poignant narrative of resilience amid Afghanistan's political instability. It chronicles the author's journey from the American University of Afghanistan to eventual relocation in the United States, highlighting the challenges faced and the resilience exhibited in pursuit of education and empowerment.

## Chapter 9: Planning the Big Escape: Lessons Learnt

Authored by Hamidullah Ahmadi and Gul Hasan Mohammadi

Amidst the chaos of Taliban rule, this chapter follows the journey of two friends from the American University of Afghanistan as they navigate challenges to pursue their studies and personal lives. Their resilience and determination serve as a testament to the human spirit amidst adversity.

## Chapter 10: Teachers' Professional Development in the Emergency Context: Resilience of Teachers in Teaching the Rohingya Children in Bangladesh

Authored by M. Mahruf C. Shohel et al.

Exploring the critical role of professional development for teachers in emergency contexts, this chapter focuses on the resilience of educators in teaching Rohingya children in Bangladesh. It highlights adaptive strategies to enhance pedagogical skills, foster socio-emotional well-being, and address the diverse needs of displaced learners.

## Chapter 11: The Extraordinary Journey of AUAF Students From Despair to Hope

Authored by Setha Fayaz and Ahmad Suhaib Majeed

This chapter narrates the remarkable journey of students from the American University of Afghanistan amid the Taliban's takeover. It explores the challenges faced in evacuation and online learning, emphasizing the resilience of students committed to education amidst adversity.

## Chapter 12: The Night I Fought for Life

Authored by Marzia Azizi

In a war-torn country, education becomes a beacon of hope amidst turmoil. This chapter recounts the author's harrowing experience surviving a Taliban attack on her university, highlighting the resilience and optimism maintained in pursuit of education and a better future.

### Chapter 13: The Russian-Ukrainian Conflict: Renewed Soviet Ambitions

Authored by Habib Al Badawi and Mohamad Zreik

Examining the complexities of the Russian-Ukrainian conflict, this chapter sheds light on historical and geopolitical factors fueling the war. It explores the implications of NATO involvement and offers insights into understanding the nature of the conflict.

### Chapter 14: When Extraordinary Circumstances Call for Mutual Aid: The Arrival of Afghan Academics in the U.S.

Authored by Stephen Rosenbaum et al.

Using narrative essays, this chapter provides insights into the experiences of Afghan scholars amid the Taliban's resurgence. It examines challenges faced by educators and students in Afghanistan and the United States, emphasizing the impact on academic freedom and human rights.

Each chapter in this volume offers a unique perspective on the resilience of educators and students in extraordinary circumstances, enriching our understanding of the power of education amidst adversity.

## IN CONCLUSION

As editors of this edited reference book, *Resilience of Educators in Extraordinary Circumstances: War, Disaster, and Emergencies*, we are profoundly moved by the narratives presented within its pages. The collective journey of educators and students navigating through the tumultuous landscapes of conflict, displacement, and political upheaval encapsulates the indomitable human spirit in pursuit of knowledge and empowerment.

From the harrowing experiences of survival recounted in chapters such as "The Night I Fought for Life" to the innovative initiatives driving education amidst crisis in chapters like "Pioneering Education for Afghan Girls in a Time of Crisis," each contribution offers a mosaic of resilience, determination, and hope.

Throughout these chapters, common threads emerge: the transformative power of education, the importance of community and collaboration, and the unwavering commitment to a better future. Amidst the chaos and uncertainty, educators and students alike stand as beacons of resilience, forging paths of progress and empowerment in the face of adversity.

As we conclude this volume, we are reminded of the immense challenges that persist in the realm of education in times of crisis. Yet, we are also inspired by the stories of courage, innovation, and perseverance that define the human spirit. It is our hope that this book will serve as a testament to the resilience of educators worldwide, igniting conversations, fostering collaboration, and driving meaningful change in the pursuit of a brighter tomorrow.

*Enakshi Sengupta*
*Borders Without Barbed Wires, India*

*Arni Thor Arnthorsson*
*American University of Afghanistan, Afghanistan*

# Chapter 1
# A Shattered, Lost, and Helpless Girl Found Her Way to a Promising Future: I Think It Will Impress You

**Somaya Salem**
*HEC Paris, France*

## ABSTRACT

*This chapter tells the narrative of a young Afghan girl determined to change the world for the better and inspire her generation and the world. However, the political circumstances of her country had a devastating impact on her life. She had to overcome numerous obstacles to reach her life milestones. Resilience, patience, and diligence enabled her to conquer life's challenges. This chapter intends to advise people from war-torn countries, particularly Afghan girls, on how to fight for their aspirations and speak up for their rights.*

## BACKGROUND

As a young girl who has grown up in a war-torn country, lived in a conservative society and gone through tough times, I am committed to taking part in the socio-economic development of Afghanistan and women's welfare. So, hopefully, fewer women will experience the suffrage and hardship that my fellow generation and I shared.

Women are the most discriminated part of Afghanistan, where girls' education is taboo. Luckily, in all the darkness, the light in our home was my father, my foremost

DOI: 10.4018/979-8-3693-1483-8.ch001

hero and supporter of girls' education. Losing him when I was only six years old was the most significant loss of my life. My mother and maternal uncle sacrificed their life to educate me and my siblings and provide for our livelihoods.

My siblings and I were the first girls to pursue an education. At first, we experienced troubles and hurdles, and some of our close relatives even cut ties with us. However, we fought tirelessly to be educated girls in a traditional community like Afghanistan. Pashtun, the ethnic group from which I am descended, was notorious for being anti-girls' education. On the other hand, my father had a different viewpoint and advocated for girls' education. I am serving as a role model for other girls in our community and working to normalise girls' education.

To ensure that my parents' sacrifices and their dream for their daughter's education are not in vain, I decided to study at a university that will bring a meaningful change in my life and help me achieve my life milestones. Therefore, I chose to look at the American University of Afghanistan (AUAF).

Getting admitted into AUAF was difficult because I had to go through different stages and, more importantly, learn English. Meanwhile, I had to overcome barriers, including cultural and financial. After many struggles, in the fall of 2017, I was awarded a U.S. Embassy Scholarship to pursue my undergraduate degree at AUAF.

In the senior year of my undergraduate program at AUAF, I started applying for postgraduate programs in different universities worldwide. While searching for doctoral programs, I heard about the French Government Scholarship for Afghan students. "Securing this scholarship is not an easy job", said a friend. Yet, I put in a lot of effort, and in January 2021, I submitted my application to the French Institute of Afghanistan for a French Government Scholarship for Afghan students. In the meantime, I applied for the Chevening Scholarship because it has been one of my goals to get my master's degree through the Chevening scholarship.

## Fifteen August 2021: I Will Never Forget It

Waiting to hear about my postgraduate application results; everything turned upside down on August 15, 2021. I will always remember 15th August; it remains like a nightmare. The image of the chaos at the airport, on roads, among people, and everything I witnessed on 15th August 2021 came to mind while writing this piece.

I well remember that my family told me not to attend office for a few days because there were rumours that most provinces had fallen to the hands of the Taliban. However, I did not hear from them and left home to go to the office even on 15th August.

Nothing felt normal as I made my way to the workplace. Everyone was concerned, and since everyone wanted to cash out their buckets, there was an increased demand at the banks. In addition, my friend and I proceeded to a nearby store to use the

ATM to cash out buckets. However, the machine was no longer operational. Next, we attempted to visit a bank branch but were not successful. We could finally locate an ATM, but we could not take out all of our money due to a lengthy queue of people waiting to cash out their buckets from a single machine.

We headed for our workplaces after cashing out our buckets. I started working on a report when I got to my office, but I found it difficult to focus. My mind was racing with so many questions. What would happen to Afghanistan? Would we have more horrific wars? Can I continue working? Can I pursue further education? Tens of queries that were similar to these were troubling me and preventing me from finishing the report. I stopped working on the report because I could not focus, and I started reading the news to stay informed about what was happening. My employer texted me around 11 a.m., telling me to head home till further notice.

I was terrified because our office was in Kabul's diplomatic area, Shashdarak. Each second, I thought that the area would be bombarded and I would lose my life. When I reached the checkpoint near our office, there was a crowd. Everyone seemed anxious and in search of a vehicle to get home. I saw everyone run for their lives and their loved ones.

I was unable to find a vehicle to get me home. Therefore, I walked for over three hours to get a rental car. While seeing the crowd of people running around for their safety, I felt like everything ended. I wanted to cry, and my chest was going to burst because I thought my struggles, my education, my dreams, and everything went in vain. I was feeling desperately disappointed. Out of nowhere, a driver stopped and picked me up at a place near our home.

## HEALING AND RESILIENCE IN POST-REPUBLIC TRANSITION

The fall of the Republic brings out a mixture of feelings, helplessness, stress and anxiety. It left me with severe mental health issues. Fifteen-August 2021 is behind me, but its pain can still be felt and hurt a lot.

I could not dare to go outside for a long time because nothing seemed like before 15 August 2021. When I was thinking about my academic and career goals and then seeing the situation, I felt shattered, telling myself that everything had gone in vain. I thought I would never be able to stand again. The Taliban ban on women's movement, employment, and every other decree restricting women were further irritating me.

I have always wanted to pursue a postgraduate degree in commercial law and establish a multinational operating law firm in Afghanistan because the country's demand for experienced commercial lawyers is more significant than ever. Almost everyone is involved in commerce in some capacity. I intended to begin working as a legal counsel for the Ministry of Industry and Commerce to establish sound regulations and implement good reforms in Afghanistan's commerce sector. However, the republic's dissolution has prevented me from fulfilling my dreams. Most institutions had specific requirements that I could not meet because of the Afghanistan situation. Thus, I was unable to apply for a Master in Commercial Law. Meanwhile, I was still looking for a source to sponsor my studies. As a result, I could not complete the significant milestones in my life.

Certain rights were taken away from citizens, particularly women, after the fall of the Republic. For example, I could not attend the office physically and thus had to work remotely for an extended period. Working remotely was difficult due to the need for more facilities in Afghanistan, such as power and internet. Nonetheless, I tried to fulfil my job responsibilities.

However, I lost my job as a result of the Taliban's restriction on women working in the public and private sectors. I will never forget the day I read the order announcing that women are no longer permitted to work in the private sector. I never imagined a day when someone or anything would take away my right to work. That, however, occurred. At first, it was difficult for me to stay at home and not go to the office. I went through difficult circumstances and survived the worst days of my life thanks to the love and support of my family and friends.

Dying is not always physical. I was just a silent observer for a long time, and it was a gradual death for me. I spent several months in depression. One day, I told myself enough is enough, "Until when should you suffer? If you give up, how will you pursue your and your parents' dreams and, more importantly, your commitment to serving your country?

I told myself I was not born to survive but to make meaningful changes to my life and others. I live a life just once. Therefore, I should stand again, even if the situation is difficult to tolerate. If I do not stand for myself today, how can I stand for others tomorrow?" I was trying to convince myself, and my family was also trying to give me hope, saying that this situation would not last forever. One day, everything will be fine, but you should stay strong.

That day, I promised that I would fight the trauma and would never let anything put me down and stop me from achieving my life milestones. The collapse of the Republic did not stop me from pursuing my goals but further strengthened me.

I tried different ways to heal, give myself hope, adjust to the circumstances and fight for my dreams. However, I was more afraid than other family members because I studied at AUAF and worked for foreign NGOs. I feared for my and my family's safety.

## MY JOURNEY TO HEC PARIS

When the Taliban retook control of the country, I had two choices: either to stay and die or try and leave. I never wanted to leave my country and to be called "REFUGEE", and I never want to have this status ever in my life. Therefore, I never applied for a settlement case; I was eligible for different cases like SIV. Since I did not find being a refugee a solution to my situation. I started once again searching for light in my darkest days of life. Therefore, I decided to stay and adapt to the problem until I could leave the country for academic purposes.

I started applying for the master's program again. I contacted the Chevening secretariat, but their reply disappointed me. They replied that they could not further support Chevening applicants inside Afghanistan. However, I could not afford to go to a third country to follow my application either. It is never easy to escape a dream, but there are times when you do not have any other option. Studying through Chevening was a dream that remained unfulfilled.

I also contacted the French Embassy in Kabul, asking about the status of my application. I was in the process of searching for a master's program. In October 2021, I received an email from HEC Paris inviting me to have a personalised meeting with them to learn about their Master in Management Program (MiM). Since I was an applicant for the French Government Scholarship, they have shared my application with HEC Paris.

In the darkest days of my life, this opportunity mesmerised me. So, I searched for HEC Paris and found it Europe's #1 business school. Then, I applied for its Master in Management program. Being in an uncertain situation, gaining admission to the #1 Business School in Europe with an enrollment rate of 8% each year was challenging. Moreover, I also had to secure a scholarship to purse my master degree. The competition was challenging because the eligible applicants for the scholarship were from other war-torn countries.

I was one of the first Afghan students who applied for this program. In March 2022, I submitted my application for HEC Imagine Fellows, a unique scholarship program for students from war-torn countries. In April 2022, I got admitted into the HEC Paris in MiM program and was chosen as an Imagine Fellowship laureate.

Once I received this news, I tried to get to France. However, the ban on solo women's travel was why I could not start my journey in HEC Paris until 2023. I have been unsuccessful and have been trying from mid-2022 till October 2023 to facilitate my travel. Yet, I did not lose my hope.

## CROSSING TORKHAM BORDER

I had difficulties from the first day I started again until I got to Paris. I was about to give up on my goals when I decided to travel to France and start studying at HEC Paris at whatever cost after remembering the hardships my mother and maternal uncle had endured for my education.

I had a terrible time organising my trip to Paris sometime in October 2023, and it seemed impossible. Given the restrictions on women travelling alone, this was the most challenging circumstance I have ever been in. I left Kabul for Pakistan on October 31, accompanied by my maternal uncle, to the Torkham border. He needed travel documents. Thus, he was unable to cross the border. The day I left Kabul for France was unlike any other. I was torn between emotions. I left Kabul and my loved ones behind to return there and fight for other girls' right to an education. I promised my family that I would return to my nation much stronger and with a larger contribution.

I want to briefly write a few words about my maternal uncle's impact on my success. Everybody has a hero, and my maternal uncle is my hero. He sacrificed his life to provide my siblings and me with an education and has supported us through life's ups and downs. I can confidently state that he is the primary cause of my current achievement, and without his help, I would not have been in my current state. He helped me so much, and I want to thank him here.

Before planning my trip to a third country, I wanted to know if it would be possible for me to cross the border without a male guardian (Mahram), so I questioned my friends and family who had travelled to Pakistan via the Torkham border. They all talked about their experiences and said, "There is no way I'm going to be able to cross the border without a Mahram." I was getting frustrated by it, and I was having a difficult time. I had all I needed—scholastic acceptance and a scholarship from the top business school in Europe—but my trip to a foreign nation without a Mahram was the only thing preventing me from beginning my studies at HEC Paris.

It was hard to say goodbye to my uncle when I got to Torkham Border. I was clueless as to what might come next. Can I get across the border, or will I not? Deciding beyond the border was essential to opening doors to more tremendous success.

I was terrified from the minute I arrived in line in front of the immigration desk until I got my passport stamped to leave Afghanistan. Every second, I worried that they would ask for my Mahram. Still, fortunately, the immigration desk officer did not inquire because the day I was leaving Afghanistan, there was a lot of rush at the border, and he was swamped. The reason for the rush was Pakistan's forced departure of Afghan refugees.

I felt free when I crossed the border and thanked Allah Almighty for making my journey possible. I felt there would be no more problems at the border as I approached my destination. Nevertheless, the Pakistani immigration desk official denied my entry into Pakistan after reviewing my passport. This was primarily because I had an Indian visa on my passport, and Pakistan and India have long-standing political issues. I tried to persuade the immigration desk officer to let me into Pakistan because I was travelling to India for medical treatment, but I found that to be unacceptable. He questioned me a lot and was not easily convinced. After receiving his responses, he eventually stamped my passport and permitted me to enter Pakistan.

In Islamabad, everything was proceeding according to plan, and I was meant to get my French visa in a week, but the procedure took longer than expected for various reasons. Due to the issues Pakistani police were causing for Afghan refugees, even the one holding the Pakistani visa, I dared not go outside.

I thought about my fellow generation members who are currently denied access to school the day I arrived at HEC Paris. Like me, I recall the women and girls who worked hard for their accomplishments, but it was all in vain. I am confident that one day, if not today, all Afghan girls will have the right to an education and will fight for every one of their rights.

## RECOMMENDATIONS

To all girls, especially Afghan girls who have lost their hopes for their future, you are the only one who can lead yourself to a bright future. You are the one who can make the final decision of victory or defeat. If I could do it, you can do it too. Thus, you should not wait for a situation to get better so that you will pursue your dreams. Stand up, fight and never give up on your dreams. I know it is easier to say than done, but believe me, I have struggled, suffered, and fought but never given up. If you badly want something, you should work to get it. Otherwise, comfort does not come with ease.

Ultimately, I would like to express my gratitude to my mother, maternal uncle, siblings, and friends for always being there for me when I needed them. I am confident that I would not have been able to overcome the mental health and trauma that I experienced after the fall of the republic. However, their encouragement propelled me to a higher level.

## CONCLUSION

Going through an unknown situation yet pursuing your goals is never an easy job. You have to push yourself to get it. The journey was never easy for me. Recalling the darkest day of my life, 15th August, makes me feel shattered. At times, I wanted to give up because the trauma never let me take an action forward, but I had to. I had no other choice.

Life is full of ups and downs. I encountered rejection after rejection academically and professionally--- before finally landing a foot in the door of success. However, when it comes to being an Afghan girl who should always suffer different life situations, especially that of her country and stand after each fall, it is never easy.

I want to end this chapter with a message to the strong Afghan girls:

*Life is full of challenges and tribulations. To attain the intended results, one must have complete faith in Allah and never submit to circumstances, no matter how hard it gets.*

Somaya Salem
December 31, 2023

# Chapter 2
# Academics in Exile:
## Navigating New Educational Landscapes Amidst Displacement and Uncertainty

**Sara Mohammed Aldous**
*Liwa International School for Gils, UAE*

**Ghaith Ismail**
*Madar International School, UAE*

## ABSTRACT

*Education is the most potent weapon that can be used to alter the world; this is a well-known adage in today's society. Individuals are increasingly interested in acquiring diverse information that they may apply in their future undertakings. Afghan professors and students faced risks to their lives and well-being, as well as academic losses. They were and still are repressed between resistance and the possibility of survival. Exiles frequently suffer from mental trauma as a result of being exposed to violent, contentious, and abusive settings. Different international NGOs and donor organisations give such services to Afghan exile professors and students so that they may cope with the mental disturbance much better. Scholarships, language development programs, recognition, accreditation, and financial assistance support such groups to deal with the challenges.*

## INTRODUCTION

Education is the most powerful weapon which can be used to change the world; this is well known saying in the present time (Abbasi & Monsutti, 2023). Individuals now are interested in gaining diversified knowledge which can be used by them in

DOI: 10.4018/979-8-3693-1483-8.ch002

their future endeavours. Different individuals have to face distinct situations which affect their academic situations. The exile is such a section of society that often has to face difficulty in accessing appropriate education, and the Afghan academic system also faced similar situations in exile (Jafari & Schuster, 2019). The concept of academics in exile reflects the human intent to resilience in the face of adversity. The narrative of such educators, scholars as well and intellectuals who have faced political conflicts or been forced to leave their land is respected (Meininghaus & Mielke, 2019).

The moment a person must leave their homeland, they are forced to deal with different kinds of challenging situations. Among such barriers, navigating the academic system can be moved towards a new landscape. The exile has to get assistance from different agencies so that they can deal with the challenges which are being faced by them (Ghosh, 2019). The students who are exiled often find themselves in such situations where the cultural gaps make it difficult for them to adjust to the host country. Afghanistan is a land where the exiled academics, as well as the students, had to face multiple disparities due to the conflicting political state. Interrupted education and barriers to resuming education are a few issues which were faced by the exiled students of Afghanistan (Alqodsi, 2023).

The essay highlights a new educational landscape that has been developed for academics in Afghanistan amidst displacement and uncertainty. The journey over the years has been discussed along with the challenges that were faced, and later on, the road to activism is also described.

## MAIN BODY

### The Perilous Journey: Escaping Conflict and Seeking Safe Havens

According to Shams, (2021) the term exile might be compelling to think about; it may be a terrible experience for the individual who is in that situation. Afghan academics and students had to face threats to their lives and well-being along with suffering a loss of academics. They were being and still are suppressed between resilience and a hope of survival. The treacherous trip these people undergo as they flee war and seek safe havens is a monument to the human spirit's tenacity. The perilous journey of the people facing such situations forced them to deal with armed conflicts, dodging minefields and even continue to cross the borders (Mäkinen, 2020). The desire to escape such speculative situations has forced the exiled population of Afganistan to leave behind the educational institutes and even their families (Abdallah & Alkaabi,2023).

Moreover, Arifani, (2021) states that education was considered significant by the Afghanistan exile, which forced them to take refuge in such regions which are not too far from their educational infrastructures. They often had been given the status of a refugee, which took a toll on their mental health, which made developing a new educational landscape important. The uncertainties which are faced by them go beyond the academic aspect, as they have to suffer different kinds of losses. On the other hand, when the Afghan exile academics shifted to a host country, they made the existing learning population of the home country vulnerable to educational gaps as they were not left with professionals who could guide them and improve their academic level (Alqodsi, Jadalhaq, & El Maknouzi,, 2023)

As per the view of Crespo & Cooperativa (2021) when a vulnerable section of society is going through certain troubles, it becomes important for other players in the community to come forwards for help; the donor agencies and non-profit organisations merged as a ray of hope to provide support to Afghan exile academics and students. Boroujerdi, (2023) provides that different organisations provided financial assistance to the exile so that they can cover up certain part of their expenses. The academics and students were provided with the cost of travel, accommodation and even expenses for basic living were covered by donor agencies (Abdallah, 2023). These actions were taken so that their lives could come a bit on track and they could focus on the academic part. Scholarships were also provided to the Afghan exile students so that they were not forced to leave their education in the middle.

Schuster & Shinwari, (2020) describes that educational resources were provided to the exile group so that they can continue with the academics and deliver quality education. When a particular section of society is weak, it needs the assistance of different supporting parties. The exiled academics of Afghanistan did receive the support of donor agencies in different ways, which provided them a reason to sustain the challenging situations. The non-profit organisations also become their advocates and voice out their perspectives so that they are not deprived of the resources which other sections of the world have (Ismail et al.,2023).

The exile academic disparities might not be known to all, but through awareness programs, light can be brought to such acts. After learning about the actual conditions of exile, academics of different Afghan agencies initiated a hand of support. This further assisted in gaining the help of other parties as well. Azami, (2021) states that such an environment provided them the support that they can deal with such challenging situations and there are people whom they can go to for seeking help. Establishing oneself in a new territory is a challenge in itself, and its difficulty level accelerates when it is done by an exile. The traditional educational landscape was hanging on the last thread, and because of that, the political conflicts in Afghanistan generated massive troublesome situations for exiled academics and students (Alqodsi & Aljahoori, 2023).

## Navigating Bureaucratic Hurdles: Immigration Laws and Academic Challenges

According to Kölbl, (2021) the exile population is obliged to take refuge in a foreign land in order to deal with the challenges and instabilities. Bureaucracies are present in all parts of the world, and their impact increases, particularly for the exile. They have to face hurdles and delays in the immigration process. There are cases the exiled students and academics have to wait years to get approvals, which further affects the entire academic process adversely. The immigration laws and the bureaucratic hurdles often contribute to such problems. The exile of Afghanistan had to go through physical hardships, and on top of that, the immigration laws also emphasised such disparities. It is argued by Ziaulhaq Rashidi, (2019) such population in this time had to take shelters of asylum as it would provide them with protection from different external conditions. This status is given to the exiled academics and students as they arrive in the host country after crossing the border. Refugee status is given to the public when they are not able to live in their home country due to conflicts or other violence. The Afghan exile students and academics were given the status of refugees as they had to leave the place for an intermediate location. Seddighi, *et al.* (2022) states that the citizenship status of the exile is often in questionable state as they had left the home country and are living in the host country, considering it their new home, and attaining its citizenship can be a permanent solution to their problem. In addition to that the women in Afghan were not provided with academic opportunities and such cultures made it difficult for the Afghan exile academics to adjust in host county's inclusive education system.

As per the view of Boroujerdi, (2023) the immigration laws are a complex and strict legal structure which may vary from nation to nation. There are stringent visa requirements which are to be followed by the exile, that makes immigration a difficult process. For example, to make visa applications, extensive paperwork would be required, and if it is for exile, then it might take even longer. This can cause delays, and the academics and students might have to wait for extensive time periods, which can hamper the quality and pace of their education (Ramadan & Ismail, 2023).

It is argued by Wardak, (2021) the application process for people with asylum and refugee status is complex and lengthy. They have to prove their creditability at different levels. They often also have to prove their status of being a refugee. This entire process can be mentally toxic for the exiled Afghan students and academics. Due to such factors, their educational continuity gets affected, and they are unable to match the calibre of other students. It is highlighted by Mielke & Meininghaus, (2021) that the exile often gets limited legal representation, which also reduces the chances of taking advantage of a situation. Navigating the complexities of such bureaucracies becomes challenging for Afghan exile students, and as a result, they often have to lose a chance to get an education in the desired host country. The disruption caused due to harsh immigration laws often results in limited access to higher education. The Afghan exile students are unable to transfer their academic credentials, and due to this, they might not receive due recognition.

## The Psychological Toll of Displacement

As stated by Ghasemi-Tafreshi & Lafortune, (2023) the mental and emotional repercussions of the exile are such that it might take years to get better. Mental trauma is often faced by the exile population as they have to experience violent, conflicting and abusive situations. The exile of Afghans also witnessed similar circumstances, and because of that, some of them also had to face post-traumatic disorders, and such conditions made focusing on the academic part difficult. They might even have to experience, nightmares, anxiety or even flashbacks which can hamper the quality of life. However, it is criticised by Dinakhel, Bakhshali & Ali, (2023) various support systems have emerged which can help the vulnerable section of society to get access to mental health services. Different international organisations and non-profit agencies provide such kind of services to the Afghan exile academics and students so that they can handle the mental disruption in a much better way.

The exile also faces loss and grief, which also causes a psychological toll on one's mental health, which is another displacement for them. GÖÇER, (2020) emphasis that the feeling of separation reduces their ability and efficiency to focus on the academic part. Uncertainties in the future life and identity dislocation are enough to disturb the mental peace of an individual. The Afghan exile students and academics did not know what life holds for them next, which made them furious, and as a result, they were not able to concentrate on the academic process. It is opposed by Khan, A., Ahmad, & Gul, (2020) if a person loses sense of belongingness, they are moved towards mental health conditions.

However, the host country did acknowledge such displacement and made arrangements for psychological support for the exile. The social well-being of such a group of people was given priority, and they were tried to be brought out of the mental trauma through different healing methods. Cultural events were also tried to be organised so that the exiled Afghans could participate and diminish their social anxieties. Isolation also emerged as a reason which contributed towards the mental health of the exile in an adverse manner (Altınay & Pető, 2022). This magnified the mental traumas, and surviving in the host country became difficult for the Afghan exile academics as well as the students. In order to cope with such mental disparities, support networks were built. Different programs such as mentorship, culturally sensitive programs, and peer introduction programs were introduced so that the feeling of isolation could be reduced.

## Transitioning to New Educational Systems

The transition that the Afghan exile academics had to make towards new educational systems brought into light uncertain academic trains. According to Nehan, (2022) an individual has to go through a multifaceted transition as critical aspects of their lives take a complete $360^0$ turn. On the other hand, it can be understood that technological advancement is happening at a rapid speed, and different sections of society are making its optimisation. As per the view, of Abawe, *et al.* (2023) it uses are also made in the academic frame where online teaching modalities are being adopted by recognised universities, which provides access to students belonging from different regions. For example, the Afghan exile professors also had to face certain challenges in dealing with the new curriculum. One of the primary reasons for such kind of barriers was pedagogical differences, as the teaching techniques were quite different in their home and host country.

Couch, (2019) criticised that a traditional approach was adopted in their home country, whereas a participative and interactive teaching methodology has become a part of the new curriculum of the host countries. Learning was the only way through which the Afghan exile professors tried to deal with such challenges as they had to forget about the old methods and learn the new ones if they had to prosper in their future endeavours. As explained by Morrice, (2022) different ways through which interactive teaching can be supported where learned, project-based learning was also focused upon along with the development of collaborative and engagement activities for the students. Moreover, collaborations were also built with international colleagues so that better participation could be made towards the new educational systems.

Language barriers emerged as another challenge for the exiled students and academics of Afghanistan (Ruhani, *et al.* 2023); due to this, they were not able to achieve the academic continuity desired by the education system. In order to bridge such gaps, they were given the opportunity to take learning outside the scope of the primary curriculum. As per the view, of Hanifi, (2021) the peer in the host country can take additional steps towards supporting the linguistic needs of the exile. They are also provided with different linguistic support services through online modes or through such mediums which are outside the scope of the curriculum. Certain experts in the host country even go out of their way to help the exile as they are well aware of the conditions they have been through.

However, it is opposed by Amiri, (2023) discrimination is still a significant part of the educational system, which can hamper the quality of education for the exile. The Afghan exile academics and students also had to face inequality challenges, which made adapting to the new education system even more difficult. On the other hand, different legislations and regulations were put to use through the code of conduct and practices so that discrimination could be reduced and the Afghan exiled academics and students could transition smoothly with the new educational system.

## Recognition and Accreditation: The Struggle for Qualification Acknowledgment

According to Abbasi & Monsutti, (2023) in the present era, the navigation of education new landscape is usually complex and having certain challenges in terms of accreditation, recognition, etc specifically for people pushed into exile because of conflicts, distractive factors, and persecution. It is a situation where they leave the professional network, study in the education universities, and even seeks refugees in different nations with various system of education, and standard (Axyonova, Kohstall & Richter, 2022). Further, transferring from one study atmosphere to other is disastrous and people are also required to adopt unique regulations so as to attempt well educational qualification in the future.

However, Ra'ees, (2023) argued there are certain challenges in acknowledgment qualification and verification of system in education is one of the major challenges because multiple nations have various systems in education. So, receiving education in one country cannot effectively align with the aspect of other host country and certificates, degrees, etc received from one nation cannot be translated in a easiest way in qualification of the other country. In addition, there are also hurdles in recognition because most of the countries lack an accurate procedure to recognize foreign credentials academics which ultimately leads to delays as well as ambiguity in acknowledging (Flamant, Fourot & Healy, 2022). It is very challenging for exile academics or people who needs their education qualification to recognize higher studies assessment as well as opportunities of employment. Figure 1 shows the total number of national Kankor participants by year.

*Figure 1. The total number of national Kankor participants by year (The Diplomat, 2020)*

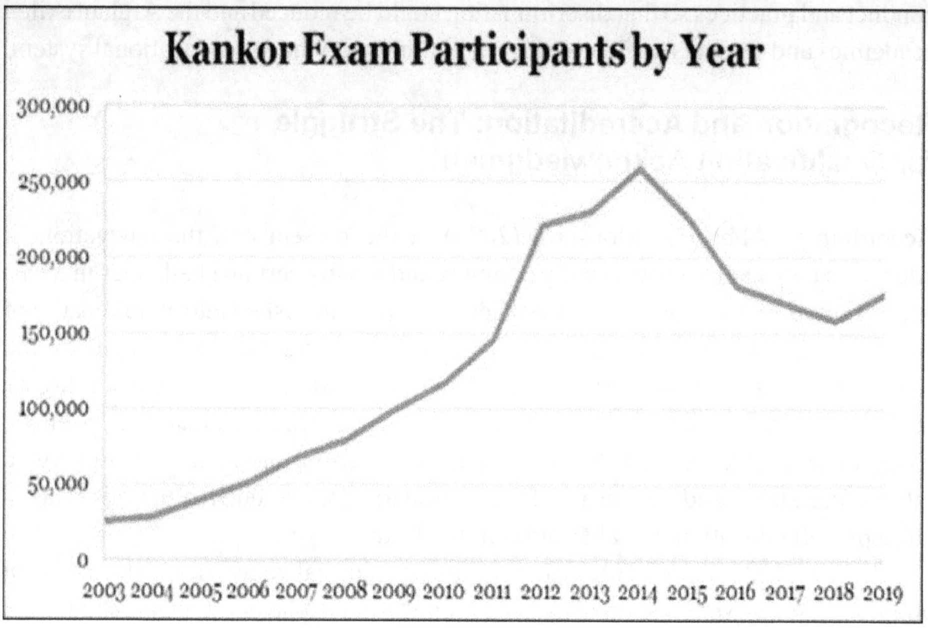


It can be analysed from the above image that the number of participants in giving exams in Kankor is decreasing the year and in the year 2019 only 175000 approximately high school graduates took part in the Kankor exam (The Diplomat, 2020).

As per the views of Konuk, (2020) there are also issues related with losing scholarly networks as exiled academics usually lose its effective networks of scholars which further makes complex for establishing connection with the entirely new environment of academics. In addition, education network are necessary to find job opportunities along with researching opportunities as well as receiving recognition in academics.

As said by Fani, (2022) it is crucial to recognize the significance of addressing several issues and even government, stakeholders, education universities, multinational firms, etc are trying to create solutions in terms of facilitating qualification acknowledging in exile academic. Standardizing process recognition is one if the best strategy for achieving foreign qualifications that provides predictability as well as clarity for exile academics. It helps to establish specific advice to evaluate as well as recognize credentials academic from unstable or conflict-impacted regions. Moreover providing cultural training is also the best way to guide educators as well as employers in terms of reducing biases and stigmatization against exiled academics (Kumar, 2023).

As opposed by Aldhahyani & Al-Eryani (2023) supporting network is also the best way to promote education in the country as it helps in displacing academics in terms of connecting with the academic communities or finding opportunities for collaboration as well as research. Recognizing academic exile qualification does not contribute in justice only but also result in potential harnessing of talented students for providing their contribution for new host nations in academic reports.

## Financial Challenges and the Role of Host Universities

According to Hayward & Karim, (2019) the issues in exile, academics are not only contributing towards administrative as well as academic hurdles which also include financial challenges but educational institutes are playing a significant role to overcome these challenges. The scholar's displacement, educators, students because of persecution, conflict as well as multiple factors which are placed in disastrous financial condition that makes it necessary for hosting institutes to offer support and set up.

As per the views of Farahi & Guggenheim, (2020) reduction in income is the main financial struggle faced by academic institutes as they were displaced and they even tried to search job in other nations despite of their secure job. There is also less scholarship assessment because financial aid is also less for exile academics and multiple scholarship programs are also restricted to permanent citizens of the host country. Further, offering scholarship to students is also limited due to less financial aid which is one of the major problems faced by the academic institutes and tuition fees is also exorbitant because it is the major issue for the student who are seeking to get an education from another country (Imran, *et al.* 2020).

However Arar, *et al.* (2021) argued the education cost in other countries is high which in turn creates a higher financial burden on students and moreover studying in another host country also increases other additional expenses such as transportation, food, accommodation, etc are very complex to face. Further, it also increases the issues in terms of making ends while meeting pursuing the educational need as well as goals of the profession for their future career and vice versa.

As explained Gutierrez, (2020) host Universities plays a significant role in assisting exile academics in research as education pursuit and through addressing as well as recognizing financial is uses faced by host institutions individuals helps in creating a more inclusive as well as supportive environment of academic in the country (Abdallah & Farhan, 2023). There are scholarships and financial aid which are offered by universities that covers tuition fees, expenses of living, etc which help in forcing students to study and gain academic knowledge to a huge extent. Moreover, providing flexibility in working opportunities is also the best way to support students by the universities along with developing opportunities for displaced academics for work in the university like positions, and research assistantships, are also the best way to provide income sources along with pursuing the goal of studies and vice versa (Webb & Paretti, 2023). Figure 2 shows the share percentage of Kankor participants in multiple regions

*Figure 2. The share percentage of Kankor participants in multiple regions (The Diplomat, 2020)*

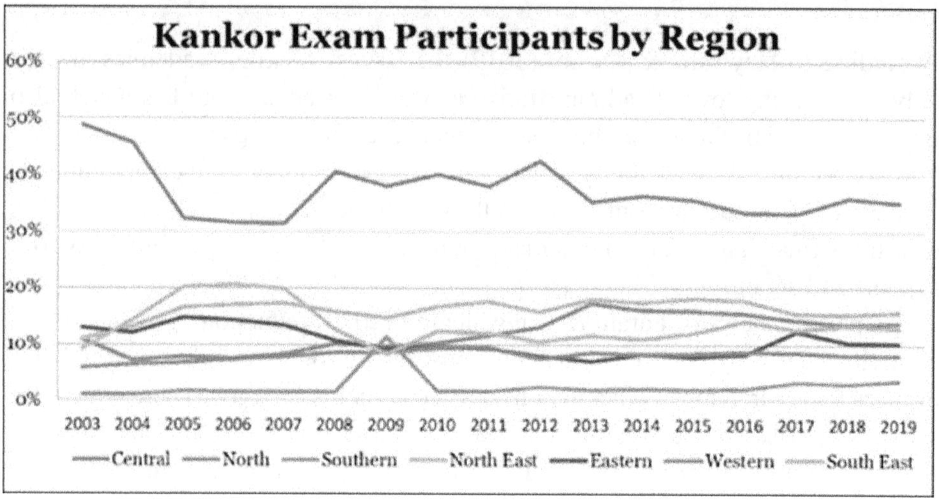


The above graph depicts that in the year 2018 from a population of 1.03 million people, only 8,616 students sat for the Kankor exam and in Helmand with a population of about 1.42 million, only 1,868 students participated in the Kankor test and so on (The Diplomat, 2020).

However, Rehman, (2023) argued providing affordable housing by the universities is also the best way to promote students for studying along with collaboration with local authorities of housing also provides the best accommodation option for students, and displaced scholars. It is also the role of universities to offer well-being as well as mental health services in terms of helping the student to cope with several psychological issues that are linked with their condition (Abdallah et al.,2023).

One of the best pieces of evidence of financial issues faced by exile academics can be considered with the Afghan academic, Dr. Aziz, who wished to fly homeland because of persecution and conflict and even did not find any significant job opportunities in other countries despite, he was renowned scholar (UNHCR, 2023). It was also difficult for him to spend living expenses regularly but Host University offered him an assistantship in research which provided him assistance and also helped in providing suggestions to live efficiently in the host country. With a financial aid combination as well as opportunities in academics Dr. Aziz highly tried to overcome challenges of exile and even made important studies contributions.

As said by Arar, *et al.* (2021) host universities played a crucial role in overcoming the challenges faced by Afghan people in terms of providing them with financial support provided them opportunities of job in terms of pursuing their professional as well as academic careers in the best possible ways.

## Scholarships, Fellowships, and Tailored Programs: A Ray of Hope

According to McNamara, (2023) fellowships, scholarship programs are emerging as hope beacons for displaced students, educators, students, etc along with offering them a lifeline in terms of continuing their journey of education despite uncertainty as well as displacement. It is designed in such a way that helps in alleviating the educational, and professional issues and all these policies also provide a hope ray by offering financial support along with accessing opportunities, and resources for professional as well as personal development. It helps in providing financial support in terms of covering living expenses, and tuition expenses, which is complex for displaced students who lose sources of income (Ehrkamp, Loyd & Secor, 2022). Moreover, resources of academics are also offered by these programs in terms of gaining library access, and academic resources in the host institutes which enable them to continue their studies and research.

As per the views of Alemi, *et al.* (2023) inclusivity as well as integration y involving displaced academics in university programs, institutions are still trying to promote inclusivity as well as diversity in academics. It ultimately led to value as well as welcome scholars in the new academic environment of academic and these programs also offers networking as well as mentorship which assist exile academics in establishing a relationship with fellow educators, and scholars which is highly valuable for making a career in the future time period and vice versa (Etzold, *et al.* 2022).

"Afghan Scholars Initiative" is one of the best evidence of exile academic program designed which was introduced by humanitarian organizations and international university consortiums and the program offers tailored scholarship which helps in covering living expenses, tuition fees, etc for academic exile in Afghan (Global Campus of Human Rights, 2023). These scholarships also assist in supporting multiple levels of academics from non-graduate to the studies of postdoctoral and it also helped in providing integration as well as language support for helping displaced in terms of increasing their skills of communication in the host nation. It adds to overcome barriers of communication along with integration boosting (Bužinkić, 2023).

However, Jackson, (2021) argued psychosocial support was also provided to participants so as to overcome emotional, and mental health issues due to such challenges in Afghanistan as well as their personal lives along with providing support and guidance to navigate the landscape of academics along with pursued their interest in research. Consequently, scholars of Afghanistan are also given the opportunity to coordinate with researchers on several Afghan-related projects in terms of rebuilding their career in academics and vice versa (Darawsheh et al.,2023).

It not only offered a hope of ray for Afghan academics but also served as a best model for increasing research and educational development in the international context. So it can be concluded that these fellowship programs are the best ways to support exile academics and even offer professional growth in period of uncertainties and it not only empowered academics but also contributed in resilience as well as diversity of the community so academic in the global world (Greenfield, Pardo & Taylor, 2021).

## Overcoming Cultural and Linguistic Barriers: Strategies for Integration

According to Axyonova, Kohstall & Richter, (2022) barriers of language is one of the most common challenge which is faced by exile academics because there are difficulties to understand articles, lectures, etc and even issues to communicate their opinion. Moreover adoption of new academic norms, culture is also problems because Afghan scholars might face issues in finding the appropriate place in the new atmosphere. They can also feel social isolation because of less networking of supporting and displaced scholars are also have a sense of belonging to other nations which in turn increases their struggle to fix or set in the host country (Ra'ees, 2023).

However Razzaque, (2023) argued there are also integration strategies in terms of solving challenges faced by displaced academics like Host University may also provide different programs for supporting language as it helps in focusing upon professional skill language that makes easy for exiled Afghans to do research as well as coursework efficiently. Providing cultural competency training is also helpful in bridging gaps in culture along with respecting as well as understanding perspectives of different cultures for promoting integration as well as harmony to a huge extent (Abdallah & Alkaabi,2023).

There are some personal stories of the success of displaced scholars also tried to integrate in host communities successfully are also a type of guidance for others and these scholars also highlight determination as well as residence of exile academics along with illustrating solutions to mitigate challenges of integration (Ehrkamp, Loyd & Secor, 2022).

## From Exile to Activism: The Broader Influence of Afghan Academics Abroad

As said by Morales, Rajmil & Andreosso-O'Callaghan, (2023) the academics exile experience is not only restricted to academic pursuits or personal struggle but it leads to realm influence as well as activity and it has been found that several Afghan scholars who went to foreign countries because of persecution, conflict, etc usually became catalysts to leverage their insights, professionalism in terms of advocating their broad global challenges. Advocacy in policy is one of the significant powers of exile academics by utilizing expertise, research in terms of informing regulations that highly affect the home country, and collaboration with multinational corporations as well as the government in terms of contributing for building peace, humanitarian efforts, etc (Konuk, 2020).

As opposed by Morales, Rajmil & Andreosso-O'Callaghan, (2023) displaced Afghans are also having opportunities for building global networking which helps to connect with policymakers, academics, etc from all over the world and this network also harnesses in mobilizing assistance to their motherland. Preservation of Afghan culture is also promoted by academic exile through participating in different cultural events by promoting literature, and art along with educating the global communities about the Afghan culture richness (Jandigulov et al.,2023).

An Afghan academic, Dr. Sahar was pushed towards exile because of conflict in Afghanistan and after the US arrival her career in academics was pursued but she was highly conscious about issues that were faced in the motherland (LSE, 2023). Her journey was embarked on due to activism and the influence of policy. But there were milestones in her journey also (Morales, Rajmil & Andreosso-O'Callaghan, 2023).

Dr. Sahar pays attention to research in academics on socio-economic effects on children as well as women conflict in Afghan and her research contributed towards recognition in the international level that leads to invitations for speaking at academic conferences along with trying to make collaborate with the institutes focusing upon achieving rights for women. Moreover, she took active participation in the implementation of policies followed by offering information to multinational institutes and agencies of government in terms of empowering females in Afghanistan along with improving the level of education to a wide extent.

The journey of Meet Dr. Sahar to activism from exile is having a crucial effect of exile academics in the international level and her leverage knowledge, and expertise along with advocating with international networks for bringing positive changes in terms of promoting the well being as well as the right of vulnerable population in the motherland. So it can be concluded that Afghanistan exile academics is having far-reaching positive impacts because their activities, working, etc served as a hope beacon that contributed towards bringing positive changes. It is even helpful in contributing towards international conflict conversation with human rights as well as peace (Morales, Rajmil & Andreosso-O'Callaghan, 2023).

## CONCLUSION

It can be concluded that the Afghan academic community had to face displacements and uncertainties, and they have exhibited remarkable adaptability and resilience. Despite the difficulties provided by violence, instability, and the necessity to seek sanctuary in foreign places, Afghan academics have displayed an unshakeable devotion to education and flourished in the face of adversity. They have to face different challenges in diverse areas, such as financial disparities, mental trauma, language barriers, cultural incompetencies, and difficulty in getting recognition and accreditation. The Afghan exile had to go beyond a limit to overcome such challenges, and they were able to attain a nominal position in it also. They reflected resilience and adaptability as they worked towards achieving language mastery by actively engaging in language learning activities so that the gaps could be bridged.

Furthermore, they also rebuild the networks so that professional connections can be built even in Afghanistan. Collaboration was also made with international entities so that the benefits of its education system could be gained. Donor organisations and non-profit organisations also played a major role in supporting the exile of Afghans as they provided financial and social assistance. A new educational system was embraced by the exile, which also contributed towards recovering from the challenges faced in the earlier state.

New ways of teaching practices were also learned through this, increasing the scope of operation for the Afghan exile academics. The desire to continue the academic journey drives the exiled population of Afghans towards the development of different strategies and approaches through which the possible challenges faced in the transition stage can be mitigated. This has also built their passion to help other exile groups deal with the displacement they are facing by providing them with support and mentorship. It is suggested that more emphasis can be paid upon providing psychological support to the exile from an initial stage so that different challenges can be handled by them graciously.

## REFERENCES

Abawe, Z., Daud, B., Daudzai, H., Jabarkhail, M., & Yousaf, F. (2023). Afghanistan and the way forward: Incorporating indigenous knowledge into policymaking. *Global Policy*, *14*(1), 192–198. doi:10.1111/1758-5899.13144

Abbasi, K., & Monsutti, A. (2023). From Muhājir to Āwāra: Figures of Migration and Exile Among Afghans. In *Migration in South Asia: IMISCOE Regional Reader* (pp. 187–200). Springer International Publishing. doi:10.1007/978-3-031-34194-6_13

Abdallah, A. K. (2023). Teacher-Led, Student-Focused, and Unleashing the Power of Teacher Empowerment for School Improvement and Success. In A. Abdallah & A. Alkaabi (Eds.), *Restructuring Leadership for School Improvement and Reform* (pp. 1–21). IGI Global. doi:10.4018/978-1-6684-7818-9.ch001

Abdallah, A. K., & Al-Kaabi, A. M. (2023). Induction Programs' Effectiveness in Boosting New Teachers' Instruction and Student Achievement: A Critical Review. International Journal of Learning. *Teaching and Educational Research*, *22*(5), 493–517.

Abdallah, A. K., & Alkaabi, A. M. (2023). Role of teachers in reinforcing students cultural and heritage awareness at Abu Dhabi schools to meet global challenge. *Cogent Social Sciences*, *9*(1), 1. doi:10.1080/23311886.2023.2194734

Abdallah, A. K., & Farhan, A. F. (2023). Breaking Barriers and Empowering Women Leaders to Drive School Improvement. In A. Abdallah & A. Alkaabi (Eds.), *Restructuring Leadership for School Improvement and Reform* (pp. 399–419). IGI Global., doi:10.4018/978-1-6684-7818-9.ch020

Abdallah, A. K., Ismail, O., Abdallah, R. K., & Alkaabi, A. M. (2023). Perceptions of Students About the Use of Webinars in Classrooms: A Case of Abu Dhabi University. *International Journal of Information and Communication Technology Education*, *19*(1), 1–17. doi:10.4018/IJICTE.322793

Aldhahyani, D. F. Q., & Al-Eryani, D. A. A. (2023). An International Peer-Reviewed Open Access Journal.

Alemi, Q., Panter-Brick, C., Oriya, S., Ahmady, M., Alimi, A. Q., Faiz, H., Hakim, N., Sami Hashemi, S. A., Manaly, M. A., Naseri, R., Parwiz, K., Sadat, S. J., Sharifi, M. Z., Shinwari, Z., Ahmadi, S. J., Amin, R., Azimi, S., Hewad, A., Musavi, Z., ... Ventevogel, P. (2023). Afghan mental health and psychosocial well-being: Thematic review of four decades of research and interventions. *BJPsych Open*, *9*(4), e125. doi:10.1192/bjo.2023.502 PMID:37424447

Alqodsi, E. (2023). Teacher Civil Liability in the Case of Breaching Educational and Control Obligations. In A. K. Alkaabi (Ed.), *Restructuring Leadership for School Improvement and Reform* (pp. 327–338). IGI Global. doi:10.4018/978-1-6684-7818-9.ch016

Alqodsi, E., & Aljahoori, S. A. (2023). Legal protection of the right to education for people with special needs: Zayed higher organization for people of determination as a model. In E. Efthymiou (Ed.), *Inclusive Phygital Learning Approaches and Strategies for Students With Special Needs* (pp. 199–213). IGI Global. doi:10.4018/978-1-6684-8504-0.ch010

Alqodsi, E. M., Jadalhaq, I. M., & El Maknouzi, M. E. (2023). *Technology-enhanced legal education: A study of its impact on student learning outcomes in the UAE*. IGI Global. doi:10.4018/978-1-6684-5518-0.ch004

Altınay, A. G., & Pető, A. (2022). Women's courageous resistance to gender apartheid in Afghanistan: A conversation with Shaharzad Akbar. *European Journal of Women's Studies*, *29*(4), 489–505. doi:10.1177/13505068221130406

Amiri, M. H. (2023). Education and Minority Groups in Afghanistan: The Case of Hazara People. *Journal of Asian Social Science Research*, *5*(1), 33–48. doi:10.15575/jassr.v5i1.70

Arar, K., Kondakci, Y., Streitwieser, B., & Saiti, A. (2021). *Higher education in the era of migration, displacement and internationalization*. Routledge. doi:10.4324/9780429345180

Arifani, D. (2021). The Taliban & Afghanistan: Conflict & Peace in International Law Perspective. *International Journal of Law Reconstruction*, *5*(2), 306–316. doi:10.26532/ijlr.v5i2.17704

Axyonova, V., Kohstall, F., & Richter, C. (2022). Academics in Exile. *The Academy in Exile Book Series*, *2*, 11–30. doi:10.14361/9783839460894-002

Azami, A. S. (2021). Warlords, the United States, and the state of anarchy in Afghanistan. *Central European Journal of Politics*, *7*(1), 46–75. doi:10.24132/cejop_2021_1

Boroujerdi, M. (2023). The Plight of Threatened Afghan and Iranian Scholars and Students. Academic Press.

Bužinkić, E. (2023). *Schooling aftershocks: Perpetual displacements of refugee youth amid multiple pandemics and social (b) orders* (Doctoral dissertation, University of Minnesota).

Couch, D. (2019). *The Policy Reassembly of Afghanistan's Higher Education Sector* (Doctoral dissertation, ResearchSpace@ Auckland).

Crespo, N. M., & Cooperativa, S. (2021). Women and Peacebuilding from the diaspora and exile in Europe. International Catalán Institute for Peace, 19, 91-96.

Darawsheh, S. R., Al-Shaar, A. S., Alshurideh, M., Alomari, N. A., Elsayed, A. M., Abdallah, A. K., & Alkhasawneh, T. (2023). The Relation Between Creative Leadership and Crisis Management Among Faculty Members at Imam Abdulrahman Bin Faisal University in Light of the Corona Pandemic from the Perspective of Department Heads, The Effect of Information Technology on Business and Marketing Intelligence Systems (Vol. 1056). Springer. doi:10.1007/978-3-031-12382-5_83

Dinakhel, M. A., Bakhshali, S. Z., & Ali, I. (2023). An Overview of Political Suppression of Legal Political Parties in Afghanistan: 1940-2010. *Khairulummah, 2*(02), 12–22.

Ehrkamp, P., Loyd, J. M., & Secor, A. J. (2022). Trauma as displacement: Observations from refugee resettlement. *Annals of the American Association of Geographers, 112*(3), 715–722. doi:10.1080/24694452.2021.1956296

Etzold, B., Isikozlu, E., Christ, S., Morosanu, L., Kraler, A., Pastore, F., . . . Al Jarrah, R. (2022). *Nothing is more permanent than the temporary: Understanding protracted displacement and people's own responses TRAFIG Synthesis Report* (Doctoral dissertation, TRAFIG).

Fani, A. (2022). Disciplining Persian Literature in Twentieth-century Afghanistan. *Iranian Studies, 55*(3), 675–695. doi:10.1017/irn.2021.14

Farahi, K., & Guggenheim, S. E. (2020). *Pathways for post-peace development in Afghanistan*. United States Institute of Peace.

Flamant, A., Fourot, A. C., & Healy, A. (2022). city network activism and the governance of migration. *Local Government Studies, 48*(6), 1017–1026. doi:10.1080/03003930.2022.2045277

Ghasemi-Tafreshi, S., & Lafortune, G. (2023). The Schooling of Forced Immigrant Afghan Youths in Iran: A Study of the Factors Leading to Exclusion. *Comparative and International Education, 52*(1). Advance online publication. doi:10.5206/cie-eci.v52i1.14714

Ghosh, A. (2019). Afghan migration and pakistan's policy response: Dynamics of continuity and change. *Public Policy Research in the Global South: A Cross-Country Perspective*, 215-230.

Global Campus of Human Rights. (2023). *Scholarship Programme for Afghan Scholars and Students at Risk*. https://gchumanrights.org/research/projects/afghan-scholarship-programme.html

Göçer, A. P. D. D. (2020). *What education builds as nation in afghanistan: a case of nation-building* (Doctoral dissertation, Middle East Technical University).

Greenfield. (2021). Afghanistan in the Era of Fentanyl. Academic Press.

Gutierrez, E. D. (2020). The paradox of illicit economies: Survival, resilience, and the limits of development and drug policy orthodoxy. *Globalizations*, *17*(6), 1008–1026. doi:10.1080/14747731.2020.1718825

Hanifi, S. M. (2021). Local Experiences of Imperial Cultures. Academic Press.

Hayward, F. M., & Karim, R. (2019). The Struggle for Higher Education Gender Equity Policy in Afghanistan: Obstacles, Challenges and Achievements. *Education Policy Analysis Archives*, *27*(139), n139. doi:10.14507/epaa.27.3036

Imran, M., Chen, Y., Wei, X. M., & Akhtar, S. (2020). Veiled courage: Inside the women's resistance against violence through their writings. *Asian Journal of Women's Studies*, *26*(1), 74–93. doi:10.1080/12259276.2020.1718392

Ismail, A. O., Alriyami, R., & Alhosani, M. (2023). The art of assuring quality education: Internal approaches and best practices. In A. K. Abdallah & A. M. Ahmed (Eds.), *Restructuring Leadership for School Improvement and Reform* (pp. 182–201). IGI Global. doi:10.4018/978-1-6684-7818-9.ch010

Jackson, A. (2021). *Negotiating survival: Civilian-insurgent relations in Afghanistan*. Oxford University Press. doi:10.1093/oso/9780197606179.001.0001

Jafari, B. A., & Schuster, L. (2019). Representations of exile in Afghan oral poetry and songs. *Crossings: Journal of Migration & Culture*, *10*(2), 183–203. doi:10.1386/cjmc_00002_1

Jandigulov, A., Abdallah, A. K., Tikhonova, Y., & Gorozhanina, E. (2023). Management and leadership in online learning. *Education and Information Technologies*, *28*(10), 13423–13437. Advance online publication. doi:10.1007/s10639-023-11699-4

Khan, A., Ahmad, A., & Gul, A. (2020). Constitution of Afghanistan: An analysis of educational and linguistic provisions. *Liberal Arts and Social Sciences International Journal*, *4*(1), 66–75. doi:10.47264/idea.lassij/4.1.7

Kölbl, M. (2021). Ethnomusicology, Fieldwork, and the Refugee Experience: Notes on Afghan Music in Austria. *Music & Minorities, 1*.

Konuk, K. (2020). *Academy in Exile* (Vol. 269). The Academy in Exile Book Series.

Kumar, R. (2023). A Study of Refugee Crisis in Mohsin Hamid's Exit West in the Backdrop of Current Afghanistan Refugee Crisis. Academic Press.

LSE. (2023). *Sahar Halaimzai*. https://www.lse.ac.uk/ideas/people/Sahar-Halaimzai

Mäkinen, A. (2020). *Memory and Exile: Spatiotemporal Narrations of Displacement in Khaled Hosseini's The Kite Runner and And the Mountains Echoed* (Master's thesis).

McNamara, E. (2023). Calling Triple A: Providing a Path From Humanitarian Parole for Afghan Refugees. Academic Press.

Meininghaus, E., & Mielke, K. (2019). Beyond Doha and Geneva: Peacemaking engagement of Afghans and Syrians in North Rhine-Westphalia and Germany. Academic Press.

Mielke, K., & Meininghaus, E. (2021). Engaging refugees and migrants as peace advocates: how Germany can contribute to peace in Syria and Afghanistan. Academic Press.

Morales, L., Rajmil, D., & Andreosso-O'Callaghan, B. (2023). The Global Economic and Political Landscape through the Lenses of Multiple and Multifaceted Crises. Academic Press.

Morrice, L. (2022). Will the war in Ukraine be a pivotal moment for refugee education in Europe? *International Journal of Lifelong Education*, *41*(3), 251–256. doi:10.1080/02601370.2022.2079260

Nehan, N. (2022). The Rise and Fall of Women's Rights in Afghanistan. *Afghanistan*, 111.

Ra'ees, W. (2023). Geostrategic and Internal Dynamics of the Afghan Conflict. *Editorial Board*, 37.

Ramadan, R. S., & Ismail, O. A. (2023). Building a better future: The role of school inspection in driving educational excellence. In *Restructuring Leadership for School Improvement and Reform* (pp. 202–221). IGI Global. doi:10.4018/978-1-6684-7818-9.ch011

Razzaque, A. (2023). *Identity Ruptures and Community Fragmentation: Afghans Resettling in the US After the Taliban Takeover* (Doctoral dissertation, Georgetown University).

Rehman, W. (2023). Afghanistan's parallel crises under the Taliban. Academic Press.

Ruhani, A., Keshavarzi, S., Kızık, B., & Çakal, H. (2023). Formation of hatred emotions toward Afghan refugees in Iran: A grounded theory study. *Peace and Conflict, 29*(4), 355–364. doi:10.1037/pac0000685

Schuster, L., & Shinwari, R. M. K. (2020). Migration in Afghan women's poetry. *Soundings, 76*(76), 111–123. doi:10.3898/SOUN.76.08.2020

Seddighi, H., Naseh, M., Rafieifar, M., & Ilea, P. (2022). Education of Afghan refugee children in Iran: A structured review of policies. Academic Press.

Shams, F. (2021). The Sensorium of Exile: The Case of Elyas Alavi and Gloria Anzaldúa. *Losing Our Minds, Coming to Our Senses: Sensory Readings of Persian Literature and Culture*, 195-214.

The Diplomat. (2020). *Afghanistan's Higher Education System Has a Stark Geographic Divide*. https://thediplomat.com/2020/07/afghanistans-higher-education-system-has-a-stark-geographic-divide/

UNHCR. (2023). *The Right to Asylum between Islamic Shari'ah and International Refugee Law*. https://www.unhcr.org/sites/default/files/legacy-pdf/4a9645646.pdf

Wardak, M. O. (2021). *Religion, Politics and Education in Afghanistan, 1950s-1980s: The Case of Habibia High School* (Doctoral dissertation).

Webb, M. E., & Paretti, M. C. (2023, June). Influences on Displaced Engineering Student Professional Identity Development: A Scoping Literature Review Across Forced Migration Contexts. *2023 ASEE Annual Conference & Exposition*. 10.18260/1-2--43696

Ziaulhaq Rashidi, D. G. U. G. (2019). Nation Building Process in Afghanistan. Academic Press.

# Chapter 3
# Capturing the Narratives of Escape and Resilience:
## A Study of Afghan Refugees

**Saad Ullah Khan**
https://orcid.org/0000-0002-3392-7844
*Amity University Rajasthan, Jaipur, India*

**Fahim Sadat**
*Independent Researcher, Germany*

**Sadaf Khan**
https://orcid.org/0009-0005-1348-0386
*IEC College of Engineering and Technology, Greater Noida, India*

## ABSTRACT

*This scholarly inquiry explores the narratives of escape and resilience among Afghan refugees in Nordhausen (Germany), employing a multidisciplinary approach. Utilizing qualitative methods, particularly open-ended structured interviews, the study illuminates the psychosocial dynamics and lived experiences of Afghan refugees. It seeks to unravel the complexities of migration, decode reasons for leaving one's country, and understand asylum-seeking processes. The research identifies adaptive mechanisms and coping strategies, shedding light on socio-cultural, economic, and political factors shaping refugee narratives. Ethically sensitive, culturally responsive engagement underscores respect for refugee voices and agency. Narratives underscore the significance of agency, resourcefulness, and community networks. The study emphasizes the role of host-country policies and social structures in shaping refugee lives. This research bears implications for academia and policy, offering insights into refugee dynamics and factors facilitating successful integration amid the global refugee crisis.*

DOI: 10.4018/979-8-3693-1483-8.ch003

## INTRODUCTION

In the landscape of contemporary global affairs, the phenomenon of forced migration has become an increasingly prevalent and complex facet, impacting millions of lives worldwide. This surge in population movements, driven by conflict, persecution, and political instability, has given rise to a pressing need for scholarly endeavours that not only chronicle the experiences of displaced individuals but also inquire into the convoluted narratives of escape and resilience that characterize their journeys (Braithwaite et al., 2018). Against this backdrop, the present research embarks on a comprehensive exploration to understand the reasons of Afghan migration, which is regularly going on for the last four decades. The focus of this study extends beyond a mere statistical or geopolitical analysis, aims to engage in a profound examination of the multifaceted narratives that emerge from the lives of Afghan refugees who have sought asylum, specifically in the context of Nordhausen, Germany. The urgency of this study is underscored by the protracted conflict and political instability in Afghanistan, which has precipitated a significant displacement of its population, making the Afghan diaspora a pertinent subject of investigation. Within the distinctive socio-cultural milieu of Nordhausen, this research unravels the diverse stories of escape and resilience that shape the lived experiences of Afghan refugees, shedding light on the multifarious challenges they encounter, the coping mechanisms they employ, and the ways in which they navigate the complexities of integration in their host society.

The conceptual underpinning of this scholarly investigation is grounded in a multidisciplinary framework, seamlessly integrating perspectives from sociology, anthropology, psychology, and migration studies. This interdisciplinary lens captures the complexity of the refugee experience, recognizing that their narratives are rich tapestries woven from various threads of personal, social, and cultural dynamics. Through an exploration of the narratives of migration, the study discerns the underlying factors that drive individuals to leave their homeland, the perilous journeys they undertake, and the coping strategies they develop in the face of adversity. Simultaneously, the focus on narratives of resilience unravels the intrinsic human capacity to adapt, overcome challenges, and rebuild shattered lives in unfamiliar environments. By delving into the stories of Afghan refugees in Nordhausen, the researchers not only contribute to academic discourse but also provide insights that can inform more effective and culturally sensitive refugee integration policies globally.

The selection of Nordhausen as the specific geographical focus of this study is deliberate, as the experiences of refugees in a particular locale are deeply entwined with the socio-cultural, economic, and political dynamics of that environment (Schuetze, 2023). Nordhausen, with its unique characteristics as a host community, presents a compelling setting for this investigation. The interactions between the

Afghan refugee population and the local community, the reception and integration policies in place, and the broader societal attitudes all shape the experiences of displaced individuals. Therefore, by situating the research within this specific context, the study generates subtle insights that can inform not only academic discussions but also contribute to the development of more effective and culturally sensitive refugee integration policies.

The ethical considerations of this research are paramount, given the sensitive nature of the topic and the potential vulnerability of the research participants. Informed consent, confidentiality, and the protection of anonymity are central tenets guiding the interactions with the Afghan refugees involved in the study. The research also adheres to the principles of cultural sensitivity, recognizing the diversity within the refugee community and avoiding essentialist assumptions about their experiences.

In the contemporary context of global challenges posed by forced migration, this study makes a significant scholarly contribution by offering a refined exploration of the narratives of escape and resilience among Afghan refugees. By amplifying the voices of those directly impacted by displacement, the research not only cultivates empathy but also challenges prevailing stereotypes, fostering informed discussions. These discussions, in turn, possess the potential to influence policies and interventions aimed at more effectively supporting and integrating refugees into their host societies. Beyond illuminating the specific experiences of Afghan refugees in Nordhausen, the narratives uncovered serve as a compelling testament to the indomitable human spirit confronting adversity. This research prompts a revaluation of societal perceptions and responses to the complex challenges posed by forced migration. Positioned at the juncture of academic inquiry and social advocacy, this study does make a meaningful contribution to both realms, thereby enriching the broader global discourse on displacement and reaffirming the profound capacity of humanity to endure and rebound from challenging circumstances.

## SIGNIFICANCE OF THE STUDY

This study systematically explores the often-neglected narratives of Afghan refugees, offering a vital forum for marginalized voices. It makes a substantial contribution to academic dialogues on refugee studies and human rights, scrutinizing the challenges and resilience within a particular host community. The insights garnered possess the capacity to influence local and national policies, informing the development of more efficient integration strategies and support services. Within the current global refugee crisis, this research assumes a timely and relevant role as a case study, augmenting our collective comprehension of refugee experiences and contributing to scholarly discourse enrichment.

## RELEVANCE OF THE STUDY

This study is highly relevant, addressing a critical humanitarian and sociopolitical issue amid the global refugee crisis and specific challenges faced by Afghan refugees. It provides a timely examination of their experiences, contributing to understanding refugee integration, social cohesion, and displaced individuals' resilience. Focused on Nordhausen, the research offers insights guiding local and national policies, bolstering support structures, fostering inclusivity, and contributing to the global discourse on human rights, migration, and refugee studies. Beyond academic value, it holds practical significance for policymakers, humanitarian entities, and host communities involved in refugee resettlement and integration complexities.

## OBJECTIVES OF THE STUDY

This scholarly pursuit explores the personal narratives of Afghan refugees in Nordhausen, with a focus on their escape circumstances and complex trajectories. Beyond a superficial analysis, the study systematically explores coping mechanisms, adaptive strategies, and remarkable resilience amid multifaceted challenges in the displacement and resettlement processes within Nordhausen's distinct context. It aims to substantially contribute to refugee studies and human rights by examining the unique challenges and strengths of the Afghan refugee population. The research extends beyond theory, offering tailored pragmatic recommendations for policymakers, humanitarian organizations, and local authorities to fortify support structures and enhance integration efforts in Nordhausen and similar host communities grappling with evolving dynamics in refugee resettlement.

## REVIEW OF LITERATURE

Four decades in the past, Afghans commenced a mass exodus precipitated by the endemic violence pervading their homeland, seeking sanctuary beyond its borders. An exodus that saw over 400,000 individuals escaping the brutalities perpetrated by the Communist-led regimes, precipitating a migratory surge into the neighbouring nations. The magnitude of displaced Afghan populace witnessed an escalating trajectory following the Soviet incursion on Christmas eve in 1979 (Dupree, 1984). By the culmination of 1980, the diaspora of Afghan refugees in nearby countries burgeoned to exceed the five million marks. Subsequently, across the ensuing four years, this demographic upheaval continued its relentless expansion, surging beyond the five-million milestone, with refugees dispersing across the borders into both Pakistan and Iran (Jazayery, 2003).

Today, the Afghan refugee demographic stands as one of the globe's most extensive displaced populations, emblematic of the enduring challenges within the realm of forced displacement (Sadiq, 2022). Within the perimeters of refugee camps, certain Afghan lineages have endured across successive generations, becoming integral components of the indigenous societal fabric. Paradoxically, despite their entrenched presence, this resilient population finds itself subjected to egregious rights deprivations, vilification, and the perpetual spectre of deportation, evoking a poignant juxtaposition of belonging and marginalization within the contours of local and global sociopolitical dynamics (Farahani et al., 2023).

Presently, the global landscape bears witness to an aggregate exceeding 2.6 million formally registered refugees originating from Afghanistan, constituting a formidable demographic that surpasses one-tenth of the total global refugee population. This staggering figure positions Afghanistan as the second most prolific source of refugees globally, second only to Syria (UNHCR Report, 2023). Beyond this enumerated count, an indeterminate multitude remains unregistered, clandestinely navigating the precarious status of asylum-seekers, their presence underscoring the complexity of the humanitarian landscape. Compounding this predicament is the substantial cohort of over a million individuals internally displaced due to the relentless throes of ongoing conflict within the nation's borders (Willner-Reid, 2016). This multifaceted scenario accentuates the manifold dimensions of displacement, necessitating a serious understanding of the diverse trajectories and challenges encountered by those compelled to seek refuge, both within and beyond Afghanistan's geopolitical boundaries.

In the year 2018, the United Nations Assistance Mission in Afghanistan systematically chronicled the apical zenith of civilian casualties ever recorded, marking an ominous milestone that included an unprecedented toll on the lives of children ensnared in the throes of conflict (Connah, 2021). The documented statistics revealed a staggering tally of nearly 11,000 casualties, comprising 3,804 fatalities and 7,189 individuals grappling with severe injuries (Afghan Civilian Deaths, 2019). Simultaneously, this period bore witness to a distressing upsurge in internal displacement, with over 360,000 individuals compelled to relocate within the nation's borders due to the prevailing conflict, as per the comprehensive findings of the United Nations Office of Coordination of Humanitarian Affairs (Afghanistan Refugees: forty years, 2019).

A subsequent report, disseminated in June 2019 by the Institute for Peace and Economics, echoed a disquieting revelation. Afghanistan, as outlined in this report, assumed the disconcerting distinction of being designated the world's 'least peaceful' country, supplanting even the strife-ridden terrain of Syria (Atakpal, 2019). This designation underscores the profound and enduring repercussions of the conflict, encapsulating the pervasive tumult that has enveloped the Afghan landscape, thus demanding a conscientious examination of the multifaceted facets contributing to this unsettling classification.

Within the spectrum of Afghan migration narratives, a subset unveils triumphant odysseys to distant realms such as the United States of America and Europe. In these distant lands, individuals have ultimately discovered the elusive realms of tranquillity and dignity that had long eluded them amid the turbulent tapestry of their homeland. Conversely, a category of less fortuitous individuals has encountered a starkly contrasting fate—forced repatriation to a homeland now fraught with heightened peril compared to their departure or detention in deplorable conditions, where they endure prolonged languishment (Naseh et al., 2018). This stark duality in outcomes underscores the profound disparities within the trajectories of Afghan migrants and compels a nuanced examination of the intricate factors shaping their divergent destinies.

Data furnished by IOM UN Migration reveals a global migration panorama, encompassing 281 million immigrants worldwide in the year 2020 (Natarajan et al., 2020). Germany, highlighted by official statistics, has become a focal point for European immigration. Over the past few years, the nation witnessed a substantial surge in immigrants, comprising 15.3 million residents with migrant backgrounds, equivalent to 18% of the population (Immigrants made up over 18%, 2023). These figures underscore Germany's pivotal role as a major recipient of diverse migration backgrounds.

Afghanistan has emerged as a notable source of immigrants in Germany, constituting the second-largest contributor after Syria, marking a distinct demographic pattern as of September 2021(European Union Agency for Asylum, 2021). Scrutiny of available data reveals a substantial Afghan presence in Germany. This demographic shift is a response to decades of instability, leading to a significant influx of Afghan migrants, solidifying Germany's position as a primary recipient for those seeking refuge. This transformation mirrors intricate geopolitical dynamics, emphasizing Germany's role as a hospitable haven solidifying the progressive contribution of its multicultural society. Germany's proactive engagement with international asylum seekers, notably in 2022, is evident in a surge of applications, by asylum seekers.

The trajectory of Afghan migration to Germany unfolds across distinct phases, tracing its roots to the mid-20 century. The initial wave saw Afghan immigrants, predominantly carpet merchants, establishing a presence in Hamburg's free port, storing their goods, and setting up branches (Ruttig, 2017). The foundation for this migration was laid in the 1950s and 1960s, fostering a robust relationship between Afghanistan and Germany. Notably, graduates from the German High School in Kabul played a pivotal role, benefiting from exchange programs facilitating enrolment in

German universities such as Berlin, Cologne, Bonn, and Bochum. This era witnessed an influx of businesspeople and students from Afghanistan, shaping the composition of newcomers during this formative period (Vadean, 2007). In the aftermath of the Soviet invasion in 1979, a subsequent wave of migration unfolded. With the establishment of communist rule in Kabul, this phase witnessed a substantial influx of immigrants, predominantly composed of Islamists. Notably, these individuals emerged as fervent adversaries to the occupying Soviet troops, marking a distinctive demographic shift in the composition of migrants during this period. Commencing in 2021, the third wave of Afghan immigrants unfolded, characterized by a mass exodus of one million individuals from Afghanistan between August and December of that year, driven by the escalating humanitarian crisis in the country. Despite being accepted by destination nations, the integration challenges faced by a significant number of these immigrants have become a source of political contention. (Albarosa & Elsner, 2023). This dynamic underscore the complex interplay between migration, humanitarian crises, and the intricate political landscape surrounding the assimilation of Afghan immigrants into their new host nations.

The cultural disparities between Kabul (Afghanistan) and Nordhausen (Germany) pose formidable challenges for new immigrants, often leading to significant difficulties in their adaptation process. One of the primary hurdles lies in the realm of language, where fluency in the host country's language plays a pivotal role in facilitating effective intercultural communication. The intricacies of navigating these cultural differences become particularly pronounced for immigrants who are not proficient in the language of their adopted nation (Nedelsky, 2023). This linguistic barrier accentuates the multifaceted nature of the challenges encountered by newcomers, emphasizing the importance of language proficiency as a key determinant in the successful integration of immigrants into the German cultural milieu (Tovar-Gálvez). Acquiring proficiency in the language of the host nation constitutes the initial and paramount step for newcomers seeking integration into the fabric of the new society. For a considerable number of recently arrived Afghan immigrants, German represents either their second or third language, posing a substantial challenge in the process of linguistic assimilation. The struggle to grasp the fundamentals of the German language emerges as a common experience for these immigrants, underscoring the significance of language acquisition as a pivotal factor influencing the successful integration of newcomers into the intricacies of German societal norms and communication patterns (Heinemann,2018).

The process of Afghan immigrants adapting to Nordhausen's culture unfolds as a multifaceted journey shaped by numerous factors, encompassing language proficiency, social norms, values, traditions, and cross-cultural interactions. Striking a delicate balance between preserving elements of Afghan culture and assimilating into the host German society becomes a nuanced imperative in the complex and evolving adaptation process. This cautious interplay of cultural dynamics highlights the significance of navigating a path that embraces diversity while fostering integration, reflecting the broader challenges and opportunities inherent in the acculturation journey.

## RESEARCH METHODOLOGY

This study employs open-ended structured interviews with seven Afghan refugees in Nordhausen, fostering unrestrained dialogue to unravel the complexities of their trajectories and resilience. A major author of this chapter, an Afghan migrant, justifies 'participant observation' for immersive community engagement. The subsequent interpretive approach empowers participants to shape their narratives authentically, elucidating resilience perplexities and elevating data authenticity. This method's value extends to informing policymaking, providing crucial insights into Afghan refugees' lived experiences in Nordhausen. The study serves as an indispensable resource for policymakers, fostering informed and empathetic decisions, and establishing a robust foundation for meaningful enhancements in support structures and integration processes within the Afghan community.

## DATA COLLECTION

The research design employed in this study is characterized by the exclusive utilization of primary data, thereby enhancing its inherent organic and qualitative nature. The study specifically involves Afghan immigrants residing in Nordhausen (Germany) since August 2021, following the Taliban's takeover of Kabul. Purposive sampling was employed to select participants, guided by the researchers' discernment of their relevance to the study's objectives. This intentional approach ensures that individuals chosen possess insights crucial for addressing the research questions and contributing valuable perspectives to the topic (Campbell, 2020). The selection criteria prioritize participants who are not only willing to share their ideas but are also deemed highly pertinent in providing comprehensive information. This sampling strategy underscores the researchers' commitment to gathering meaningful and relevant data from individuals who can offer exclusive insights into the subject matter (Bigsby, 2018).

For this research, interviews were conducted with seven Afghan immigrants (males and females) to elicit their opinion on diverse points. The respondents are chosen in a way that they represent major geo-political zones and represents diversity. Individuals falling within the subaltern category are also included in the study cohort (Khan, Khan & Mukherjee, 2024). Interviews are conducted using an interview guide (Rapley, 2001), which consisted of open-ended questions, broadly focusing on the following themes: 1) Demographics. 2) Respondent's Life in Afghanistan 3) Escape and journey to Nordhausen, Germany. 4) Process of resettlement 5) Coping Mechanisms and Adaptive Strategies 6) Support and Integration in new country 7) Expectation from this research study.

**Data Limitations**: Given the nature of this study, obtaining responses from refugees proved challenging, primarily due to a linguistic barrier and apparent hesitancy among respondents to disclose comprehensive information at once. Consequently, the study relies on insights from seven participants representing diverse backgrounds. Despite these challenges, this research stands out as a valuable source, offering essential information crucial for comprehending the refugee crisis, its challenges, and the rehabilitation process. The unique perspectives shared by the participants contribute to a mature understanding of the complexities surrounding refugee experiences.

In consideration of the significant information elicited from the data, respondents' names are deliberately withheld and replaced with a unique 'code.' This method aligns with ethical standards, fostering a secure environment that encourages participants to candidly share their thoughts and experiences, assured of the protection of their privacy.

## DATA ANALYSIS AND INTERPRETATION

This study employs a qualitative data analysis approach, precisely scrutinizing the interview data to glean comprehensive insights. The process encompasses discerning underlying themes, patterns, and pivotal elements intrinsic to refugee narratives. Following this identification, the researcher constructs a cohesive narrative elucidating the discovered patterns and establishes connections between the identified key elements and the overarching research topic. In tandem with in-depth analysis, the researchers employ a data visualization method, employing tables, graphs, or charts, to distil and transform extensive and intricate information into a visual format. This strategic use of visualization enhances the accessibility and interpretability of the data, providing a comprehensive and effective understanding of the complex facets under scrutiny.

*Table 1. Demographic details of the participants*

| Participants | Age | Gender | Domicile Province in Afghanistan | Profession | Marital Status | Children | Educational Qualifications |
|---|---|---|---|---|---|---|---|
| Participant 'A' | 30 | Male | Kabul | Translator (Government Job) | Married | Yes | Master's degree (MA) |
| Participant 'B' | 33 | Male | Paktika | Security (Government Job) | Married | Yes | School Education |
| Participant 'C' | 34 | Male | Kunduz | University Teacher (Private Job) | Married | Yes | Master's degree (MA) |
| Participant 'D' | 34 | Male | Nangarhar | University Teacher (Private Job) | Married | Yes | Master's Degree (MA) |
| Participant 'E' | 27 | Female | Khost | Nurse (Private Job) | Married | Yes | Undergraduate Degree (Medicine) |
| Participant 'F' | 29 | Male | Kandahar | Security (Government Job) | Married | Yes | Master's Degree (MBA) |
| Participant 'G' | 33 | Female | Herat | School Teacher (Private Job) | Married | Yes | Undergraduate Degree (BA) |

(1) **Demographics**: Examination of demographic particulars uncovers intriguing insights. The respondents boast an average age of 31.4 years, underscoring a youthful and labour-capable cohort, promising long-term benefits for Germany. Notably, all participants possessed prior employment and educational qualifications in their homeland. A noteworthy gender contrast emerges, with Afghan men constituting most migrants. Compellingly, all migrant respondents are married with children, portraying distinctive familial dimensions within the migration dynamic.

(2) **Life in Afghanistan**: Regarding life in Afghanistan, a subset of respondents characterizes it as humble, while participants from Kabul and northern areas deem it decent, albeit within an enduring backdrop of fear and insecurity. Intriguingly, female respondents express a perception of greater happiness in their lives compared to their male counterparts. This divergence reflects the varied subjective experiences, underscoring regional disparities and gender-specific perspectives within the respondents' reflections on life in Afghanistan. Unanimously, all participants profess a belief in God, interpreting suffering as a test of faith. However, a noteworthy observation emerges as a minority, particularly among women, ascribe significance to the influence of fate and destiny in assessing life's events. This subtle divergence underscores the complex interplay of religious beliefs and personal philosophies, illuminating the distinctive perspectives within the participants' worldview.

(3) **From Afghanistan to Nordhausen**: The respondents have narrated their due course of migration as follows:

**Participant 'A'**: Participant 'A' experienced a relatively smooth migration, facilitated by his employment with a German NGO, which prompted his family's relocation to Germany just before the fall of Kabul. The immigration process was seamlessly supported by German authorities. However, challenges arise as he and his family grapple with linguistic barriers, navigating the complexities of adapting to their new cultural milieu.

**Participant 'B'**: Participant 'B' encountered significant migration hurdles due to family and financial constraints. Overcoming challenges at checkpoints controlled by the Taliban, he faced difficulties reaching the airport. Evacuated by Germany, he struggled to learn the German language but eventually found a sense of belonging. Germany has become his new reality, making Afghanistan a distant memory.

**Participant 'C'**: Following the collapse of the republic on August 15, 2021, widespread insecurity prevailed in Afghanistan. In the context of humanitarian evacuations driven by various political motives, Germany approached individuals for evacuation. The participant was part of Germany's humanitarian evacuation policy, leading him to Germany via a 24-hour stop in Saudi Arabia and a 15-day stay in Spain at the Rota Naval Base. This abrupt relocation induced psychological trauma and marked an unplanned journey toward an uncertain future in Europe. He perceives migration, after a decade of striving in his hometown, as a painful tragedy. Restarting life in a wholly different environment poses challenges comparable to adapting to the new life itself.

**Participant 'D'**: Fleeing Afghanistan due to threats from the Taliban, the participant, a vocal critic on national and international TV, face extreme challenges in securing visas for family members and escaping. Choosing secrecy, he navigated Kabul's checkpoints under the guise of heavy snow, reaching the airport before Taliban reinforcement. Luck played a role, and despite flight delays, he along with his family finally departed, evading stringent checks. In Islamabad, concerns of Pakistani interference proved unfounded, perhaps due to a simultaneous focus on COVID-19 and the evolving Afghanistan's situation. From Islamabad, he along with his family landed in Germany after ten hours of non-stop flight.

**Participant 'E'**: The immigration process proved straightforward due to respondent's husband's prior employment with the German organization GIZ. Following the Taliban takeover, GIZ contacted her husband, and the family's passport and personal details were submitted. Approval came on September 25, 2021, with Pakistan visas and flight tickets provided by GIZ's Islamabad office. She spent 45 days in Islamabad, where the German Embassy finalized her visa process. Charter flights facilitated her journey to Germany, with GIZ Islamabad ensuring accommodation and minimizing challenges, although surviving in Pakistan for 45 days was a bigger challenge.

**Participant 'F'**: Participant 'F,' in a pivotal role with Afghan security forces, faced upheaval during Nimruz's fall, prompting relocation to Kabul for reinforcements. Despite the Taliban's formidable backing from Pakistan, Participant 'F' anticipated a presidential strategy. Disillusioned with President Ghani's NSA, he strategically moved to Herat on August 10, 2021. Crossing into Iran, he found refuge with colleagues planning relocation. With precision, Participant 'F' aimed for Europe, navigating Turkey's challenges to reach Italy in November 2021. Communicating his background, International Refugee Agencies facilitated his relocation to Germany by April 2022. Six months of bureaucratic navigation secured official refugee status. Transitioning from Hamburg to Nordhausen, Participant 'F' leveraged German language skills for integration. Now concluding his language course, Participant 'F' eyes entry into German government services, embodying resilience amid challenges. His journey reflects adaptability and determination, showcasing the human spirit's pursuit of a better life against odds.

**Participant 'G'**: Participant 'G's' husband served as a translator for a German non-governmental organization, while she independently managed diverse German academic projects as a freelance professional, besides teaching at a private school. The familial decision to relocate to Peshawar in May 2021 was motivated by their astute assessment of the imminent downfall of the democratic republic. In proactive response, Participant 'G's' husband-initiated contact with the German embassy. Remarkably, within a brief period of 90 days, systematic diplomatic processes unfolded, culminating in the issuance of a visa that granted Participant 'G' the opportunity to embark on a journey to Germany. She is learning German language now and aspire to teach at German Universities after earning her PhD. She misses Afghanistan and her hometown every single day, wondering whether she might be able to return someday.

*Figure 1. A graphical depiction illustrating the magnitude of the migration journey*

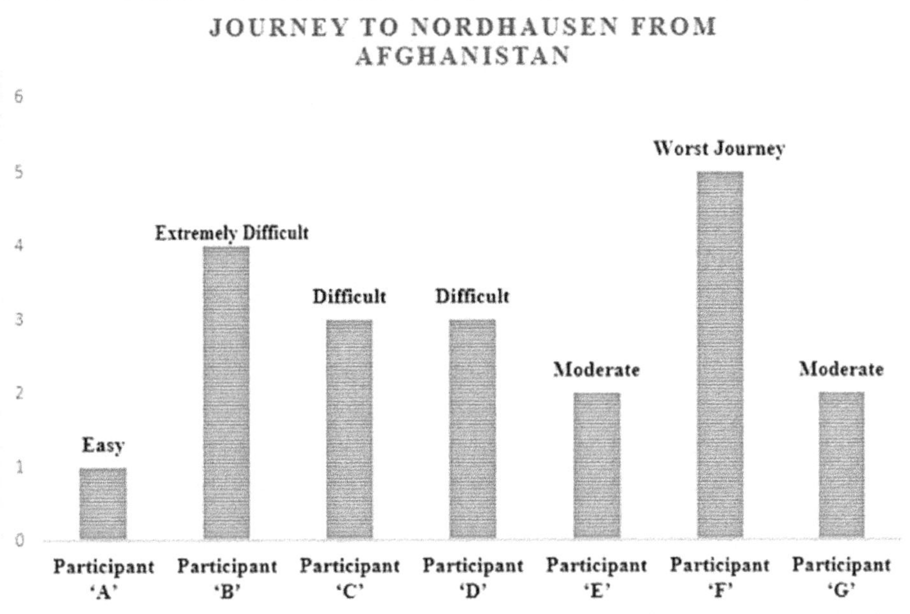


(4) **Process of Resettlement**: Expressing profound gratitude, all participants extend their appreciation to the German government, German aid agencies, and UN relief agencies for the unwavering support, warmth, and security extended to them. Notably, all respondents, except for Participant 'F,' have easily obtained settlement permits, a testament to the efficiency of the evacuation process orchestrated by Germany. A paramount challenge faced by these refugees revolves around acclimating to the complexities of the German language and seamlessly integrating into the cultural mainstream. This linguistic and sociocultural adaptation is pivotal for their holistic assimilation into the German society. Despite the success in securing refuge in Germany, the arduous journey is compounded by the plight of family members still stranded in Afghanistan. The respondents earnestly strive to navigate bureaucratic hurdles and ensure the rehabilitation of their kin within the German borders. Nordhausen has garnered unanimous favour among the respondents, attributed to its tranquil environment and rich historical significance. This affinity for the city underscores its appeal as a conducive setting for the refugees to rebuild their lives, fostering a sense of belonging and stability in the face of tumultuous pasts.

(5) **Coping Mechanisms and Adaptive Strategies**: Coping mechanisms and adaptive strategies are lifelines for refugees navigating the tumultuous terrain

of displacement. Forced to flee their homes due to conflict or persecution, refugees confront a myriad of challenges, both psychological and practical. Coping mechanisms become essential for mental and emotional resilience, enabling individuals to grapple with trauma and uncertainty. Meanwhile, adaptive strategies are crucial for survival, facilitating the integration into new societies by addressing language barriers, cultural differences, and the complex legal landscapes of host countries. These mechanisms not only empower refugees to rebuild shattered lives but also contribute to the fabric of diverse communities, showcasing the remarkable resilience of individuals who, in the face of adversity, forge pathways to recovery and renewal.

**Participant 'A'**: Participant 'A' underscores the importance of language acquisition as a key strategy for integration into the German community. In addition to mastering the language, he actively cultivates an appreciation for German culture by exploring its culinary delights, indulging in German films, and immersing himself in the country's music. When confronted with moments of melancholy, he seeks solace in the companionship of fellow Afghan refugees, engaging in heartfelt conversations with family members back in Afghanistan. Notably, his evolving interests extend to a newfound love for reading, a pursuit that was not prominent in his past.

**Participant 'B'**: Participant 'B' places a profound belief in harnessing inner strength and fostering connections within the community as indispensable tools for stress management. A distinctive aspect of his coping strategy involves a dedicated commitment to understanding and adhering to the legal framework, aspiring to be a law-abiding citizen. This commitment extends beyond a mere compliance with rules; Participant 'B' actively idolizes the principles of law, viewing them as a guiding force in shaping his identity within the community. Moreover, the acquisition of the German language stands as a cornerstone in his journey, not only instilling a sense of confidence but also fostering a profound sense of belonging.

**Participant 'C'**: Participant 'C' believes in engaging in a German language course, undertaking self-directed study, participating in sports activities, and forging new friendships as highly effective coping mechanisms. The acquisition of the German language not only facilitates communication but also enhances a sense of belonging in the host culture. Additionally, self-study endeavours contribute to personal growth and skill development, fostering adaptability in the face of new challenges. In tandem, involvement in sports serves as a holistic approach, promoting physical well-being and providing an avenue for social interaction. Equally crucial is the recognition that letting go of the past and embracing the current reality serves as a paramount adaptive mechanism. Acceptance of the present circumstances enables a more resilient response to the challenges of displacement, paving the way for a smoother integration into the host society.

**Participant 'D'**: Participant 'D' places paramount importance on communication with friends and family as a significant stress-relief strategy. In addition, he advocates embracing various subtle coping mechanisms, such as engaging in sight-seeing, family shopping outings, and dining outside, which collectively contribute to a sense of normalcy and well-being. Intriguingly, he finds an unconventional yet effective stress outlet in watching Hindi Cinema and Afghan TV, highlighting the diverse sources that individuals draw upon for comfort. Moreover, Participant 'D' underscores the pivotal role of language as a cornerstone in the adaptive process, recognizing its influence in fostering a connection with the new environment.

**Participant 'E'**: Participant 'E' underscores the paramount significance of 'spirituality' as a cornerstone in her coping arsenal. Delving into the realms of faith and spiritual practices serves as a profound source of solace and resilience. In tandem, she ardently advocates for the acquisition of language and understanding of the host culture as essential components of adaptive mechanisms.

**Participant 'F'**: Participant 'F' has skilfully navigated his passion for music and literature, channelling these interests into avenues of personal growth. In addition to cultivating these artistic pursuits, he has actively enrolled in an academy to hone the intricacies of acting, demonstrating a commitment to continuous learning and skill development. Beyond these creative endeavours, Participant 'F' firmly believes in embracing the host culture as the most effective adaptive mechanism. By immersing himself in the nuances of the local customs, traditions, and societal dynamics, he not only facilitates a smoother integration but also fosters a profound sense of belonging.

**Participant 'G'**: Participant 'G' holds a profound conviction in the transformative power of worship and cooking as pivotal stress-relief mechanisms and coping strategies. Recognizing the intrinsic link between culture and cuisine, she ardently believes that delving into the culinary traditions of a culture is a key element in the process of cultural assimilation. In alignment with this belief, Participant 'G' has actively undertaken a workshop focused on German Bakery products, demonstrating her commitment to not only mastering culinary skills but also immersing herself in the gastronomic aspects of German culture. By embracing these practices, she not only finds solace and resilience in her daily life but also actively contributes to the nuanced process of adopting and integrating into the cultural fabric of Germany.

*Figure 2. A graphical depiction illustrating coping and adaptive strategies*

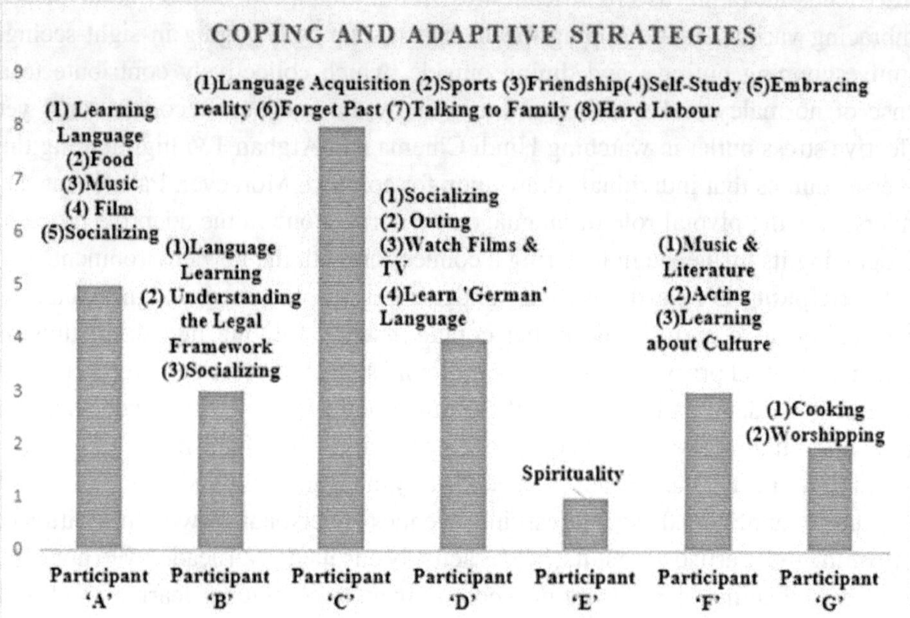


(6) **Recommendations for Support and Integration**: For effective integration and employment of Afghan refugees in Nordhausen (Germany), all the participants unanimously opined that a concerted effort between the German government and non-governmental organizations (NGOs) is necessary. The German government can initiate tailored language programs that go beyond basic language skills, incorporating vocational and technical training to enhance employability. Cultural orientation programs, supported by NGOs, can facilitate a smoother transition by offering insights into societal norms, values, and workplace dynamics. Financial assistance for educational and skill development initiatives, provided in collaboration with NGOs, would empower refugees to overcome barriers to entry into the job market.

NGOs can play a crucial role in establishing mentorship programs, connecting Afghan refugees with professionals in relevant industries. These mentorship initiatives provide not only career guidance but also valuable networking opportunities, contributing to a more seamless integration into the German workforce. Moreover, NGOs can offer targeted job placement services, linking refugees with employers seeking diverse talent. To address the potential challenges of credential recognition, the German government, in tandem with NGOs, can create pathways for recognizing foreign qualifications and experience, ensuring that the skills and expertise of Afghan refugees are appropriately acknowledged.

Collaborative awareness campaigns can combat stereotypes and prejudices, fostering a more inclusive environment. Legal support services provided by NGOs can ensure that refugees are informed about their rights, helping them navigate the complexities of the job market. By combining these measures, the German government and NGOs can establish a comprehensive support system, empowering Afghan refugees not only to integrate successfully into German society but also to contribute meaningfully to the nation's workforce and cultural tapestry.

Moreover, it is imperative for the German government to implement substantive reforms aimed at optimizing the asylum-seeking process, recognizing the pragmatic constraints faced by refugees who, upon arrival, may lack essential documentation. The compounded ramifications of the Ukrainian and Afghan conflicts have exerted considerable strain on both governmental structures and societal resources in Germany, precipitating a perceptible shift in focus following the influx of Ukrainian refugees. Nevertheless, the predicament of Afghan refugees, particularly in the realm of family reunification, remains a formidable challenge. The instigation of a more empathetic and all-encompassing policy framework in this domain would not only ameliorate bureaucratic impediments but also provide a substantial psychosocial support network for Afghan individuals contending with the ongoing separation from loved ones in Afghanistan. While Nordhausen assumes a relatively modest scale in comparison to urban centres like Munich, Berlin, and Hamburg, it is incumbent upon the authorities to ensure that the services rendered to refugees within its precincts adhere to uniform standards. These integrated approaches emphasize the exigency of a comprehensive and all-encompassing strategy to effectively address the multifaceted needs of Afghan refugees as they seek sanctuary and opportunities for resettlement in Germany.

(7) **Interrogating Pedagogical Hurdles: Exploring Educators' Challenges and the Evolving Landscape of Education:** The Taliban's resurgence in Kabul has precipitated a crisis in education, with scholars compelled to flee and curricula reshaped to align with Taliban's interpretation of Shariah law. This upheaval disproportionately impacts girls, who are systematically denied access to higher education. In this turbulent environment, educators face unprecedented challenges in upholding academic standards and fostering intellectual inquiry. The evolving educational landscape necessitates a rigorous interrogation of pedagogical hurdles, highlighting the urgent need for international solidarity to support educators, protect academic freedom, and advocate for inclusive education policies. Only through collective action, Afghans can mitigate the

detrimental effects of Taliban rule and safeguard the fundamental right to education for all.

**Participant 'A':** Participant 'A' stated that educators in Afghanistan faced insurmountable challenges under Taliban rule, including limited academic freedom, censorship, and threats to their safety. These oppressive conditions forced many to flee to seek opportunities elsewhere. Under the Taliban's regime, the status of education is bleak, characterized by gender disparities, curriculum restrictions, and a lack of resources. Such conditions not only impede educational progress but also perpetuate a cycle of ignorance and repression, exacerbating the exodus of skilled professionals.

**Participant 'B':** Participant 'B' believes that the Taliban's imposition of strict Islamic teachings limits the curriculum, undermining efforts to provide a diverse and comprehensive education. This restricts critical thinking and stifles intellectual development among students. Being an ex-security personnel, Participant 'B' believes that there are broader concerns about the implications of Taliban-controlled education for global security. A lack of education or the propagation of extremist ideologies within the educational system could contribute to radicalization and the spread of terrorism beyond Afghanistan's borders.

**Participant 'C'**: Participant 'C' being a university teacher himself, highlighted the detrimental impact of Taliban's policies on education. The imposition of a rigid curriculum by Taliban authorities has eroded educational quality, hampering critical thinking and stifling innovation. Furthermore, an atmosphere of pervasive fear and intimidation has curtailed academic freedom, discouraging educators from presenting diverse perspectives. Insufficient resources and inadequate infrastructure have compounded these challenges, perpetuating cycles of poverty and ignorance. Taliban's prioritization of ideological indoctrination over education poses significant threats to intellectual development and societal progress. As a result, the educational sector is particularly vulnerable, facing immense obstacles under Taliban rule. The urgent need to address these challenges is paramount to safeguarding the future of education in Afghanistan and promoting a more inclusive and prosperous society.

**Participant 'D'**: Participant 'D' reflects on the tragic events of August 15, 2021, marking the fall of Kabul, a day etched in infamy within Afghanistan's historical narrative. In the aftermath, the once revered profession of teaching, often perceived as humble and undervalued within society, now faces an ominous future under Taliban rule. Participant 'D' finds himself haunted by profound psychological distress, grappling with the devastating collapse of educational institutions. Even amidst seeking solace in Nordhausen through therapy sessions, his resilience remains steadfast, buoyed by a singular determination: to establish an 'Online

Open University for Afghans,' officially registered under the aegis of the Ministry of Education in Germany.

**Participant 'E'**: Participant 'E', being an erstwhile nurse, stated that under Taliban's regime, nursing education in Afghanistan has encountered formidable challenges, particularly with restrictions on female education and the limited curriculum enforced by the Taliban's stringent interpretation of Islamic law. Female nursing students and teachers face threats to their safety, inhibiting their ability to pursue education and contribute to the nursing workforce. This situation exacerbates the shortage of skilled nurses and impedes efforts to improve healthcare infrastructure. Concurrently, female doctors endure suffering due to limited employment opportunities, threats to personal safety, and obstacles to continuing education and professional development. The scarcity of female healthcare professionals exacerbates disparities in healthcare access and quality, especially for women in conservative communities who may be reluctant to seek medical care from male providers, thus amplifying the healthcare challenges faced by Afghan women under Taliban rule.

**Participant 'F'**: Participant 'F', a former security officer in Afghanistan before the Taliban's assumption of power, offers a profound assessment of the implications of this regime change, specifically focusing on the Ministry of Education. His analysis highlights a deep-seated concern regarding the potential trajectory towards a curriculum excessively imbued with religious teachings, potentially overshadowing critical areas of scientific exploration and technological innovation within the educational framework. Furthermore, Participant 'F' identifies a conspicuous absence of a comprehensive strategy within the Taliban administration to cater to the educational requirements of women, minorities, and other marginalised categories. The widespread departure of skilled educators, driven by a combination of unpaid salaries and the growing influence of the Taliban within educational institutions, foreshadows a bleak future characterized by a scarcity of capable instructors and a subsequent decline in the quality of education provided.

**Participant 'G':** Participant 'G', herself a teacher, emphasized the dire consequences of Taliban rule on education. With the closure of all girls' schools, 50% of the population is deprived of education, leaving female teachers unemployed, some of whom were sole breadwinners. Many have fled to seek refuge in Europe and the US, while those remaining endure depression and trauma, with reports of suicide attempts. Boys' education also suffers due to Taliban's substandard curriculum. Participant 'G' urges UNICEF to finance Afghan schools, potentially pressuring Taliban to reconsider their stance on education, particularly for girls. Since Taliban deems education non-productive, UNICEF's financial commitment could incentivize a change in policy, offering hope for education in Afghanistan and alleviating the plight of teachers and students alike.

(8) **Expectation From This Research Study**: Collectively, all participants express a unanimous consensus on the imperative continuity of research concerning refugees and their challenges. They advocate for a direct, firsthand engagement between researchers and refugees, emphasizing the reliance on primary data over secondary sources to authentically capture the intricacies of their experiences. Furthermore, the participants assert that the outcomes of such studies should be disseminated extensively, urging the German government's active attention. They emphasize that researchers must comprehensively analyse and comprehend their struggles, underscoring the transformative potential of this research in reshaping the lives of those residing in Nordhausen. The participants contend that this research, anchored in firsthand experiences and disseminated widely, holds the potential to act as a catalyst for positive change and draw the attention of relevant authorities toward addressing the unique challenges faced by refugees in Nordhausen.

## CONCLUSION

In the culmination of this exhaustive research endeavour, encapsulating the subtleties of escape and resilience among Afghan refugees in Nordhausen, a profound narrative mosaic emerges, delineating the profound depths of human experience in the face of displacement. The methodological precision employed in this study, anchored in an intricate blend of qualitative methodologies, has unravelled narratives that extend beyond mere anecdotal recollections, transcending into the realm of sociological and psychological inquiry. The refugees' stories unveil a multifaceted panorama, navigating the treacherous terrain of escape, adaptation, and resilience. The persistent echoes of trauma, the indomitable spirit of survival, and the complex interplay between personal agency and external constraints are systematically delineated. The integration of first-hand accounts and primary data sources not only bolsters the credibility of the findings but also amplifies the voices of those often marginalized in mainstream discourse. As the narratives unfold, a nuanced understanding of the myriad challenges faced by Afghan refugees crystallizes, elucidating the socio-cultural, economic, and psychological dimensions of their journey

Moreover, this research illuminates the pivotal role of resilience as both a psychological construct and a tangible force shaping refugees' adaptive strategies. The study discerns how personal agency, community networks, and cultural preservation act as salient facets of resilience, influencing the trajectory of refugee experiences. The narratives underscore the dynamic nature of resilience, revealing its transformative potential in mitigating the adverse effects of displacement and fostering empowerment. Furthermore, the research underscores the necessity of elaborative and context-specific interventions that acknowledge the intersectionality of refugee identities and experiences, as well as the multifaceted nature of resilience.

As this research endeavours to transcend the academic realm, its implications resonate on a broader societal and policy-oriented spectrum. The documented narratives, rich in both tribulation and triumph, constitute a compelling advocacy tool, imploring policymakers, humanitarian agencies, and advocacy groups to reassess existing frameworks and cultivate more comprehensive, empathetic responses to the multifaceted needs of Afghan refugees in Nordhausen. By amplifying these narratives, this research acts as a catalyst for social change, challenging prevailing stereotypes and fostering a more informed, compassionate understanding of the refugee experience.

In essence, this study extends beyond the confines of a conventional book chapter. It serves as a testimony to the resilience of the human spirit, an exploration of the human condition amidst adversity, and a clarion call for a paradigm shift in the approach towards refugee integration. The meticulousness of the research design, the richness of the narratives uncovered, and the theoretical contributions to the fields of migration studies and resilience research collectively position this study as a seminal exploration that not only deepens academic understanding but also propels a transformative discourse on the experiences and agency of Afghan refugees in Nordhausen.

## REFERENCES

*Afghanistan civilian deaths hit record high in 2018: UN.* (2019, February 24). Aljazeera. https://www.aljazeera.com/news/2019/2/24/afghanistan-civilian-deaths-hit-record-high-in-2018-un

*Afghanistan security situation update: country of origin information report: September 2021.* (2021). European Union Agency for Asylum. https://euaa.europa.eu/publications/coi-report-afghanistan-security-situation-update-september-2021

*Afghanistan's refugees: forty years of dispossession.* (2019, June 20). Amnesty International. https://www.amnesty.org/en/latest/news/2019/06/afghanistan-refugees-forty-years/

AlbarosaE.ElsnerB. (2023, January 06). *Forced Migration and Social Cohesion: Evidence from the 2015/16 Mass Inflow in Germany.* Social Science Research Network. https://doi.org/ doi:10.2139/ssrn.4319003

Atakpal, H. (2019, June 14). *Afghanistan Ranked World's Least Peaceful Nation*. Tolo News. https://tolonews.com/afghanistan/afghanistan-ranked-world%E2%80%99s-least-peaceful-nation

Bigsby, E. (2018). *The SAGE Encyclopedia of Communication Research Methods*. SAGE Publications, Inc. doi:10.4135/9781483381411

Braithwaite, A., Salehyan, I., & Savun, B. (2019). Refugees, forced migration, and conflict: Introduction to the special issue. *Journal of Peace Research*, *56*(1), 5–11. doi:10.1177/0022343318814128

Campbell, S., Greenwood, M., Prior, S., Shearer, T., Walkem, K., Young, S., Bywaters, D., & Walker, K. (2020). Purposive sampling: Complex or simple? Research case examples. *Journal of Research in Nursing*, *25*(8), 652–661. doi:10.1177/1744987120927206 PMID:34394687

Connah, L. (2021). US intervention in Afghanistan: Justifying the Unjustifiable? *South Asia Research*, *41*(1), 70–86. doi:10.1177/0262728020964609

Dupree, L. (1984). Afghanistan in 1983: And Still no Solution. *Asian Survey*, *24*(2), 229–239. doi:10.2307/2644442

Farahani, H., Langari, M. N. M., Eliasi, L. G., Tavakol, M., & Toikko, T. (2023, April 18). "How Can I Trust People When They Know Exactly What My Weakness Is?" Daily Life Experiences, and Resilience Strategies of Stateless Afghans in Iran. *Journal of Immigrant & Refugee Studies*, 1–16. doi:10.1080/15562948.2023.2199252

*Germany: Immigrants made up over 18% of 2022 population.* (2023, April 20). Deutsche Welle.https://www.dw.com/en/germany-immigrants-made-up-over-18-of-2022-population/a-65383249

Heinemann, A. M. B. (2018, March 22). The making of 'good citizens': German courses for migrants and refugees. *Studies in the Education of Adults*, *49*(2), 177–195. doi:10.1080/02660830.2018.1453115

Jazayery, L. (2003). The Migration-Development Nexus: Afghanistan Case Study. *International Migration (Geneva, Switzerland)*, *5*(40), 231–254. doi:10.1111/1468-2435.00218

Khan, S. U., Khan, S., & Mukherjee, T. (2024). The "Subaltern" Will Speak: Investigating Portrayals in the Acclaimed OTT Series of India and Pakistan. In N. Kalorth (Ed.), *Exploring the Impact of OTT Media on Global Societies* (pp. 1–24). IGI Global. doi:10.4018/979-8-3693-3526-0.ch001

Naseh, M., Potocky, M., Stuart, P. H., & Pezeshk, S. (2018, October 09). Repatriation of Afghan refugees from Iran: A shelter profile study. *Journal of International Humanitarian Action, 3*(13), 1–12. doi:10.1186/s41018-018-0041-8

Natarajan, A., Moslimani, M., & Lopez, M. H. (2022, December 16). *Key facts about recent trends in global migration.* Pew Research Centre. https://www.pewresearch.org/short-reads/2022/12/16/key-facts-about-recent-trends-in-global-migration/

Nedelsky, N. R. (2023). A New German 'We'? Everyday Perspectives on Germanness and its Boundaries. *Nationalities Papers, 51*(6), 1215–1234. doi:10.1017/nps.2022.106

Rapley, T. J. (2001). The art(fulness) of open-ended interviewing: Some considerations on analysing interviews. *Qualitative Research, 1*(3), 303–323. doi:10.1177/146879410100100303

Ruttig, T. (2017, February 17). *Afghan Exodus: Afghan asylum seekers in Europe (3) – case study Germany.* reliefweb. https://reliefweb.int/report/germany/afghan-exodus-afghan-asylum-seekers-europe-3-case-study-germany

Sadiq, A. (2022). Leading Literate Lives: Afghan Refugee Children in a First-Asylum Country. *Journal of Literacy Research, 54*(1), 28–50. doi:10.1177/1086296X221076431

Schuetze, C. F. (2023, September). A Small German City Staves Off a Threat From the Far Right. *The New York Times.* https://www.nytimes.com/2023/09/24/world/europe/germany-afd-mayor-election.html

Tovar-Gálvez, J. C. (2022). Acculturation in the Integration Course for Immigrants in Germany: A Study from the Cultural Bridge Approach. *Journal of Ethnic and Cultural Studies, 9*(3), 72–91. https://www.jstor.org/stable/48710351. doi:10.29333/ejecs/1061

*UNHCR Projected Global Settlement Needs 2023.* (2023). UNHCR. https://www.unhcr.org/sites/default/files/legacy-pdf/62b18e714.pdf

VadeanF. P. (2007). *Skills and remittances: The case of Afghan, Egyptian, and Serbian immigrants in Germany.* Social Science Research Network. https://dx.doi.org/ doi:10.2139/ssrn.1317135

Willner-Reid, M. (2016). IDPs in Afghanistan: A Confused National Glossary. *Refugee Survey Quarterly, 35*(4), 79–96. doi:10.1093/rsq/hdw013

## ADDITIONAL READING

Aikins, M. (2022, February 15). *Naked Don't Fear Water: An Underground Journey with Afghan Refugees*. Harper.

Alimia, S. (2022). *Refugee Cities: How Afghans Changed Urban Pakistan*. University of Pennsylvania Press.

Chatty, D. (2010). *Deterritorialized Youth: Sahrawi and Afghan Refugees at the Margins of the Middle East: 29 (Forced Migration)*. Berghahn Books.

Ghufran, N. (2010). *Politics of Afghan Refugee Repatriation from Pakistan*. VDM Verlag.

Jehangir, A. (2023). *Afghan Refugees, Pakistani Media and the State: The Missing Peace (Routledge Research in Journalism)*. Routledge. doi:10.4324/9781003325291

Kuspanov, A. (2016). *UN programs for Afghan refugees in Kazakhstan*. LAP Lambert Academic Publishing.

Mahfouz, S., & Kapoor, M. (2023). *Defiant Dreams: The Journey of an Afghan Girl Who Risked Everything for Education*. Doubleday.

Miller, C., & Shawali, L. (2024). *The Alterations Lady: An American, an Afghan Refugee, and the Stories that Define Us*. Apollo Publishers.

Nasiri, A. (2022). *Forced to Flee: A Tale of Two Afghan Refugees*. Page Publishing Inc.

Ulrich, A. M. (2012). *Lifting the Chaderi: My Life as an Afghan Refugee*. Two Worlds Press.

## KEY TERMS AND DEFINITIONS

**Asylum:** Asylum, an essential humanitarian concept, signifies the refuge and protection granted by a nation to individuals fleeing persecution. It embodies the sanctity of shelter, embodying the profound principle of international solidarity.

**Deportation:** Deportation is the lawful expulsion of individuals from a country, typically due to visa violations or criminal activities. It involves the state's formal removal of non-citizens, enforcing legal immigration protocols.

**Internally Displaced People:** Internally Displaced People (IDPs) are individuals forcibly uprooted within their own country due to conflict, persecution, or natural disasters. Their plight reflects the poignant narrative of resilience amid societal upheavals.

**Lingual Barrier:** Lingual barrier signifies the impediments hindering effective communication between individuals due to linguistic differences. It encompasses challenges related to grammar, vocabulary, and pronunciation, obstructing seamless understanding and expression in diverse language contexts.

**Nation:** A nation is a complex, culturally interconnected community bound by shared history, values, language, and often a common territory. It fosters a collective identity, with citizens expressing a sense of solidarity.

**Refugee:** A refugee is an individual forcibly displaced from their homeland due to persecution, conflict, or distressing circumstances, seeking sanctuary in a foreign land, often marked by profound human resilience and hope for a secure future.

**Social Assimilation:** Social assimilation embodies the intricate process wherein individuals from diverse cultural backgrounds integrate seamlessly into a host society. It encompasses the harmonious blending of values, customs, and social interactions, fostering inclusive cohesion.

**UNAMA:** UNAMA, the United Nations Assistance Mission in Afghanistan, is an organization facilitating peace, stability, and development in Afghanistan. It provides diplomatic support, monitors human rights, and aids in humanitarian efforts.

**UNHCR:** The UNHCR, or United Nations High Commissioner for Refugees, epitomizes global commitment to safeguarding displaced populations. As a beacon of compassion, it champions the rights, protection, and well-being of refugees worldwide.

**Welfare State:** A welfare state is a societal system where the government assumes responsibility for citizens' well-being, providing services like healthcare, education, and unemployment benefits, aiming to ensure social and economic equity.

# Chapter 4
# Changing and Adapting Curriculum While Teaching:
## Narratives From Afghan Academics and Students

**Sara Mohammed Aldous**
*Liwa International School for Gils, UAE*

**Ghaith Ahmad Ismail**
*Madar International School, UAE*

## ABSTRACT

*Education is vital for professional development and is a fundamental human necessity. Unfortunately, individuals in many nations lack access to high-quality education. This study explores curriculum adjustments in Afghan education, emphasizing the narratives of academics and students. To meet evolving needs, institutions must prioritize ongoing curriculum modifications for improved educational quality. Afghan public universities can benefit from incorporating outcome-based education and student-centered learning. Local governments must understand displaced people's needs, adapting the curriculum accordingly. Implementing changes may pose psychological and cultural challenges for students and teachers. International collaboration is crucial for successful curriculum adoption. Decision-makers should focus on developing flexible curricula to address displaced individuals' unique needs. The resilience of Afghan educators and students showcases the human spirit's power. Collaborative efforts can break barriers, fostering a globally accessible education landscape.*

DOI: 10.4018/979-8-3693-1483-8.ch004

# 1. INTRODUCTION

## 1.1 Brief Overview of the Challenges Faced by Afghan Academics and Students in the Context of War and Displacement

Education is utmost important for the professional growth of the individuals and also it is being considered as a basic human need. But in some of the countries the individuals are not able to receive good quality education due to multiple reasons. This essay elaborates about the changing and adapting curriculum while teaching: Narratives from Afghan Academics and Students". There are different challenges which are being faced by the Afghan academics and students due to the wars and the displacements as well (Gindeel and Etim, 2014). Due to the long term of conflicts in Afghanistan the education institutions has faced issues related to funding and destruction. When it comes to displacement then it has affected the overall education aspirations of millions of students due to the regular conflicts and disruptions (Arar *et. al.* 2020). Along with this the refugee students also suffer a lot in getting quality education due to not so effective education system for Afghan academics. Adding to this the refugee students also face countless issues related to language barriers, discrimination and several other issues which is affecting their opportunities to have a good professional career.

## 1.2 The Importance of Curriculum Adaptation in Ensuring Continuity of Education

For adapting the changing environment it is important for the education institutions to focus on curriculum adaptation for improving the quality of education continuously. Along with this curriculum adaptation helps in modifying the teaching materials, methods and content for meeting the learning needs of different students (Bannova and Mayorova, 2019). Moreover it can be said that regular curriculum adaptation is also important for offering up to date knowledge to the students, adapting the latest technologies and to adjust with the uncertainties such as Covid-19 pandemic (Gelmez-Burakgazi, 2020). Thus for offering the high quality education it is necessary to focus on curriculum adaptation.

# 2. THE AFGHAN EDUCATIONAL LANDSCAPE: A HISTORICAL CONTEXT

## 2.1 A Brief History of the Afghan Educational System

As per Alamyar (2018) there was no such education system in Afghanistan until 1875. The first two formal schools have being built in Afghanistan under the ruling of Amir Shir Ali Khan. Adding to this more schools was being built under the leadership of King Habibullah Khan. Following this the King Amanulla Khan (1919-1929) has made the primary education compulsory for all which has bought great awareness related to education among the Afghan people. Under this period only the Ministry of Education has been set in Afghanistan and has started paying attention over offering better and good quality education to all. In later part of the report the author also stated about the situation of education system of Afghanistan during Taliban regime (1996-2001) during this time education is banned for females except the Kabul Medical Faculty. But later on in 2001 the Taliban regime has issued an education law under which everyone is having the right to education and for women education special laws were being developed in accordance with Sharia.

In opposite of this Hedayati, (2023) has argued that no Sharia barrier is there for education for women. Also the author has argued that the Islamic texts promote the necessity of women education and due to this reason it should not be banned as per Taliban regime. For having a modern civilized society in Afghanistan it is very important to have well educated people and the education system should promote education to all (Alqodsi, 2023).Khwajamir (2016) has argued that after the collapse of Taliban the government and the international communities are making continuous efforts to reform the education system in Afghanistan. After all this the concept of modern education with well-defined system is being set up in the Afghanistan education system.

## 2.2 The Impact of War on Education and Curriculum Development

Slattery (2012) has stated that the situations related to war in the country always had an adverse influence on the education system and the overall development of the children of the country. In the case of Afghanistan the war has always affected the education due to multiple reasons. According to Deo (2014) the conflict leads by Taliban regime has affected the education of females in Afghanistan to a greater extent. As a result of the education ban on women has created the gender disparities in education system of Afghanistan. Not only this there is great politics related to education in Afghanistan and due to this reason less attention is being paid over prioritising education for all. Karlsson and Mansory (2008) has added that due to war situations in Afghanistan brain drain is the major issue due to shortage of expertise and qualified educators. As a result of ongoing wars and conflicts the people especially the teaches, academics or professors had flee the country which leads to the overall shortage of qualified educators (Alqodsi et al, 2023).

In the view of Spink (2005) the war related situations has affected the overall development of the modern education system in Afghanistan. War has led a major loss to the education infrastructure which created more challenges to the education system. Following this Burde (2014) has argued that the curriculum also not get updated on time due to the situations related to war. Along with this the manipulated ideology related to education and political pressure has limited the students exposure to get high quality education. But for the rebuilding of nation it is very important for the Afghan government to pay attention over better education. Welch and Wahidyar, (2020) has mentioned in their report that chronic underfunding is one of the major challenges for the Afghan academics as a result of which they are not able to get better higher education in Afghanistan. Along with this the author has highlighted that the limited funding, limited capacity of the ministry of education and increasing level of insecurities has affected the overall quality education in Afghanistan. In addition to this the on-going wars has also lead issues related to the infrastructure of the universities as the campuses are continuous being used for the military purpose.

## 3. NARRATIVES OF DISPLACEMENT: THE IMMEDIATE CHALLENGES

### 3.1 Personal Accounts From Afghan Students and Faculty on the Initial Challenges Faced During Displacement

According to Brenner and Kia-Keating, (2016) the refugees and especially the youth under the age of 18 face major challenges related to education due to displacement. Here it is important to understand that the students and their parents face issues related to the linguistic and cultural differences between their own country and the host country as a result of which students face a lot of challenges. Shakya *et. al.* (2010) has stated that the major challenge which is being faced by the Afghan student is related to the unfamiliar hostile environment. Along with this the students has faced issues related to discrimination, language barriers and find difficulties in getting education in the host countries (Alqodsi & Aljahoori, 2023). The refugee students of Afghanistan has faced great discrimination by other students while getting higher education in Toronto which has become a major challenge for them to adjust into a host country.

Teachers and faculty like all the other Afghans has faced the challenges related to the displacements, insecurity and instability. As per Grawert and Mielke, (2018) the Afghans has faced major issues in securing their livelihood in the host countries like Iran and Pakistan. Following this displacements the faculty has also lost their jobs and has faced a lot of challenges in getting new jobs in the host countries initially. Along with this during the initial time of displacement also the faculty member got least support from the education institutions and has loss their professional identity as they are not able to use their knowledge and expertise.

### 3.2 The Urgency of Adapting to New Educational Environments

Due to the conflict and instability many people have being displaced to the host countries as a result of which they are facing major educational challenges. Smith (2009) has argued that for the Afghan refugees it is necessary to rebuild their lives as per the new environment of the host country. Especially for the students it is necessary to adapt the new environment as education can help them in having a sense of normalcy in the new country (Marshall *et. al.* 2016). Along with this education is an fundamental human right and the Afghan students need to get quality education in the host country so that they can add to the host communities.

According to Pincock *et. al.* (2021) there are different international communities which are working hard for offering and supporting education to the displaced population including the Afghan refugees as education for all is a basic human right. Thus it can be said that adjusting to the new educational environment is paramount importance for Afghan academics and students for their well-being and realisation of their full potential as well. In support to this Cerna, (2019) has mentioned that the education institutions in different countries need to have more inclusive and flexible language support for helping the Afghan refugees and also the refugees from around the world to adapt to their new educational environment more easily.

## 4. CURRICULUM ADAPTATION: NECESSITY AND STRATEGY

### 4.1 The Reasons Behind the Need to Adapt the Curriculum

Based on the views of Oates *et. al.* (2019) adapting and developing the curriculum within Afghanistan is a crucial doing for educators and pupils alike, provided the disputed past of the nation as well as the constantly shifting socio-political landscape. However, there are multiple factors that make this requirement necessary. The requirement for curriculum adaptation arises as a critical component in the search of relevant and successful education in the wake of persistent wars, political upheavals, and educational obstacles. On the critical note, Easar *et. al.* (2023) has mentioned that the current curriculum, created under various conditions and systems, may no longer be pertinent and align to modern educational needs. This discrepancy can affect the development of critical abilities and expertise between students. One of the primary reasons beneath changing the curriculum lies in the need to address current educational needs and international norms.

Corrales *et. al.* (2020) stated the fact that curriculum must also meet the requirements of pupils with various origins and goals while showcasing the variety of Afghan society. Although the current curriculum may be viewed as being overly strict, a revised curriculum can include multicultural viewpoints, languages, and a wider range of courses, promoting inclusion and enhancing the educational process. On the contradictory note, Alariqi *et. al.* (2019) defined that the process of adapting is not devoid its difficulties. The political environment and resource limitations can make it difficult to execute adjustments in a timely and efficient manner. The procedure can become more difficult if there aren't enough trained teachers to teach the updated curriculum or if there aren't enough suitable teaching resources. On the other side, some would contend that changing the curriculum could result in the loss of historical or cultural legacy. Creating a balance between maintaining a sense of culture and embracing development is challenging for ministry of education in Afghanistan (Corrales *et. al.* 2020). By giving students an educational experience that both honours the past and equips them for the future, the curriculum modification can be a chance to close the gap between modernity and the past.

## 4.2 Strategies Employed by Afghan Academics to Modify Curriculum Content to Suit New Environments and Student Needs

As mentioned by Nazari *et. al.* (2021) the difficult burden of adapting curricular material to different circumstances and changing student requirements has been placed before Afghan educators. Contextualization is a key tactic used by Afghan scholars. They are aware that the curriculum has to be modified to reflect the particular social context of Afghanistan. Finding a balance between adhering to international educational norms and including material that is pertinent to Afghan society is not so easy. In opposite to this, Katawazai (2021) defined that is a difficulty for Afghanistan to adapt the modern education system and to make changes in curriculum. With the aid of this technique, students may learn more about their own history and culture while still learning knowledge that is useful across borders. The use of regional expertise and customs is a crucial additional strategy. Afghan scholars have discovered methods to include indigenous viewpoints and traditional knowledge into the curriculum. This enhances learning and contributes to the preservation of cultural heritage. However, on a more serious note, there is a danger of supporting outmoded or superstitious notions, necessitating a careful consideration of the curriculum's content.

According to the views of Pillai (2020) the academic community in Afghanistan has also embraced technology to modify its curriculum but still looking forward towards tech-solutions. They have been able to overcome some of the difficulties brought on by limited resources and unstable political environments because to access to internet resources and digital platforms. Students' perspectives have been broadened by online classes or educational resources, which provide them access to a wider variety of disciplines and enable distance study. However, Mahdi and Wani, (2021) argued that issues including digital devices and insufficient infrastructure still need to be resolved. Collaboration is another tactic used by academics in Afghanistan. They have been able to access a variety of knowledge and resources by cooperating with foreign partners, non-governmental organizations, and educational institutions. These partnerships have produced greater access to educational resources, teacher training programmes, and curricular improvements. However, people worry that these alliances might undermine national sovereignty and have an unfavorable impact on the curriculum according to Afghan goals.

## 5. CASE STUDY: A UNIVERSITY'S APPROACH TO CURRICULUM ADAPTATION

### 5.1 Empirical Data Showcasing a Specific University's Approach to Curriculum Adaptation for Afghan Students

A study of Katawazai, (2021) has stated about the adaption of outcome based educations and student centred learning in the curriculum of the Afghan Public Universities. Also the study has highlighted the attitude of the teachers and the major challenges which are being faced by them while adapting the changes related to OBE-SCL learning. The results of the study have showed that teachers exhibit positive attitudes and are inclined to integrate this approach into their teaching and learning methods (Abdallah et al, 2023). Teachers who responded positively were those who had undergone training but those who have not supported this approach are the one who has not received any kind of training related to this. Based on this it can be said that there are challenges related to lack of training and lack of resources when it comes to adaptation of new approaches (Amouzeshi *et. al.* 2021). Here it is necessary to understand that the universities need to make more efforts for implementing the changes and adapting something new in the curriculum.

While it's evident that OBE-SCL is still in its initial phases within the Afghan context, there is a clear need for pay attention and dedication at the governmental and organizational levels as it progresses towards subsequent implementation stages. In support to this Otukile-Mongwaketse *et. al.* (2016) has stated that for implementing any of the change in the curriculum it is necessary to have effective support and appropriate resources as well. Not only are this collaborative efforts being required by the government, educational institutions and the ministry of education for making it more successful.

### 5.2 Challenges Faced and Solutions Implemented

Some of the major challenges which were being identified while making changes in the curriculum and adaptation are related to lack of efforts by the ministry of Afghanistan and lack of effective resources as well (Himat *et. al.* 2021). For the implementation of the OBE-SCL approach in the universities the solution which is being proposed in the study that. For a more effective implementation of OBE-SCL in Afghanistan, the Ministry of Higher Education needs to initiate the accurate definition of teaching duties and responsibilities for the academic staff. This entails determining the appropriate number of classes each teaching staff should handle (Akramy, 2021). Moreover, in cases where universities need to alleviate the teaching workload of lecturers, they should be encouraged to announce teaching vacancies. Simultaneously, the Ministry should issue an official directive to all public universities, compelling them to adopt OBE-SCL without exceptions and to underscore its significance in the educational context.

## 6. THE ROLE OF PARTNERSHIPS: LOCAL GOVERNMENTS AND HUMANITARIAN ACTORS

### 6.1 How Collaborations Between Educational Institutions, Local Governments, and Humanitarian Actors Facilitated Curriculum Adaptation

The regions which have faced the issues related to conflict and displacement had to redefine their education system with better curriculum. According to Mentes and Talas, (2021) for adapting the new curriculum it is necessary to have better collaboration between the education institutions and local government. Local government need to pay attention over understanding the needs of the affected people during the displacements and then having more responsive curriculum. In addition to this better collaboration with the local government can help the educational institutions in setting up the curriculum which is more aligned to the cultural and contextual factors which is important for the countries like Afghanistan (Bird, 2009).

Rubin and Rudeforth, (2016) has stated that effective support from the humanitarian organisation help the educational institutions and local government in resource mobilization. For Ministry of Education of Afghanistan lack of resources and expertise is a major challenge which is limiting the potential to adapt the new curriculum with the changing educational landscape around the world (Burde, 2014). In addition to this the global educational institutions aligned with the humanitarian groups can help a lot in supporting with the expertise required for changing and adapting the curriculum in Afghanistan after displacement and crisis situation. Not only this the humanitarian organisation can help with effective funding through different nations and international organisations which is necessary for Afghanistan to have better education system and updated curriculum as well. As per Welch and Wahidyar, (2020) for making successful changes for quality assurance in Afghan higher education it is important to have better funding from the local government and the international humanitarian groups as well. In support to this Miszak and Motwani, (2021) has stated that most funding in Afghan for education purpose still comes from the foreign source as there is a better collaboration between the local government, educational institutions and the humanitarian agents across the world.

## 6.2 Best Practices From Around the World in Curriculum Adaptation for Displaced Populations

According to Gonchar *et. al.* (2019) for supporting the refugees in their learning and education it is necessary to have effective curriculum adaptation strategies. Having effective inclusive curriculum with attractive materials, interactive activities, blended learning and effective use of digital tools can really help in easy adaptation of the new curriculum. In support to this Miller *et. al.* (2022) has stated that having highly trained educators especially for the displaced students can really help them in adjusting to the new educational environment and new curriculum in an effective manner. For this the teachers should be specially trained so that better support can be offered to them. It can be said that there are diverse method and practices which are being adapted around the world for curriculum adaptation for displaced populations.

# 7. PSYCHOLOGICAL AND CULTURAL IMPLICATIONS OF CURRICULUM CHANGES

## 7.1 The Emotional and Cultural Challenges Faced by Students and Faculty Due to Curriculum Changes

In accordance with the views of Alam, (2022) for both students and professors, the adoption of a curriculum can result in a wide range of psychological and cultural difficulties. These difficulties, though frequently disregarded, have a significant impact on the educational setting. They may also be viewed as chances for development and adaptability, though. The anxiety brought on by adjusting to an entirely novel curriculum is one of the major emotional difficulties that pupils must overcome. Anxiety and uncertainty might be brought due to a drastic change in the subject matter, instructional strategies, or expectations. However, Thomas and Bigatti, (2020) argued that as kids learn to manage change and grow more robust in the face of difficulties, this mental health issue can promote personal development. As students successfully acclimatize to the new instructional framework, it may help promote a sense of accomplishment.

Jenkins (2020) described when a curriculum fails to effectively reflect the different backgrounds of pupils, cultural problems occur. Students from disadvantaged or marginalized groups may find it difficult to find their place in the curriculum, which can cause feelings of isolation. Nevertheless, addressing these cultural differences gives educators a chance to encourage inclusion and diversity. The curriculum may be used as a tool for cultivating empathy and understanding among students from varied cultural origins by embracing different points of view. When putting a new curriculum into practice, faculty members also experience emotional difficulties. If they enjoyed a close relationship to the prior curriculum, they might experience a feeling of loss. It may be emotionally exhausting to live in fear of shifts and unpredictability. On the critical note, Clemens *et. al.* (2020) said that these difficulties can serve as a springboard for professional development by motivating teachers to modify and improve their teaching approaches.

Hamedani and Markus (2019) ascertained the fact that faculty members encounter cultural difficulties related to the requirement for cultural competency. Teachers may feel unprepared to manage cultural preferences in the classroom when the curriculum fails to adequately recognize the variety of the pupils. Recognizing these difficulties offers faculty members the chance to take part in courses on cultural competence and better equip themselves to deal with cultural concerns in the classroom (Abdallah & Farhan, 2023).

## 7.2 Strategies Employed to Ensure Cultural Relevance and Sensitivity

Gibson *et. al.* (2020) stated that several initiatives that attempt to establish a welcoming atmosphere for learning have been developed as a result of the desire for relevance to culture and sensitivity in education. These tactics do not come without difficulties, though, and their efficacy varies. Including varied viewpoints and views in the curriculum is a popular tactic. This strategy aims to give pupils resources and information that reflect a diverse variety of cultures, situations, and backgrounds. Striking a balance between inclusiveness and an overrepresentation may be difficult, though, since a strong emphasis on one culture may unintentionally marginalize others. Another common tactic is to train educators in cultural competence (Zhan *et. al.* 2021). Giving educators the information and abilities necessary to comprehend and successfully negotiate cultural differences is part of this. The effectiveness of these training programmes might vary, so it's important to provide continuous support and assessment to make sure that educators are still using what they've learnt (Abdallah, & Musah, 2023).

However, Adams and Ecker (2021) argued that a key tactic for ensuring relevance to culture and sensitivity is community engagement. By adding regional customs, information, and viewpoints into the curriculum, this strategy enables educational institutions to interact with the cultural setting in which they operate. However, community participation can pose logistic and administrative difficulties and may call for a sizable time and resource commitment. Corrales *et. al.* (2020) mentioned in their study that an increasingly popular method is curriculum collaboration with pupils and community members. It is easier to guarantee that curriculum content is culturally appropriate and current when these stakeholders are involved in its creation. To leverage the aggregate expertise of various voices while maintaining the authenticity of the curriculum, an organized and effective procedure must be developed (Abdallah & Abdallah,2023).

# 8. FUTURE IMPLICATIONS AND RECOMMENDATIONS

## 8.1 The Long-Term Implications of Curriculum Adaptations for Afghan Students and the Broader Educational Community

It is important to understand that with the implication of curriculum adaptation the Afghan students are much prepared for complex situations and all this also sharpened their skills and knowledge to a greater extent (Taheryar, 2017). It is necessary to understand that with the changing educational landscape across the globe the curriculum adaptation has also helped the Afghan students to improve their overall quality of education and to meet the global standard of education (Abdallah et al, 2023).

International collaboration plays a significant role in adapting the new curriculum in an effective manner. Better collaboration between the Afghan institutions and the international firms can help in better resource sharing which helps in having more connected global educational community (Salih and Taniwall, 2020). In addition to this when better technologies are being used with the modern teaching methods then it becomes easy for students to adapt the changes and learn the new curriculum in a better way.

## 8.2 Recommendations for Institutions and Policymakers on How to Effectively Adapt Curricula for Displaced Populations in the Future

Adapting curricula for refugees, particularly in Afghanistan, is a difficult but crucial endeavour. Institutions and decision-makers should place a high priority on creating adaptable curriculum that is simple to adapt to meet the unique demands of displaced people. However, it is crucial to make sure that the quality and scope of learning are not jeopardized by these adjustments. The curriculum's material should continue to be pertinent to students' academic and career goals (Orfan *et. al.* 2021). The curriculum need to be inclusive, taking into account the linguistic and cultural variety of Afghanistan's displaced communities. The existence of different ethnic communities and their particular requirements must be acknowledged. Achieving cultural sensitivity may be difficult, though, as it calls for a thorough comprehension of regional traditions, beliefs, and practices (Leufer *et. al.* 2019). To do this, local communities' participation can be quite helpful. Institutions and politicians should prioritize digital learning solutions given Afghanistan's unstable climate. These can give learners remote access to learning materials, which is crucial for people who are relocated or residing in unsafe places. However, the digital gap continues to be an issue, making fair access to technology essential (Darawsheh et al, 2023).

It is crucial to train teachers to work to displaced people. Institutions should fund initiatives for teacher preparation that concentrate on meeting the particular requirements of relocated students, such as using trauma-informed teaching techniques. Policymakers must provide enough financing and support for these projects because this demands significant financial and human resources (Arar *et. al.* 2020). Collaboration among schools, non-governmental organizations, and international organizations may be a great help in adjusting curriculum for people who have been displaced. These collaborations may provide resources, knowledge, and a broader viewpoint on best practices. However, it is crucial to keep a balance among national sovereignty and global collaboration in concerns of education Abdallah & Alkaabi,2023).

# 9. CONCLUSION

## 9.1 Reflecting on the Resilience and Adaptability of Afghan Students and Faculty

The adaptation and resiliency of Afghan teaching members and students serve as a monument to the strength of the human spirit. They have persevered in their pursuit of knowledge and education despite having to deal with decades of conflict, political unrest, and economic uncertainty. These people have persevered despite insurmountable obstacles, such as dangers to their safety and a lack of resources, and they are thriving today. Their persistent dedication to growth, learning, and innovation is an inspiration to the entire world. Afghan educators and students are living examples of how education can be a source of hope and a way forward even in the face of hardship (Jandigulov et al, 2023). From the above essay it is evident that Afghan academics have confronted displacement and instability with extraordinary adaptability and resilience. Despite the numerous obstacles posed by violence and the need to seek refuge in foreign lands, they have demonstrated effective dedication to education and thrived to make effective changes in curriculum with time. Different challenges faced by the Afghan academics and faculty, includes financial disparities, psychological trauma, language barriers, cultural differences, adapting the new educational environment (Ramadan & Ismail, 2023).

## 9.2 The Importance of Documenting and Learning From These Narratives for Future Educational Crises

It is important to understand that how people around the world are getting educated after the displacements. Documentation of the narratives of Afghan academic and students is important for understanding that how people get education during the crisis. Both the Afghan academics and the students have faced multiple issues due to the crisis and the displacements. From this it can be said that in the future crisis education needs to be priorities like food and shelter as it is also an important human right. Based on this the educational institutions across the world and the local government should develop effective strategies and practices for ensuring that even during or after the crisis situation the students or the refugees might get the best possible high quality education.

## 10. PERSONAL NARRATIVES

The different study mentioned in the essay narrates the issues and challenges which are being faced by the Afghan students and faculty due to the displacement. The linguistic issues are being faced by them while adjusting to the new educational environment. Along with this the major issues related to discrimination is also being faced by them while getting education in the host country. When it comes to the faculty members then they have mentioned about their regular efforts for making modification in their teaching methods with the development of education system of Afghanistan and the new things being added to the curriculum (Abdallah, 2023). Also the faculty members have helped the refugees by offering them better psychological support to adapt the new educational environment. Not only this it is being observed from this report that Ministry of education in Afghanistan has paid less attention towards women education and making adaptation in the curriculum with the changing world. Although now the educational landscape is changing in Afghanistan due to regular efforts being made for offering best possible quality education to all. The local government, international organisations, educational institutions and the humanitarian agents all are working in a collaborate manner for improving the quality of education in Afghanistan.

## REFERENCES

Abdallah, A. K. (2023). Teacher-Led, Student-Focused, and Unleashing the Power of Teacher Empowerment for School Improvement and Success. In A. Abdallah & A. Alkaabi (Eds.), *Restructuring Leadership for School Improvement and Reform* (pp. 1–21). IGI Global. doi:10.4018/978-1-6684-7818-9.ch001

Abdallah, A. K., & Abdallah, R. K. (2023). Achieving academic excellence: The intersection of teacher development, quality education, and entrepreneurship. In S. Chakravarti (Ed.), *Innovations in Teacher Development, Personalized Learning, and Upskilling the Workforce* (pp. 136–158). IGI Global. doi:10.4018/978-1-6684-5518-0.ch007

Abdallah, A. K., & Al-Kaabi, A. M. (2023a). Induction Programs' Effectiveness in Boosting New Teachers' Instruction and Student Achievement: A Critical Review. International Journal of Learning. *Teaching and Educational Research*, *22*(5), 493–517.

Abdallah, A. K., & Alkaabi, A. M. (2023b). Role of teachers in reinforcing students cultural and heritage awareness at Abu Dhabi schools to meet global challenge. *Cogent Social Sciences*, *9*(1), 1. doi:10.1080/23311886.2023.2194734

Abdallah, A. K., AlKaabi, A. M., & Ramadan, R. S. (2023). The critical role of principals in leading effective inclusive schools. In E. Efthymiou (Ed.), *Inclusive Phygital Learning Approaches and Strategies for Students With Special Needs* (pp. 256–281). IGI Global. doi:10.4018/978-1-6684-8504-0.ch012

Abdallah, A. K., & Farhan, A. F. (2023). Breaking Barriers and Empowering Women Leaders to Drive School Improvement. In A. Abdallah & A. Alkaabi (Eds.), *Restructuring Leadership for School Improvement and Reform* (pp. 399–419). IGI Global., doi:10.4018/978-1-6684-7818-9.ch020

Abdallah, A. K., Ismail, O., Abdallah, R. K., & Alkaabi, A. M. (2023). Perceptions of Students About the Use of Webinars in Classrooms: A Case of Abu Dhabi University. *International Journal of Information and Communication Technology Education*, *19*(1), 1–17. doi:10.4018/IJICTE.322793

Abdallah, A. K., & Musah, M. B. (2023). Principal and Teacher Licensing as a Tool to School Improvement. In A. Abdallah & A. Alkaabi (Eds.), *Restructuring Leadership for School Improvement and Reform* (pp. 278–300). IGI Global. doi:10.4018/978-1-6684-7818-9.ch014

Akramy, S. A. (2021). Implementation of outcome-based education (obe) in Afghan universities: Lecturers' voices. *International Journal of Quality in Education*, *5*(2), 27–47.

Alam, A. (2022). Investigating sustainable education and positive psychology interventions in schools towards achievement of sustainable happiness and wellbeing for 21st century pedagogy and curriculum. *ECS Transactions*, *107*(1), 19481–19494. doi:10.1149/10701.19481ecst

Alamyar, M. (2018). Education in Afghanistan: A Historical Review and Diagnosis. *College and University*, *93*(2), 55–60.

Alariqi, A. A., Najafi, M., Abdulrab, M., Murray, C., & Slimanzai, H. (2019, October). Factors affecting e-learning effectiveness in a higher learning institution in Afghanistan. *Proceedings of the 11th International Conference on Education Technology and Computers* (pp. 176-181). 10.1145/3369255.3372275

Alqodsi, E. (2023). Teacher Civil Liability in the Case of Breaching Educational and Control Obligations. In A. K. Alkaabi (Ed.), *Restructuring Leadership for School Improvement and Reform* (pp. 327–338). IGI Global. doi:10.4018/978-1-6684-7818-9.ch016

Alqodsi, E., & Aljahoori, S. A. (2023). Legal protection of the right to education for people with special needs: Zayed higher organization for people of determination as a model. In E. Efthymiou (Ed.), *Inclusive Phygital Learning Approaches and Strategies for Students With Special Needs* (pp. 199–213). IGI Global. doi:10.4018/978-1-6684-8504-0.ch010

Alqodsi, E. M., Jadalhaq, I. M., & El Maknouzi, M. E. (2023). *Technology-enhanced legal education: A study of its impact on student learning outcomes in the UAE.* IGI Global. doi:10.4018/978-1-6684-5518-0.ch004

Amouzeshi, A., Abedi, F., Amiri, M., Amouzeshi, Z., Jamshidi, A., & Salehi, F. (2021). Afghan students' willingness to continue their studies abroad and the related factors: A study of Afghan students studying at selected Afghan universities of medical sciences in the academic year 2019-2020. *Educational Development of Judishapur, 12*(1), 248–254.

Arar, K., Kondakci, Y., Kaya Kasikci, S., & Erberk, E. (2020). Higher education policy for displaced people: Implications of Turkey's higher education policy for Syrian migrants. *Higher Education Policy, 33*(2), 265–285. doi:10.1057/s41307-020-00181-2

Arar, K., Kondakci, Y., & Streitwieser, B. (2020). Higher education for forcibly displaced migrants, refugees and asylum seekers. *Higher Education Policy, 33*(2), 195–202. doi:10.1057/s41307-020-00184-z

Bannova, O., & Mayorova, V. 2(019, November). Building trans-national and multi-disciplinary academic curricula through adaptation of a project-based approach. In AIP Conference Proceedings (Vol. 2171, No. 1). AIP Publishing. doi:10.1063/1.5133292

Bird, L. (2009). Promoting resilience: developing capacity within education systems affected by conflict. *Think piece commissioned for the EFA Global Monitoring Report 2011: The Hidden Crisis–Armed Conflict and Education.*

Brenner, M.E. & Kia-Keating, M. (2016). Psychosocial and academic adjustment among resettled refugee youth. *Annual review of comparative and international education 2016*, 221-249.

Burde, D. (2014). *Schools for Conflict or for Peace in Afghanistan.* Columbia University Press.

Cerna, L. (2019). *Refugee education: Integration models and practices in OECD countries.* Academic Press.

Clemens, V., von Hirschhausen, E., & Fegert, J. M. (2020). Report of the intergovernmental panel on climate change: Implications for the mental health policy of children and adolescents in Europe—a scoping review. *European Child & Adolescent Psychiatry*, 1–13. PMID:32845381

Corrales, A., Goldberg, F., Price, E., & Turpen, C. (2020). Faculty persistence with research-based instructional strategies: A case study of participation in a faculty online learning community. *International Journal of STEM Education*, *7*(1), 1–15. doi:10.1186/s40594-020-00221-8

Darawsheh, S. R., Al-Shaar, A. S., Alshurideh, M., Alomari, N. A., Elsayed, A. M., Abdallah, A. K., & Alkhasawneh, T. (2023). The Relation Between Creative Leadership and Crisis Management Among Faculty Members at Imam Abdulrahman Bin Faisal University in Light of the Corona Pandemic from the Perspective of Department Heads, The Effect of Information Technology on Business and Marketing Intelligence Systems (Vol. 1056). Springer. doi:10.1007/978-3-031-12382-5_83

Deo, N. (2014). The politics of education in Afghanistan. *Pakistan Journal of Women's Studies: Alam-e-Niswan*, *21*(1).

Easar, F., Azizi, H., Rahmani, K., Moradi, M., Taieb, R., & Faqiryar, W. N. (2023). *Education in Afghanistan since 2001: Evolutions and Rollbacks*. MISC.

Gelmez-Burakgazi, S. (2020). Curriculum adaptation and fidelity: A qualitative study on elementary teachers' classroom practices. *Issues in Educational Research*, *30*(3), 920–942.

Gibson, K. E., Barnett, J., Haslam, N., & Kaplan, I. (2020). The mental health impacts of climate change: Findings from a Pacific Island atoll nation. *Journal of Anxiety Disorders*, *73*, 102237. doi:10.1016/j.janxdis.2020.102237 PMID:32485590

Gindeel, R., & Etim, J. S. (2014). Education in War Affected Areas (Internally Displaced Persons communities) in Eastern Sudan. *Journal of Studies in Education*, *4*(4), 108–121. doi:10.5296/jse.v4i4.6540

Gonchar, O., Kryvuts, S., Petukhova, O., & Tokar, M. (2019). Organization of Pedagogical Relationships between Learners and Teachers in the Framework of Inclusive Education. *Romanian Journal for Multidimensional Education/Revista Romaneasca pentru Educatie Multidimensionala,* 11(2).

Grawert, E., & Mielke, K. (2018). *Coping with protracted displacement: How Afghans secure their livelihood in Afghanistan, Iran and Pakistan. Academic Press.*

Hamedani, M. Y. G., & Markus, H. R. (2019). Understanding culture clashes and catalyzing change: A culture cycle approach. *Frontiers in Psychology*, *10*, 700. doi:10.3389/fpsyg.2019.00700 PMID:31031669

Hedayati, A. (2023). Women's education from the perspective of Islam and consequences of restrictions on women education in Afghanistan. *Sprin Journal of Arts. Humanities and Social Sciences*, *2*(08), 1–9.

Himat, A. N., Takal, M. H., & Hakimi, M. F. (2021). Afghan students' satisfaction from online learning during COVID-19 at Kandahar University, Kandahar, Afghanistan. *American International Journal of Social Science Research*, *6*(1), 16–29. doi:10.46281/aijssr.v6i1.931

Jandigulov, A., Abdallah, A. K., Tikhonova, Y., & Gorozhanina, E. (2023). Management and leadership in online learning. *Education and Information Technologies*, *28*(10), 13423–13437. Advance online publication. doi:10.1007/s10639-023-11699-4

Jenkins, G. (2020). Teacher agency: The effects of active and passive responses to curriculum change. *Australian Educational Researcher*, *47*(1), 167–181. doi:10.1007/s13384-019-00334-2

Karlsson, P., & Mansory, A. (2008). *Islamic and modern education in Afghanistan: Conflictual or complementary*. Institute of International Education, Stockholm University.

Katawazai, R. (2021). Implementing outcome-based education and student-centered learning in Afghan public universities: The current practices and challenges. *Heliyon*, *7*(5), e07076. doi:10.1016/j.heliyon.2021.e07076 PMID:34095574

Khwajamir, M. (2016). History and problems of education in Afghanistan. *In SHS Web of Conferences (Vol. 26, p. 01124)*. EDP Sciences.

Leufer, N., Prediger, S., Mahns, P., & Kortenkamp, U. (2019). Facilitators' adaptation practices of curriculum material resources for professional development courses. *International Journal of STEM Education*, *6*(1), 1–18. doi:10.1186/s40594-019-0177-0

Mahdi, M. I., & Wani, N. U. H. (2021). Factors affecting e-learning adoption in Afghanistan: Empirical evidence from technical and vocational education and training authority. *Kardan Journal of Economics and Management Sciences*, *4*(2), 1–22. doi:10.31841/KJEMS.2021.1

Marshall, E. A., Butler, K., Roche, T., Cumming, J., & Taknint, J. T. (2016). Refugee youth: A review of mental health counselling issues and practices. *Canadian Psychology*, *57*(4), 308–319. doi:10.1037/cap0000068

Mentes, O. & Talas, M. (2021). The international humanitarian aid delivered for afghanistan on primary and secondary education after 2001. *Journal of World of Turks/Zeitschrift für die Welt der Türken,* 13(1).

Miller, E., Ziaian, T., de Anstiss, H., & Baak, M. (2022). Practices for inclusion, structures of marginalisation: Experiences of refugee background students in Australian secondary schools. *Australian Educational Researcher*, *49*(5), 1063–1084. doi:10.1007/s13384-021-00475-3

Miszak, N. & Motwani, N. (2021). *Afghanistan Case Study*. Academic Press.

Nazari, M., Miri, M. A., & Golzar, J. (2021). Challenges of second language teachers' professional identity construction: Voices from Afghanistan. *TESOL Journal*, *12*(3), e587. doi:10.1002/tesj.587

Oates, L., Goger, L.K., Hashimi, J. & Farahmand, M. (2019). *An early stage impact study of localised OER in Afghanistan.* Academic Press.

Orfan, S. N., Akramy, S. A., Noori, A. Q., & Hashemi, A. (2021). Afghan lecturers' perception of problem-based learning: A case study of takhar university. *Journal of Problem-Based Learning*, *8*(2), 62–68. doi:10.24313/jpbl.2021.00059

Otukile-Mongwaketse, M., Mangope, B., & Kuyini, A. B. (2016). Teachers' understandings of curriculum adaptations for learners with learning difficulties in primary schools in B otswana: Issues and challenges of inclusive education. *Journal of Research in Special Educational Needs*, *16*(3), 169–177. doi:10.1111/1471-3802.12069

Pillai, V. S. (2020). No-tech solutions to contınue learning: An example from Afghanistan. *Education*.

Pincock, K., Betts, A., & Easton-Calabria, E. (2021). The rhetoric and reality of localisation: Refugee-led organisations in humanitarian governance. *The Journal of Development Studies*, *57*(5), 719–734. doi:10.1080/00220388.2020.1802010

Ramadan, R. S., & Ismail, O. A. (2023). Building a better future: The role of school inspection in driving educational excellence. In *Restructuring Leadership for School Improvement and Reform* (pp. 202–221). IGI Global. doi:10.4018/978-1-6684-7818-9.ch011

Rubin, B., & Rudeforth, C. (2016). Enhancing access to education: Challenges and opportunities in Afghanistan. Center in International Cooperation, New York University.

Salih, S.K. & Taniwall, N.J. (2020). Issues and challenges of E-learning system Adoption in a public university of Afghanistan: a case study of shaikh zayed university. *J. Human. Soc. Sci.(IOSR-JHSS),* 25(1), 63-69.

Shakya, Y. B., Guruge, S., Hynie, M., Akbari, A., Malik, M., Htoo, S., Khogali, A., Mona, S. A., Murtaza, R., & Alley, S. (2010). Aspirations for higher education among newcomer refugee youth in Toronto: Expectations, challenges, and strategies. *Refuge: Canada's Periodical on Refugees*, *27*(2), 65–78. doi:10.25071/1920-7336.34723

Slattery, P. (2012). *Curriculum development in the postmodern era: Teaching and learning in an age of accountability*. Routledge.

Smith, V. J. (2009). Ethical and effective ethnographic research methods: A case study with Afghan refugees in California. *Journal of Empirical Research on Human Research Ethics; JERHRE*, *4*(3), 59–72. doi:10.1525/jer.2009.4.3.59 PMID:19754236

Spink, J. (2005). Education and politics in Afghanistan: The importance of an education system in peacebuilding and reconstruction. *Journal of Peace Education*, *2*(2), 195–207. doi:10.1080/17400200500185794

Taheryar, H. (2017). Perceptions of quality in higher education in Afghanistan: A case study of Shaheed Rabbani Education University. Academic Press.

Thomas, M., & Bigatti, S. (2020). Perfectionism, impostor phenomenon, and mental health in medicine: A literature review. *International Journal of Medical Education*, *11*, 201–213. doi:10.5116/ijme.5f54.c8f8 PMID:32996466

Welch, A., & Wahidyar, A. (2020). Quality assurance in Afghan higher education: Achievements and challenges. *Asian Education and Development Studies*, *9*(4), 479–493. doi:10.1108/AEDS-09-2018-0146

Zhan, R., Liu, X., Wong, D. F., & Chao, L. S. (2021, May). Meta-curriculum learning for domain adaptation in neural machine translation. *Proceedings of the AAAI Conference on Artificial Intelligence*, *35*(16), 14310–14318. doi:10.1609/aaai.v35i16.17683

# Chapter 5
# Disrupted Educational Pathways in Afghanistan:
## Evaluating the Disparities in Education Under the Taliban Affecting Women's Educational Access

**Nassir Ul Haq Wani**
https://orcid.org/0000-0003-3871-9857
*Kardan University, India*

**Farha Naz Khan**
https://orcid.org/0000-0001-6223-2286
*Anil Surendra Modi School of Commerce, SVKM's Narsee Monjee Institute of Management, India*

**Shukriya Sadaat Hashmi**
*Kardan University, Kabul, Afghanistan*

## ABSTRACT

*The chapter explores the impact of education disparities under the Taliban on women's educational access and capabilities development in Afghanistan, a country on the brink of collapse due to ongoing conflicts and political upheaval. The study uses a case study technique to analyze the current state of women's education and its impact on women's capability development. The chapter argues that women's potential can revolutionize Afghan society and secure the country's existence as a modern nation-state. Women's education can help Afghan women enhance their capacities and contribute to the country's progress. The chapter concludes with reflections on the implications for government, development partners, communities, non-governmental organizations, and educational institutions in strengthening education services and building resilience in fragile contexts. It also highlights the need for linkages with social protection, justice, and psychosocial support services.*

DOI: 10.4018/979-8-3693-1483-8.ch005

## INTRODUCTION

The Universal Declaration of Human Rights states that everyone has the right to education, starting at the most basic levels (Eriksson et al., 2010). However, educational discrimination in Afghanistan violates this declaration, particularly regarding gender. There is a significant research gap regarding gender discrimination in education, especially for Afghan women. Afghanistan, a country plagued by poverty and significant social and economic challenges, is particularly vulnerable to this prejudice. Therefore, the country must address this issue and harness the potential of its human capital through education. This would ensure equal access to education for all individuals, regardless of their background or circumstances. Afghanistan has emerged as a focal point in the formulation of international policies. Massive financial investments and human lives have been shed by international powers to revolutionise Afghan society. The nation has emerged as a central focus for policymaking among the main global powers. The primary focus has been on advocating for educational initiatives to instigate a transformative societal shift within this tribal community.

Considerable endeavours have been devoted to the education of Afghan women. Consequently, female literacy increased substantially before the Taliban regained control of the country. With the Taliban's consolidation of power and obstruction of educational opportunities for women, this matter assumes a profoundly significant nature. Women's education extends beyond female literacy, with implications for geopolitics and international security. The education for women was non-existent during the initial Taliban regime of 1996–2001 (New York Times, 2021). However, after the Taliban's downfall, female literacy increased to 30 percent by 2018, as reported by UNESCO in 2021. Since the withdrawal of the United States, women have been prohibited from resuming their studies, restricting feminist education once again (Walt, 2021). Comparatively, only 33% of adolescent females possess fundamental literacy skills, while 67% of boys in the same age group do so (Human Rights Watch, 2022). The disparity in female literacy between rural and urban areas is particularly pronounced, as evidenced by the fact that metropolitan regions have 70% of enrolled girls in primary education, while rural areas have only 40% (Centre for Global Development, 2022; Muram and Wani, 2020). It appears that no amount of pressure from the West can halt the Taliban's consolidation of power in Afghanistan.

The education of women in Afghanistan during the Taliban era provides ample material for research. Regarding Afghanistan during the Taliban regime, the education of women faces numerous obstacles. The history, society, economy, and culture of the nation ingrained these manifold impediments to female literacy.

Without discrimination, access to education alters societal norms and advances meritocracy concerning societal transformations (Dore, 1984). The promotion of merit renders all forms of discrimination associated with gender, lineage, ethnicity, or religion void. Furthermore, within the realm of business, a greater gender disparity in employment reduces the pool of qualified individuals available to companies and organisations, which harms the economy's overall productivity and competitiveness (Klasen and Lamana, 2000). The worldwide financial burden resulting from gender-based discrimination in education amounts to an estimated \$15–\$30 trillion in lost lifetime earnings and productivity (World Bank, 2022; Wani, Dhami, and Sidana, 2022).

The matter of gender discrimination in education has become increasingly urgent and significant in Afghanistan with the return of the Taliban to power. The tenuous progress towards female literacy in Afghanistan before the Taliban's resurgence is under significant pressure as the nation transitions towards a quasi-religious tribal theocracy. As a fundamental human right, access to education serves as the foundation for all other rights that are associated with the development of capabilities and education. Furthermore, through an examination of past and current research in the global realm of gender discrimination, this study endeavours to identify any deficiencies in the field that warrant additional investigation.

This study's objective is to examine the obstacles that the Taliban regime in Afghanistan presents to the illiteracy of women. Through an exploration of gender discrimination in education in Afghanistan, this study seeks to illuminate the myriad challenges that lie ahead by drawing on the perspectives and experiences of Afghan women. This study also aims to show how gender bias in education could hurt the country by limiting Afghan women's views and goals to a certain range of skills and progress made possible by education. Furthermore, it will emphasise the potential tangible and intangible advantages that education may offer in Afghanistan at the tertiary and secondary levels because of women's education. These research findings can also provide valuable policy guidelines for any multilateral or bilateral assistance aimed at advancing female literacy in Afghanistan.

Furthermore, through an examination of past and current research in the global realm of gender discrimination, this study endeavours to identify any deficiencies in the field that warrant additional investigation. Furthermore, it endeavours to expand upon prior investigations in the field and conduct a comparative analysis of prior discoveries. By considering the perspectives and understandings of Afghan women, the study's findings could potentially contribute to our theoretical comprehension of the subject matter and yield valuable policy implications. Without discrimination, access to education alters societal norms and advances meritocracy concerning societal transformations (Dore, 1984). The promotion of merit renders all forms of discrimination associated with gender, lineage, ethnicity, or religion void.

The social, economic, and cultural development of a nation demands an unharnessed demographic dividend for the marginalised section of society (Klasen and Lamana, 2000; Dore, 1984). Women's empowerment in all spheres of life can improve the present state of Afghanistan, characterised by poverty, social disarray, and economic growth. Education paves the way for the foundation of women's empowerment. Klasen and Lamana (2000), in their study, found a correlation between the gender gaps in education and regional economic growth and prosperity.

The primary religious sources of Islam, especially the Holy Qur'an, focus on the importance of education in human life, regardless of gender discrimination. For example, Allah (SWT) says in Sura Al-Alaq, "Read in the name of your Lord who created; created man from a clinging substance" (The Qur'an, 96:1-2). Also, it says in Sura Az-Zumar, "Say: 'Are they equal, those who know and those who do not know?' Only those with minds remember" (The Qur'an, 39:9). Furthermore, Anas bin Malik narrated that Prophet Mohammad (PBUH) stated, "Every Muslim must seek knowledge" (Sunan Ibn Majah, n.d.). Consequently, key leaders in the Islamic world called education a fundamental right for all human beings, including women (Sarwari & Hameidi, 2023).

Thus, another dimension of development focused on in the article is 'capability development." Capacity building goes beyond task completion and is concerned with changing attitudes and mindsets. Capacity Development addresses Sustainable: Revitalising the Global Partnership for Sustainable Development and is committed to transformation from within. It entails the adoption of innovation and technology in the least developed nations of the world. Thus, the present article focuses on the impact of the Taliban on women's educational access and capability development. We are conducting this study to explore the perceptions and understandings of the first-hand sufferers, i.e., Afghan women, to foster policy implications.

Gaining a comprehension of the problem of gender discrimination in education in Afghanistan during the current Taliban regime not only provides valuable policy insights but also illuminates the inherent difficulties that this issue presents. These research findings can also provide valuable policy guidelines for any multilateral or bilateral assistance aimed at advancing female literacy in Afghanistan.

## 1. CURRENT STATUS OF WOMEN HIGHER EDUCATION UNDER TALIBAN 2.0

Afghanistan has been a challenging place for women, with the Taliban's rise in power in the 1990s and 2021 putting higher education at risk. This has led to increased restrictions on women's education, employment, freedom of movement, and expression, further damaging gender equality and development in the country.

The pre-2001 Taliban attitudes towards education have changed since the 1990s, with the Taliban now being a very different entity and its approach to governance substantially changing. During the 2000–2001 academic year, primary school enrolment rates were as low as 29% for boys and less than 1% for girls. The Taliban imposed severe restrictions on women's ability to participate in public life and access basic services such as education and healthcare.

The pre-2001 Taliban attitudes towards education have changed since the 1990s. There are several ways to answer this. One part of the answer is that the Taliban is a very different entity now, and its approach to governance has largely changed substantially. Taliban spokesperson Zabiullah Mujahed said in 2012 that the Taliban government 'tried to contact many foreign NGOs for funds [for girls' education], but none were ready to provide actual assistance. There was 'no money, no help, except some of the Gulf countries funding only Wahabi-type madrasas, and the only books we could get were from Pakistan'.

During the 2000–2001 academic year, primary school enrolment rates were as low as 29% for boys and less than 1% for girls. Taliban officials typically argue that they simply could not support female education during that period.

Segregating male and female students would be necessary to permit female education. They would need female teachers, separate schools for girls, boundary walls, and transport so that girls could commute and study discreetly. The Taliban claim they could not obtain the necessary resources for all of this. Regardless of the rationale they present now, the fact is that most girls' schools were shut down under the Taliban, and the movement imposed severe restrictions on women's ability to participate in public life (including teaching) and access basic services such as education and healthcare. Finally, it is important to emphasise that Afghanistan itself has dramatically changed since the 1990s. There have been enormous advances in education in the country. School enrolment has increased by a factor of nine since 2001 (Amiri & Jackson, 2021).

*Figure 1. School enrolment in Afghanistan (World Bank Development Indicators, 2022)*

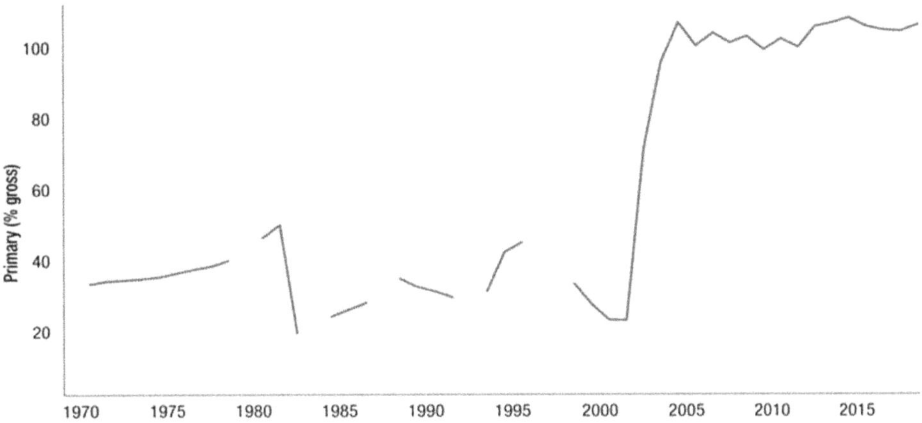


## 2. LITERATURE REVIEW

The literature review section provides a theoretical background for the study, focusing on higher education and women's perceptions of challenges and opportunities. It will highlight previous research and serve as a foundation for the study's research. The literature is classified into three categories: women's educational access under the Taliban, women's capability development, and disparities in education under the Taliban that affect women's educational access and capability development.

### 2.1: Women's Educational Access Under the Taliban Government

Education contributes to the betterment of societies, and it benefits people for a better life. Education's aim is the integral development of individuals and groups. According to the study by the International Institute for Applied Systems Analysis (IIASA, 2018), it is worth investing in education because the processes of educating and learning are a "natural and defining characteristic of human society." Education is an essential mechanism for the developmental level of communities at the national and international levels. The policy brief by IIASA (2018) states, "There is a positive relationship between education and other dimensions of development," and it adds that progress in education and educational sectors is "in everyone's interest." Thus, the demand for quality education is high on a global level. Education as a human right has been a priority in international frameworks and practices. The 2030 Agenda for Sustainable Development Goals (SDGs) reflects on the importance of education for all. Goal 4 of the SDGs aims to “ensure inclusive and equitable quality education and promote lifelong learning opportunities for all,” while Millennium Development Goal (MDG) 2 is only focused on “achieving universal primary education” (Ahmadi, 2022; Wani and Dhami, 2021).

According to Rodriguez (2019), around 129 million girls are out of school (UNICEF, 2022). Around 39 million girls are the direct victims of conflict when it comes to education. There are many barriers to girls’ education, but conflict is a major one. Ahmadi (2022) claims that the education system in Afghanistan faces many challenges, including a lack of quality education and resources, social and traditional obstacles to girls’ enrolment, insecurity, poverty, limited seats in governmental universities, gender-based discrimination, early marriage, a lack of higher education institutions and facilities in provinces, and other problems. Women in Afghanistan face social, economic, and cultural barriers that contribute to gender inequality and hinder their advancement. According to a 2009 report, more than 90% of females are illiterate in Afghanistan.

Furthermore, there are significant differences between governmental and non-governmental universities in terms of structure, quality of education, and resources. Kankor examinations in governmental universities are very competitive, and the Kankor candidates are much higher than in nongovernmental universities. The governmental universities accept only students who pass the Kankor examination (Noora and Wani, 2020). While non-governmental universities have their own Kankor examination, which is helpful for students who did not get into government universities and students who work in the daytime, the quality of education in government universities is better than that in private universities (Ahmadi, 2022).

Moreover, there are huge differences between universities in the capital and provinces, as there are more public and private universities in the capital and big cities than in provinces and remote regions. Based on Burridge et al. (2016), there are only limited opportunities for older women who missed the opportunity to attend school during the Taliban era to receive adult education or literacy training. Poverty, early marriage, insecurity, lack of family support, lack of female teachers, lack of proximity to schools, the poor quality of education, a lack of girls’ only schools, and community attitudes are major obstacles to the education of girls. Also, most of the students expressed concerns about the costs involved in higher education as a barrier to their continued participation. A number also indicated that they felt additional financial pressure on them to obtain good employment at the end of their studies (Burridge et al., 2016). Access to public universities in Kabul is free, but the number of spaces available is very limited.

Another peculiar barrier to education for women has a more psychological nature than the socio-economic ones. Prevalent gender stereotypes in less developed societies negatively impact women’s educational opportunities by devaluing their educational and professional potential (Inayatullah, 2022). Furthermore, institutions of higher learning and universities estimate that female enrolment in higher education is below 20 percent. Education levels also vary considerably by region. In some districts, the percentage of eligible girls attending school is in the single digits, and female literacy has barely budged. In many of these low-enrolment areas, the Taliban exercises effective control over education, prohibiting girls from attending school or allowing them to do so only under certain conditions, such as using a Taliban-vetted curriculum. If the Taliban can reassert control over larger parts of the country, its control over education will increase, further threatening girls' access to education (Powell,2014).

## 2.2: Women's Capabilities Development

The capabilities approach "focuses on the ability of human beings to lead lives they have reason to value and to enhance the substantive choices they have." The capability approach of Amartya Sen and Martha Nussbaum centres on the well-being and agency of people. Sen and Nussbaum believe that education enhances women's capabilities (Moravej,2022)

The issue of women's education and capability development in a less-developed, rural, and tribal context is a contentious issue when education and capability development are narrowly defined and viewed solely through an economic growth perspective. Women's capabilities and development through education can contribute to national prosperity in two important ways. Firstly, through their participation in productive activities, they can contribute to the labour markets, export manufacturing, service sector, and food supply. Secondly, women's education can reduce fertility rates, minimise child and mother mortality rates, and improve the overall family health system. Furthermore, future human capital enrichment, which correlates with women's education, positively impacts technological change, the single most influential factor for global growth and development (Inayatullah, 2022).

Many view education as a primary pathway to achieving empowerment for girls and women. Studies supported the positive role of education in empowering women, arguing that education leads to an empowerment process for women if it expands women's knowledge and understanding, improves self-esteem, and raises awareness of gender equity (Moravej, 2022). Despite the disruptions caused by war, Afghan women have sought education and knowledge as a means of empowerment. They have fought against the denial of education, established underground schools, and pursued learning opportunities to equip themselves with the skills necessary to rebuild their lives and contribute to society (Heamatzai & Rafiqzad, 2023).

According to Inayatullah (2022), educating women burnishes their decision-making and social communication skills, thereby opening potential political opportunities for them to engage in national decision-making. Such capability development via education is not only beneficial for women as it improves their individual lives but can have a positive influence on their local and national politics by expanding the talent pool and through diversification. Such women also bring experience and knowledge with a visible effect on public policies.

## 2.3: Disparities in Education Under the Taliban Affecting Women's Educational Access and Capability Development

The Fourth World Conference on Women, held in Beijing in 1995, recognised that women's education, specifically literacy, is key to empowering women's participation in decision-making in society and improving families' well-being. Ahmadi (2022) claims that higher education has helped a lot of women in their lives. Higher education helped women to be financially independent, employed, and have dignity in the community. Even the Millennium Development Report 2013 shows a 10% increase in the ratio of girls to boys' enrolment in all levels of education since 2008. This progress is particularly admirable when compared to the Taliban regime, which did not allow girls to go to school. However, more recent data highlights the challenges that significantly hinder the progress and sustainability of the stated MDG accomplishments. Afghanistan's Ministry of Education affirmed that 7% of girls drop out each year due to issues related to poverty, early marriages, transportation, a lack of sufficient facilities, and security. This leaves us with only a 3% improvement in girls' enrolment. Hence, we can infer from these data that the MDG's enrolment ratios are merely conducted at the beginning of the academic year and do not represent the accurate number of students present throughout the academic year. Moreover, since the goals were implemented in 2004, by 2013, the MDG was supposed to increase the ratio of girls to boys' education by 50%, which has not happened. The data suggest that the MDG has not achieved its goal of reducing gender inequality, even by its measure of "increasing the ratio of girls to boys at schools. (Yousofi, 2016)

The story of Farkhunda reveals how the systematic aggression, societal oppression, and injustices of patriarchal society pose the biggest challenge to women's empowerment. Furthermore, it exposes how gender inequality is deeply ingrained in the thoughts and actions of certain young, ordinary citizens. The 2008 Global Rights Report found that 87% of Afghan women had experienced some level of violence in their lives. In the Taliban-populated areas like Kandahar and Helmand, the figure was above 90%. In the past fifteen years, many women's empowerment and gender equality projects have been implemented, but the statistics indicate that the trajectory of Afghan women's rights remains the same. The UN's Millennium Development Goals (MDG), which are one of the largest development projects, identified women's empowerment as central to global poverty reduction. Hence, the third goal of the MDG is dedicated to women's empowerment, and it has been implemented worldwide (Yousofi, 2016; Seddiqi, 2022).

One of the most touted positive statistics that is often repeated in reports like MDG regarding Afghan women's political representation is the presence of women parliament members in Afghanistan. In 2005, for the first time in Afghan history, women entered both houses of the Afghan parliament in large numbers. Since then, women's presence in parliament has been consistently around 27–28%. However, these statistics measure only the presence of women in Parliament, not their political effectiveness. The reality of the power dynamic between male and female MPs in parliament is quite different. Fleschenberg in Afghanistan's Parliament in the Making: Gendered Understandings and Practices of Politics in a Transitional Country points out how, in Afghanistan's male-dominating society with influential conservative veto actors, a lack of support for women among male MPs leaves the female MPs with limited agency and autonomy.

The Afghanistan Independent Human Rights Commission reported 4250 cases of violence in the first nine months of 2015. Many acts of violence remain unreported due to cultural issues such as fear of social stigmatisation and social exclusion. In the most recent MDG report, no data related to the progress of women's access to justice has been reported. The consistently staggering reports of violence against women suggest that MDG has had little influence in increasing women's access to justice. Farkhunda's case was a prime example of how the judicial system failed people's pleas for sustaining justice. The initial outrage of the demonstrators led to the judicial system declaring death sentences for the four killers. However, the judicial system reviewed and reduced their later trial sentences to only ten or twenty years of imprisonment. The judicial system sentenced the policemen who failed to protect Farkhunda to one year in prison, while eight were acquitted. The result outraged many Afghans. (Yousofi,2016)

In 2012, government offices reported a female-to-male ratio of 27% in employment. The World Bank reported the female-to-male ratio at 16% in 2014. Additionally, Afghan women also earn 49% less than their male counterparts. 18 Given the current unemployment rate of 40%, no recent data has been released illuminating the details of its implications for the lives of women. However, one can only imagine the struggles of unemployed women. 19 Based on the trend of the job market, we can infer that MDG's effort to reduce the employment disparity and alleviate women's poverty has not been effective. (Yousofi,2016)

To sum up this section, one of the common patterns in the MDG reports is the underlying assumption that an increase in the ratio of girls and women, be it in the educational, political, or economic sectors, equates with empowerment and increased gender equality. Given all the problems mentioned above, we know that these assumptions are flawed and that the MDG's approach to tackling the challenges facing women in Afghanistan is very narrow-sighted. Empowerment is dependent on the development of capabilities. Insufficient development of individuals' capabilities and functioning hinders them from reaching their full potential. The complicated nature of gender requires a much more robust analysis and goal-setting process in which the important variables that have limited women's participation in the social, economic, and political spheres of their society are considered. In addition, despite their narrow goals, the Millennium Development Goals have also failed to achieve their stated goals. (Yousofi,2016)

Gender equality and women's empowerment are the cornerstones of the fifth Sustainable Development Goal (SDG). Efforts to promote women's empowerment and gender equality are critical to fostering health and human development globally. Empowerment promotes the potential and capability of underprivileged populations by removing existing obstacles to individual decision-making and autonomous action, improving overall well-being. Gender equality could be achieved through equal rights, opportunities, and access to education, health care, decent work, and political and economic privileges for both men and women. This could be effectively achieved by empowering women across three main categories: agency that indicates the decision-making abilities regardless of the existing power structure; resources that are described as channels through which one exercises agency, such as education, health, and physical assets; and achievements that are the product of agency, such as economic and socio-political gains. It is well known that women's empowerment not only benefits women but also contributes to the development of the whole society (4). Promoting women's empowerment has been linked to improved men's and women's health and reduced child mortality and morbidity. Empowered women are more likely to use modern contraception and have access to antenatal care, institutional delivery, and skilled birth attendance. Besides, it has been shown that the children of empowered women are less likely to suffer from malnutrition, and their daughters are more likely to spend longer time in education and receive equal treatment as their sons in inheritance (Dadras, 2022).

Finally, Powell (2014) suggests that the government of Afghanistan's accountability to all citizens and respect for its constitutional and international commitments, including gender equality, are crucial to overcoming the systemic gender discrimination that thrived under the Taliban's rule. Improving the status of Afghan women and girls is also important for the nation's political development and stabilization. Investing in girls' education and women's economic opportunities has a multiplier effect; it enhances development outcomes, including the health and education of children and the economic status of families and communities. 1 Strengthening the status of women also correlates with reduced rates of conflict and violence. 2 The relationship between gender equality and security creates a virtuous circle: greater security is critical for women and girls to participate on equal footing in public life; greater equality for women and girls correlates to a more secure and economically viable environment for the community as a whole. Promoting gender equality is, therefore, not only a significant human rights issue but also an economic and security matter.

## 3. RESEARCH METHODOLOGY, OPERATIONALIZATION, AND CONCEPTUAL FRAMEWORK

A qualitative research methodology was used, with structured questionnaires and focus group discussions being used to gather primary data. Six provinces were identified in Afghanistan for the study purposes, 289 Afghan girls filled out the questionnaire and a pool of 60 Afghan girls were interviewed. Security issues and social sensitivity towards women interviewing caused the researcher to reach fewer respondents.

Semi-structured interviews were used to examine and investigate the perceptions, aspirations, and understanding of the participants about the issue of gender discrimination in education in Afghanistan. The study aimed to ease participants into the interview with a few pre-determined questions and confront them with deeper unplanned and spontaneous questions to delve deeper into their understanding of the issue at hand. The deductive approach was used, which is coordinated with the qualitative nature of the study. The findings from the study within the context of Afghanistan under a peculiar set of endogenous and exogenous conditions render the findings of this specific study pertinent and valid only within the context of this specific case.

The study involved fifty-nine interviews with Afghan women, twenty-one of whom were school-going, and the rest were university-level students. The participants were chosen based on their level of education and accessibility, as conducting research in Afghanistan is risky for foreigners. The study aimed to select Afghan women living abroad with some level of education attained inside Afghanistan and who were easy to access.

Validity and reliability are essential features of a qualitative research study, as they enhance the quality of the study by enabling it to be replicable and valid across different settings. Reliability ensures repeatability of the study, while validity enhances the study's integrity by making the conclusions valid. To ensure the validity and reliability of the study, the interview participants were Afghan women with first-hand educational and capability development experience within Afghanistan. Their views and opinions can be relied upon and are valid within the context of women's education and capabilities development.

The study bases its theoretical assumptions on Nussbaum's capabilities theory, which is based on seven elements: life, bodily health, practical reason, affiliation, sense, imagination and thought emotions, and control over one's environment. The research design was based on a pre-existing longitudinal sample with whom the research team had already developed rapport in previous data collection rounds.

All interviews were recorded, transcribed, and translated into English by bilingual transcribers. A subset of each transcriber's interviews was checked by a research supervisor to ensure quality translations. A detailed debriefing discussion was held with the interviewer team to discuss emerging context-specific issues and themes, and to inform the development of a thematic codebook developed in line with the conceptual framework. Data was then analysed thematically using a thematic analysis approach.

## 4. DATA ANALYSIS AND FINDINGS

This section includes the analysis of the primary data that was obtained from the study participants through questionnaires and interviews, as well as its ensuing analysis. Moreover, it is worth briefly alluding to the composition of the project participants and their current significance to the research to provide readers of the study with an overview of the data generated and the context for the data analysis. On August 15, 2021, the Taliban took over the Islamic Republic government of Afghanistan, leading to an uncertain future for Afghans, especially women. After taking control of the entire country, the Taliban announced their interim government, including a male cabinet. The Taliban issued a statement demanding women stay home until further instruction from them; however, several protests erupted around the country to demand women's rights and their access to education because the Taliban had replaced the Ministry of Women's Affairs with their Ministry of the Propagation of virtue and the Prevention of vice. On the other hand, the Taliban Minister of Education, in an announcement, declared the reopening of boys' schools from grades six to twelve, but no comment was made regarding the fate of girls' schools. However, the Taliban Minister of Higher Education said that girls would be given equal access to education, albeit in gender-segregated classes. In addition, only private higher education institutes and universities are currently running under the new segregation policy but due to financial and logistic reasons, public universities are closed now (Kayen,2022). A total of 289 responses were collected through Google form questionnaires as well and 60 women in higher education were interviewed in person and through WhatsApp calls from Kabul, Herat, Mazar-e-Sharif, Bamyan, Khost, Logar and Parwan provinces.

*Table 1. Geographical representation of the respondents*

| | | Frequency | Percent | Valid Percent | Cumulative Percent |
|---|---|---|---|---|---|
| Valid | | 100 | 25.7 | 25.7 | 25.7 |
| | Bamyan | 19 | 4.9 | 4.9 | 30.6 |
| | Parwan | 8 | 2.1 | 2.1 | 32.6 |
| | Khost | 77 | 19.8 | 19.8 | 52.4 |
| | Kabul | 48 | 12.3 | 12.3 | 64.8 |
| | Logar | 74 | 19.0 | 19.0 | 83.8 |
| | Mazar-e- sharif | 50 | 12.9 | 12.9 | 96.7 |
| | Herat | 13 | 3.3 | 3.3 | 100.0 |
| | Total | 389 | 100.0 | 100.0 | |

Source: Data output from SPSS v 22.0 based on the data from the fieldwork

*Table 2. Age distribution of respondents*

| | | Frequency | Percent | Valid Percent | Cumulative Percent |
|---|---|---|---|---|---|
| Valid | | 100 | 25.7 | 25.7 | 25.7 |
| | Old Girls (16-21)-Married | 90 | 23.1 | 23.1 | 48.8 |
| | Old Girls (16-21)-Single | 127 | 32.6 | 32.6 | 81.5 |
| | Young Girls (12-15)-Married | 20 | 5.1 | 5.1 | 86.6 |
| | Young Girls (12-15)-Single | 52 | 13.4 | 13.4 | 100.0 |
| | Total | 389 | 100.0 | 100.0 | |

Source: Data output from SPSS v 22.0 based on the data from the fieldwork

*Table 3. Level of education of respondents*

| | | Frequency | Percent | Valid Percent | Cumulative Percent |
|---|---|---|---|---|---|
| Valid | | 100 | 25.7 | 25.7 | 25.7 |
| | Higher Education | 123 | 31.6 | 31.6 | 57.3 |
| | Primary School | 66 | 17.0 | 17.0 | 74.3 |
| | Secondary School | 100 | 25.7 | 25.7 | 100.0 |
| | Total | 389 | 100.0 | 100.0 | |

Source: Data output from SPSS v 22.0 based on the data from the fieldwork

Generally, all interviewees strongly believe that there is an indirect relationship between women's education and the Taliban government, as education was easily accessible before the second regime of the Taliban, but in the past two years, girls above the age of six could not pursue their education. There are some responses from interviewees about education and the Taliban.

Ms. Madina_Herat,

*"It was easy for us to gain education before the Taliban regime, but now it is impossible, and we want the present government to allow all girls to get educated and take an active part in society."*

Ms. Manizha_Mazar-e- Sharif,

*"There was no barrier to girls' education before the Taliban era, but after that, all education doors have been closed to women, and there is no education opportunity for women right now."*

Ms. Nazo_Khost,

*"The Taliban must provide education opportunities for women so women can take an active part in society."*

*Table 4. Responses about women education and Taliban*

| | | Frequency | Percent | Valid Percent | Cumulative Percent |
|---|---|---|---|---|---|
| Valid | | 100 | 25.7 | 25.7 | 25.7 |
| | Negative | 231 | 59.4 | 59.4 | 85.1 |
| | Positive | 58 | 14.9 | 14.9 | 100.0 |
| | Total | 389 | 100.0 | 100.0 | |

Source: Data output from SPSS v 22.0 based on the data from the fieldwork

## 4.1 Physical/Logistical Barriers and Women's Education

Ms. Hadia-engineering Student_Kabul,

*" Financial and economic situations have a direct impact on women's education. Low financial facilities limit student's access to education as required".*

Ms. Zarmina_Khost,

*"There are no facilities in schools and universities, and schools and universities are far from residence areas; therefore, reaching women there takes a lot of time and trouble and even sometimes negatively affects families' decisions on allowing their female members to get educated."*

One of the women from Parwan province says,

*"There are fewer facilities like chairs, buildings, books, and other required materials in remote areas of schools and universities."*

*Table 5. Responses about physical/logistical barriers & women's education*

| | | Frequency | Percent | Valid Percent | Cumulative Percent |
|---|---|---|---|---|---|
| Valid | | 100 | 25.7 | 25.7 | 25.7 |
| | Direct relationship | 195 | 50.1 | 50.1 | 75.8 |
| | Indirect relationship | 94 | 24.2 | 24.2 | 100.0 |
| | Total | 389 | 100.0 | 100.0 | |

Source: Data output from SPSS v 22.0 based on the data from the fieldwork

## 4.2 Social/Cultural Barriers and Women's Education

One of the biggest barriers to women's education is the Afghan culture and social norms. In most cases, the families are not supportive enough of women's education because of their cultural values.

Sharara_Herat,

*''My family is supportive enough for my education, but my in-laws and my husband have different mentalities about women going to university, especially governmental universities, which sometimes puts me in trouble."*

One of the interviewees_Khost,

*"Since men are the breadwinners of the families, they think that there is no need for girls to get educated. And most of the time, in most cases, families are against women's education."*

Other interviewees_Logar,

*"Women pursuing education, especially in provinces like Logar, have to fight against people's mentalities about girls' education, and working women start from family to relatives and society because of the culture that puts extra pressure on them."*

However, the situation is a bit different in Parwan, Bamyan, Mazar, and Kabul provinces, as most people are open-minded and supportive enough of women's education.

*Table 6. Responses about the social/cultural barriers and women's education*

| | | Frequency | Percent | Valid Percent | Cumulative Percent |
|---|---|---|---|---|---|
| Valid | | 100 | 25.7 | 25.7 | 25.7 |
| | Direct relationship | 213 | 54.8 | 54.8 | 80.5 |
| | Indirect relationship | 76 | 19.5 | 19.5 | 100.0 |
| | Total | 389 | 100.0 | 100.0 | |

Source: Data output from SPSS v 22.0 based on the data from the fieldwork

## 4.3 Poverty and Women Education

Economic stability is important while getting educated. There is an indirect relationship between poverty and women's education. The interviewees have the same ideas and views regarding poverty and women's education.

*"I am the only breadwinner in my family since my dad and brother had government jobs and lost their jobs after the Taliban took over Afghanistan. Now I must think of my family's survival, not my education. I can pursue my education, which may be some years later, but now I need to think of my family".*

Since we are living in a male-dominated society, and in all cases, males have the first chance over females even when it comes to education. In the Taliban regime, the economic situation of the families is not good, and they can't afford for all members to pursue their education, so they prefer the male members to pursue their education rather than the females.

Participant 1:

*"It is hard to survive daily. Everything is so expensive, and then there are no jobs. It is very hard for families to pay the school fees or the school bus fare".*

Participant 2:

*"Things are hard, but women's education is important, and many families, including mine, would cut on many things but will support my education".*

Participant 3:

*"How can I ask my parents for money for school or the bus when I see them struggle daily for food? The government has no support for women going to school".*

Participant 4,

*"We can't live like that. We can't send kids to school because we don't have money. Everyone wants to get out of the country".*

*Table 7. Responses to poverty and women's education*

| | | Frequency | Percent | Valid Percent | Cumulative Percent |
|---|---|---|---|---|---|
| Valid | | 100 | 25.7 | 25.7 | 25.7 |
| | Direct relationship | 227 | 58.4 | 58.4 | 84.1 |
| | Indirect relationship | 62 | 15.9 | 15.9 | 100.0 |
| | Total | 389 | 100.0 | 100.0 | |

Source: Data output from SPSS v 22.0 based on the data from the fieldwork

## 4.4 Security and Women Education

As mentioned above, the instance's security has a direct relationship with women's education. Ms. Beheshta from Kabul thinks.

*"During these two years, the security situation became worse, and even if the Taliban allowed girls to pursue their education, just because of the security issues, they would not go to schools and universities."*

To one of the interviewees_Mazar-e-Sharif,

*"Security is really important when it comes to women's education inside Afghanistan. Women must be secure enough while going out of their homes to get an education, especially those who are far from the city."*

The interviewee from Herat says,

*"There are fewer security issues in Herat province."*

*Table 8. Responses about the security and women education*

| | | Frequency | Percent | Valid Percent | Cumulative Percent |
|---|---|---|---|---|---|
| Valid | | 100 | 25.7 | 25.7 | 25.7 |
| | Direct relationship | 229 | 58.9 | 58.9 | 84.6 |
| | Indirect relationship | 60 | 15.4 | 15.4 | 100.0 |
| | Total | 389 | 100.0 | 100.0 | |

Source: Data output from SPSS v 22.0 based on the data from the fieldwork

## 4.5 Education and Capabilities Development

Living a normal and useful life devoid of any physical limitations might reduce its worth or cut it short (Nussbaum, 2000). Nussbaum's capability list begins with life. Life encompasses all capabilities and exhibits the essence of all human rights. Not only is it celebrated for its functionality, but also for its true value and worth. Life on Nussbaum's list does not represent it in its mere existence but instead is cherished for its potential and capabilities. Its relevance to the study cannot be exaggerated, as women, through education, can truly understand and find meaning in their lives. It is through capability development via education that women can experience an internal and external transformation concerning the true meaning and worth of life. In its basic raw form in a tribal society and a developing country like Afghanistan under the Taliban, it means uplifting the plight of women through capability development and ensuring the decency and respect their existence demands (Inayatullah, 2022).

Overall view of interviewees on education concerning capabilities,

*"Education enables women to have a better life; educated women can make a better family that makes a better society. Education helps women to be independent and more aware of their rights because, if a woman is educated, she knows what her rights are in the community and how to use them. In addition, education helps women to have the position in society that they deserve. Education makes women stronger.*

*Educated women can be the best role models for every other woman. By using education rightly, women never let others impose something on them. Generally, education makes the world a better place to live for women".*

To other interviewees, the responses were as:

*"Education enabled me to overcome my life problems and understand my rights, helps us to have a good life in the future."*

One of the participants responded.

*"It is believed that education adds value to women's lives. For example, in my family, my mom is a gynaecologist, and she has a good personality and is active in society. The same is true for my aunts. One of them is a gynaecologist and the head of Azizullah Safar Governmental Hospital, and my other aunt was a lawyer in the previous government in Afghanistan. They are respected in our society, and other people respect them because of their knowledge. They never get flabbergasted when dealing with their problems. Since educated women are more powerful than uneducated women, therefore educated women are more knowledgeable and powerful." "They will become independent with the help of education."*

*Table 9. Responses about education and capabilities development in terms of life*

| | | Frequency | Percent | Valid Percent | Cumulative Percent |
|---|---|---|---|---|---|
| Valid | | 100 | 25.7 | 25.7 | 25.7 |
| | Negative | 39 | 10.0 | 10.0 | 35.7 |
| | Positive | 250 | 64.3 | 64.3 | 100.0 |
| | Total | 389 | 100.0 | 100.0 | |

Source: Data output from SPSS v 22.0 based on the data from the fieldwork

## 4.6 Bodily Health

Enjoying a well-nourished and well-sheltered healthy life, which includes reproductive health as well (Nussbaum, 2000). According to Inayaltullah (2022), bodily health is the second most fundamental issue plaguing Afghan women, especially the illiterate ones. Instead of enjoying a well-nourished and well-sheltered healthy life they at best are the recipients of meagre meals and scanty shelter devoid of any basic facilities. Their response is either muted or absent given this blatant violation of their basic human right. However, the existence of such a state and its survival is the result of a lack of education amongst Afghan women and by extension the absence of capabilities development.

The interviewees believe,

*"An educated women have a financially stable life and can take care of her own and her children's health. A healthy mother makes a healthy family and a healthy society, an educated woman is aware of birth control which directly impacts mother and child health, and when it comes to bodily health during the Taliban era, even women want to think of their health they cannot work for. The recent bans on women's sports and gyms negatively impacted women's health."*

*Table 10. Responses about education and capabilities development in terms of bodily health*

| | | Frequency | Percent | Valid Percent | Cumulative Percent |
|---|---|---|---|---|---|
| Valid | | 100 | 25.7 | 25.7 | 25.7 |
| | Negative | 45 | 11.6 | 11.6 | 37.3 |
| | Positive | 244 | 62.7 | 62.7 | 100.0 |
| | Total | 389 | 100.0 | 100.0 | |

Source: Data output from SPSS v 22.0 based on the data from the fieldwork

## 4.7 Practical Reason

Another significant capability developed via education is the ability to have a conscience. Only through education and learning can one be able to breathe some life into one's conscience as one becomes aware of their minuscule existence in the grander scheme of things. To have a living conscience one needs to be elevated from an individualistic standpoint and be placed in more pluralistic settings to evaluate their presence about other humans and the rest of the existence. Without education, the development of such a capability among Afghan women is nothing.

In the interviewee's words,

*"Education enabled me to overcome my life problems and understand my right, the educated and they got a job and then they became a teacher with a great salary".*

*Table 11. Responses about education and capabilities development in terms of practical reason*

| | | Frequency | Percent | Valid Percent | Cumulative Percent |
|---|---|---|---|---|---|
| Valid | | 100 | 25.7 | 25.7 | 25.7 |
| | Negative | 62 | 15.9 | 15.9 | 41.6 |
| | Positive | 227 | 58.4 | 58.4 | 100.0 |
| | Total | 389 | 100.0 | 100.0 | |

Source: Data output from SPSS v 22.0 based on the data from the fieldwork

## 4.8 Affiliation

Affiliation here reflects the mindful nature of other human concerns and is non-discriminatory to others in any way, form, or shape. In the case of Afghanistan in general and Afghanistan under the Taliban in particular, affiliation as a capability takes centre stage for several reasons.

Interviewees opine,

*"An educated woman is well aware of her own and other rights, and responsibilities against the people and the society, an educated woman not only defends and stands for her right but all other women's rights in the society and an educated woman knows to respect people and their opinions and creates and safe and peaceful work and home environment".*

*Table 12. Responses about education and capabilities development in terms of affiliation*

| | | Frequency | Percent | Valid Percent | Cumulative Percent |
|---|---|---|---|---|---|
| Valid | | 100 | 25.7 | 25.7 | 25.7 |
| | Negative | 52 | 13.4 | 13.4 | 39.1 |
| | Positive | 237 | 60.9 | 60.9 | 100.0 |
| | Total | 389 | 100.0 | 100.0 | |

Source: Data output from SPSS v 22.0 based on the data from the fieldwork

## 4.9 Sense, Imagination, and Thought

The sense, imagination and thought represent the ability to use all human senses to do or experience life in a truly human way. This capability encompasses numerous human traits that are a direct result of human imagination, thought, and reasoning. It is rather the use of this very capability that truly distinguishes humans from every other living mammal (Nussbaum, 2000).

In the interviewee's words,

*"An educated woman has better control over her emotional intelligence and can handle any situation properly, empowers and enables women to be self-sufficient, improved my decision-making ability and helped me gain independence financially and I overcame my financial and economic problems".*

Other participants opined that:

*"Education enabled me to take an active part in society and I am aware of my rights and responsibilities, enables women to know their rights in society and the knowledge on differentiation of good and bad deeds, Education enabled me to have a better life with great facilities that other uneducated women, Education make women familiar of modern life. And the educated woman has more self-esteem, and she is more confident in her life. An educated woman is more responsible about her own her family and children's future which has a significant impact on future of the society, Education makes a woman follow her rights and dreams".*

*Table 13. Responses about capabilities development in terms of emotions*

| | | Frequency | Percent | Valid Percent | Cumulative Percent |
|---|---|---|---|---|---|
| Valid | | 100 | 25.7 | 25.7 | 25.7 |
| | Negative | 65 | 16.7 | 16.7 | 42.4 |
| | Positive | 224 | 57.6 | 57.6 | 100.0 |
| | Total | 389 | 100.0 | 100.0 | |

Source: Data output from SPSS v 22.0 based on the data from the fieldwork

## 4.10 Control Over One's Environment

By control over one's environment, we understand to be part of a political process; and have full property rights as of anyone else (Nussbaum, 2000). To develop such capabilities, one needs to be fully aware of their surroundings and whatever transpires in them. Afghan women can only truly have control over their environment if they are educated. Through reading and learning can they inquire and search for answers and alternative narratives and challenge the existing status quo so bent on marginalizing them and discriminating against them. Through education and learning can they break through the glass ceiling and convince and persuade others to be cognizant of their interests and rights. They can even aspire to be policymakers and lobbyists to help improve the lives of other women and add their share to their country's development.

Overall interviewee's view,

*"Providing education related to political rights and participation in political activities for women can help increase their awareness and empowerment by making them understand their rights and showing the way to take their rights. Considering that there are always a lot of rights for women to take part in governmental issues, just by learning knowledge and using their education rights, it's possible to enjoy their political rights because education makes women more aware of their rights; otherwise, there is no way to get them. If a woman is not educated, she will lose respect and rights, whether they are political or social. By educating women, they can defend their rights. One of the most important points about boosting women's knowledge is that women become able to learn about their educational rights, political rights, and the right to freedom of speech. If women have a high level of education, it means that they can improve their skills and defend their rights against oppressors. They can advocate for their rights, and they can freely participate in the politics of their country and know better the rules and laws."*

*Table 14. Responses about education and capabilities development in terms of control over one's environment*

| | | Frequency | Percent | Valid Percent | Cumulative Percent |
|---|---|---|---|---|---|
| Valid | | 100 | 25.7 | 25.7 | 25.7 |
| | Negative | 56 | 14.4 | 14.4 | 40.1 |
| | Positive | 233 | 59.9 | 59.9 | 100.0 |
| | Total | 389 | 100.0 | 100.0 | |

Source: Data output from SPSS v 22.0 based on the data from the fieldwork

## DISCUSSION OF THE RESULTS

Development in Afghanistan suffers from the lack of capabilities of women. As the nation fails to reap the demographic dividend from this source of diversity. The present state calls for revolution, driven by the mindset of educated youths, particularly the male section of society. Until and unless the minds of men are revolutionised, gender sensitivity and women's development cannot be achieved. This diversity benefits society as well as the workplace, which calls for equality. Financial independence, freedom for decision-making, supporting the family financially, and having a good standard of living If a woman gets educated, she educates the entire family. The multiplicative effect of many women getting an education can prosper the entire nation. Exposure and education will open avenues for various professional and vocational courses. Civilization in society needs a paradigm shift in the lifestyle of the people, driving modification and updating in prevailing cultural and traditional roles. This nourishing of the nurturer (the so-called weaker section) of society can bring wealth to the entire nation. The present government must leverage the benefits of the international fund and help the citizens, which will be reflected in the growth of the nation. Thus, enabling the government to rebrand itself.

Female education in Afghanistan, which is largely attributed to traditional norms, has always been a challenge. Despite substantial measures taken by governments to provide education for girls, it remains trivial and limited. Cultural barriers are one of the main obstacles to the growth and spread of girls' education in Afghanistan. In Afghanistan, poverty, a lack of supporting infrastructure, and international aid for girls' education are only superficial aspects of the challenge. The root causes are predominant rigid cultural norms such as child marriage, disagreement for co-ed schools and male instructors for girls, sexual abuse, gender stereotypes, and dominant patriarchal codes. Afghan families tend to give the boy child better prospects, even though giving equal social, economic, and political opportunities will lead to national prosperity and promise a better quality of life to the citizens. Thus, to understand the root of what impedes girls' education, we need to delve more into cultural norms deriving from the country's multi-ethnic, multilingual, traditional society, which has been governed along tribal lines and by weak states. Different interpretations of Islam, tribal laws, and cultural practices determine these cultural norms and gender roles. Throughout history, society has perceived women as the guardians of "honour," and they have been expected to confine themselves to domestic responsibilities.

The issue of gender discrimination in education attains enormous significance and urgency as the country falls back into Taliban control. Access to education, a human right, is the centre of all human rights stemming from education and capability development. The Taliban portrays its attitudes and policies toward education as an insurgency committed to eventually taking over the education system. The Taliban has pursued a strategy of gradually controlling and coercing the education system, co-opting, and shaping the delivery of education on the ground, as well as determining who can access education and under what conditions. This is already impacting the education of millions of Afghans. While the Taliban is aware of its limited finances and human resources, it is not clear that it truly appreciates the level of funding and capacity required to maintain current access to education. The Taliban's approach to education is dependent on the government and donors, but its policy documents avoid conceding this reality. Current Taliban policies governing education have brought the country and its higher education system to the brink of collapse. The hope that the Taliban has transformed into a more moderate force over the past two decades, originating from the group's professional appearance and polished public relations, is gradually dissolving. The great concern is that the Taliban cannot govern and lead the higher education sector. They do not believe in democracy, pluralism, human rights and individual freedoms, or progress and modernity, which are the foundation of a flourishing education system. These tragic facts leave little hope for a better higher education under the Taliban in the 21st century. Even the base-case scenario with the Taliban in power appears grim and rife with uncertainty (Ahmadi, 2022). Further, there is a mismatch between the Taliban's education policies and practices, and this ambiguity helps the Taliban manipulate implementation. Education in Afghanistan is heavily reliant on international aid, which the Taliban does not readily acknowledge. The Taliban showed a commitment to pass education law so every individual in Afghanistan, including women, could have access to education under a special law by the Islamic Sharia, yet that special law was never issued. National and international NGOs play a crucial role in supporting girls to get access to higher education. Some NGOs, such as BRAC and CW4WA, sponsor female students from poor backgrounds and those from remote areas. The Taliban has issued policies supporting education as a "basic human need," including for girls "within the scope of Sharia and Afghan traditions." However, the Taliban continue to deny and disrupt service provision, including education, in many parts of the country. A gap exists between Taliban policy and practice, questioning the leadership's will and ability to address that gap.

## 5. CONCLUSION AND RECOMMENDATIONS

In the conclusion and recommendation section of the study, the researcher concludes the answers drawn from the research questions and in the context of these conclusions we may be able to form an understanding of these issues as well as identify areas for further research. To begin with, it is important to restate the research questions for the study to prove a marker and a reference point around from which conclusions can be made and assessed. The study research questions thus are: What are the barriers to female education in Afghanistan under the Taliban? How can the promotion of women's education in Afghanistan enhance women's capabilities development and empowerment? In this given context of the research question, the conclusions have been drawn from the data generated via interviews and questionnaires with school and university-going Afghan women from six provinces of Afghanistan (Kabul, Herat, Bamyan, Mazar-e-sharif, Khost, Logar, and Parwan). Therefore, there are physical, cultural, and psychological barriers exist to women's education in Afghanistan under the Taliban. The cultural barrier preventing women from pursuing higher education is not the same for all Afghani women. To a greater extent than other factors, one's race, tribe, and geographic area determine how strict this barrier is.

The issue of poverty, which is in and of itself a barrier to women's education, differs depending on how a family views and interprets the idea that education directly benefits their women's lives. All participants agreed that education is directly related to the capability's development, which directly affects human rights. Capabilities development through the education of Afghani women can be simply separated into two groups: primary and secondary.

The primary capabilities development is those that have a direct bearing on fundamental/basic human rights like life and health. The secondary capabilities are like Maslow's requirements for hierarchy and are those that are linked to a higher order of human rights and needs. These capabilities are based on an expansion of the core capabilities.

## 5.1 Recommendations and Future Scope of the Study

The current study recommends that there are many hindrances in women's education during the Taliban that directly impact women's educational access and their capabilities development. Hence, following are some of the policy enunciations:

- Examining the growth of Afghan women's skills in the workforce about their educational level.
- International communities should negotiate with the Taliban and work for a policy on women's education.
- Awareness programs about the importance of women's education must be conducted for male family members in Afghanistan.
- Further research can be made considering a larger number of women from all provinces of Afghanistan or at least one province from each zone.

## 5.2 Limitations and Future Scope of the Study

The study faced several limitations in data collection and research, including a lack of resources for face-to-face interviews with province females, time constraints, and the respondents' local language being Dari and Pashto. The study also struggled to conduct interviews in English, which would have improved data quality. Access to Afghan women for face-to-face interviews was limited, leading to interviews being conducted through WhatsApp calls. Additionally, the sensitive topic of women's education in Afghanistan led to many females being afraid to participate. A larger sample size was needed to represent various racial groups, tribes, and geographic areas of Afghani women. Despite these limitations, the study relied on Skype interviews due to logistics, time, and cost constraints. The original interviews were conducted in Dari, and translation and transcripts were formed in English. The study also faced challenges in capturing meanings and thoughts lost in translation.

## REFERENCES

Ahmadi, F. (2022). Higher education opportunities and challenges in Afghanistan: Afghan women's perspective (Doctoral dissertation, BRAC University).

Ahmadi, M. H. (2022). Higher Education of Afghanistan under the Taliban Rule. Academic Press.

Amiri, R., & Jackson, A. (2021). Taliban Attitudes & Policies Towards Education. Working Paper 601.

Amiri, R., & Jackson, A. (2021). Taliban Attitudes & Policies Towards Education. Working Paper 601. Available at: https://cdn.odi.org/media/documents/taliban_attitudes_towards_education.pdf

Burridge, N., Maree Payne, A., & Rahmani, N. (2016). 'Education is as important for me as water is to sustaining life': Perspectives on the higher education of women in Afghanistan. *Gender and Education*, *28*(1), 128–147. doi:10.1080/09540253.2015.1096922

Dore, R. (1984). *Education in Tokugawa Japan*. Athlone Press, Center for Japanese Studies. The University of Michigan.

Heamatzai, M. N., & Rafiqzad, A. J. (2023). Unleashing the Potential of Afghan Women: Strides, Challenges, and Pathways to Prosperity in Economic Development. Challenges, and Pathways to Prosperity in Economic Development.

Inayatullah, S. (2022). Women's Education in Afghanistan Women's Education in Afghanistan: Disparities in Education Under the Taliban Affecting Women's Capabilities Development.

International Institute for Applied Systems Analysis (IIASA). (2018). Education and the sustainable development agenda. International Institute for Applied Systems Analysis (IIASA). https://www.jstor.org/stable/resrep24554

Kayen, H. S. (2022). Improvements and Setbacks in Women's Access to Education: A Case Study of Afghanistan. *Muslim Education Review*, *1*(1), 19–36. doi:10.56529/mer.v1i1.5

Klasen, S., & Lamana, F. (2009). 'The impact of gender inequality in education and employment on economic growth: New evidence for a panel of countries. *Feminist Economics*, *15*(3), 91–132. doi:10.1080/13545700902893106

Meera, M., & Yekta, K. (2021). The Challenges to Political Participation of Women in Afghanistan: Historical and Cultural Perspectives. *Asian Studies (Quezon City, Philippines)*, *9*(1), 65–91. doi:10.4312/as.2021.9.1.65-91

Moravej, M. (2022). Higher Education, Gender, and Empowerment in Afghanistan: A Nation in Conflict (Doctoral dissertation, Lehigh University).

Muram, S., & Wani, N. U. H. (2020). Linkage between International political relations and foreign direct investment: A case study of Afghanistan, *Kardan. Journal of Social Sciences and Humanities*, *3*(1), 1–32.

Noora, M., & Wani, N. U. H. (2021). Linkage between imposed career study, career planning, and employability perceptions among higher education students: An empirical investigation of Afghanistan, *Kardan. Journal of Social Sciences and Humanities*, *4*(2), 1–18.

Nussbaum, C. (2000). Women's Capabilities & Social Justice. *Journal of Human Development*, *1*(2), 219–247. doi:10.1080/713678045

Powell, C. (2014). Women and girls in the Afghanistan transition. Council on Foreign Relations® Incorporated.

Rodriguez, L. (2019). 7 Obstacles to Girls' Education & How to Overcome them. Global Citizen. Available at: https://www.globalcitizen.org/en/content/barriers-togirls-education-around-the-world/

The Qur'an. 39:9. https://quran.com/39?startingVerse=9

The Qur'an. 96:1-2. https://quran.com/96

UNICEF. (2022). Girls' Education: Gender Equality in Education Benefits Every Child.

Viczko, M., & Matsumoto, R. (2022). Problematizing access to higher education for refugee and globally displaced students: What's the problem represented to be in Canadian university responses to Syrian, Afghan and Ukrainian crises? *Journal of Contemporary Issues in Education*, *17*(1), 40–56. doi:10.20355/jcie29504

Wani, N. H., Dhami, J. K., & Sidana, N. (2022). Impact of education on poverty alleviation in Afghanistan: An empirical evaluation. *Kardan Journal of Social Sciences and Humanities*, *5*(1), 1–14. doi:10.31841/KJSSH.2022.47

Yousofi, Z. (2016). The Millennium Development Goals: Women's Empowerment and Gender Equality in Afghanistan. Academic Press.

# Chapter 6
# Educational Challenges in Refugee Camps and Conflict Zones

**Mustafa Kayyali**
https://orcid.org/0000-0003-3300-262X
*HE Higher Education Ranking, Syria*

## ABSTRACT

*This chapter embarks on an intricate exploration of the manifold educational challenges that confront individuals in the tumultuous and often perilous realms of refugee camps and conflict zones. In these settings, education takes on a profound and multifaceted significance, serving not only as a means of acquiring knowledge but as a beacon of hope and a cornerstone of resilience. The authors delve into the distinct and daunting obstacles faced by educators, students, and entire communities as they navigate the turbulent waters of such environments. Through a comprehensive lens, they examine the critical roles educators play in these challenging landscapes, where they often find themselves contending with limited resources, security concerns, and the immense responsibility of nurturing the minds of the next generation. It calls upon educators, policymakers, and the global community to recognize the significance of addressing these educational challenges and nurturing the human spirit's enduring pursuit of knowledge, irrespective of the adversity it faces.*

DOI: 10.4018/979-8-3693-1483-8.ch006

## INTRODUCTION

In the vast tapestry of human history, there exist narratives that shimmer with resilience, tenacity and hope, casting their radiant light even in the darkest of hours. These stories are often born in the crucible of adversity, where tumultuous landscapes, characterized by the discordant symphony of refugee camps and conflict zones, seem to be dominated by chaos and despair. Yet, within this turbulence, a profound and deeply human tale emerges—a story of unwavering dedication that transcends the boundaries of ordinary existence. In the heart of these troubled regions, where the consequences of conflict and displacement weigh heavily on individuals and communities, a unique breed of heroes emerges—educators. They stand as sentinels of enlightenment, upholding the torch of education in the most extraordinary circumstances. It is through these educators that we can glimpse the indomitable human spirit, which refuses to bow to the adversity that swirls around them. Their stories are a testament to the power of education and the belief that knowledge can break the shackles of despair.

Our journey into the heart of "Educational Challenges in Refugee Camps and Conflict Zones" immerses us in the world of these remarkable educators. As we delve deeper into their lives, we bear witness to the remarkable struggles and triumphs of both instructors and students. This narrative unfolds against a backdrop of instability, violence, and constant relocation. It is a world where ordinary concepts of teaching and learning are stretched to their limits, where the traditional classroom is often replaced by makeshift spaces under tattered tents, and where the sound of gunfire can interrupt the sound of chalk on a blackboard. These individuals and organizations, frequently forced into the heart of armed conflict, natural disasters, and emergencies, embark on a transforming journey fueled by an unyielding dedication. Their mission goes beyond the sheer transfer of knowledge; it is about bringing solace, information, and a semblance of normalcy to individuals who find themselves caught in the harsh realities of war and displacement. They are the unsung heroes who understand that education is not only a luxury but a lifeline—a means to escape the cycle of violence, poverty, and despair. The problems confronted by these educators are enormous and complex. In refugee camps and crisis zones, they must navigate a dangerous path through the cyclone of violence, scarcity, and uncertainty. Security issues, lack of resources, and the emotional toll of witnessing the suffering of their kids are just some of the hurdles they tackle daily. Yet, their tenacity serves as both a monument to and an example of the human spirit's ability to overcome adversity.

As we investigate this tale, we discover the genuine core of their commitment, a commitment that goes well beyond the classroom and the usual areas of teaching. It is a dedication to built-in empathy, in the notion that education is not just about teaching knowledge but also about nourishing the human spirit and delivering a glimmer of hope. These educators often become surrogate parents, mentors, and role models to their kids, offering emotional support and direction in a world that is often empty of both. This chapter depicts the indomitable human spirit in the face of enormous catastrophe. It goes beyond being a mere collection of anecdotes; it is a monument to the ongoing effect of education and its unwavering ability to throw light on even the darkest corners of the human experience. It is an investigation into the depths of sacrifice, persistence, and courage that differentiate individuals who dare to stand at the nexus between combat and education. In the chapters that follow, we will embark on a journey alongside these educators, students, and the communities they serve. We will look into the various techniques they take to overcome the particular problems of their locations. We will learn about the transforming power of education in mending shattered lives and communities, and we will witness the resilience of the human spirit in the face of seemingly insurmountable difficulties. As we turn the pages, we welcome you to join us in celebrating the indomitable human spirit and the enduring power of education in the most challenging of situations. The story of educational struggles in refugee camps and combat zones is a story of optimism, of the victory of the human spirit over adversity, and a reminder that in the darkest hours, the light of knowledge may still shine brightly.

## The Crucible of Conflict Zones

In this extensive exploration, we embark on a profound journey through the heart of the crucible of conflict zones, where the complexities of adversity are laid bare for examination. These regions, often mired in a perpetual state of turmoil and despair, become the backdrop against which we endeavor to untangle the multifaceted web of forces that perpetuate their instability (Smith, J. A. 2018). The origins of this instability are as diverse as the regions themselves, and in this inaugural section, we will dive into the essence of the matter. The crucible of conflict is a place where the very foundations of society seem to crumble under the relentless pressure of political ideologies clashing and power struggles intensifying. In these regions, ethnic and religious tensions become incendiary sparks that threaten to ignite a devastating conflagration. Scarce resources take on an ever-increasing value, amplifying economic disparities that serve to exacerbate the grievances of marginalized communities. The social fabric slowly unravels, leaving in its wake a trail of trauma and shattered identities, and an indelible mark on the human spirit. Amid this intricate and often chaotic backdrop (Williams, L. M. 2020), educators find themselves facing monumental challenges. They navigate treacherous paths through landscapes where violence is a constant companion, resources are scarce, and the specter of insecurity looms large. Yet, in the face of these adversities, they remain unwavering, determined to impart knowledge and provide a glimmer of hope in a place where hope is in short supply. The crucible of conflict presents a grim reality: education becomes a battleground within these turbulent regions. Institutions, students, and educators unwillingly find themselves thrust into a harrowing drama, ensnared in the crossfire of ideological and physical warfare (Anderson, R. E. 2019).

The deliberate targeting of education infrastructure and personnel is a sinister tactic employed by those seeking to further their agendas in many conflict zones. Schools are bombed, occupied, or otherwise rendered unusable. Teachers are threatened, and students are forcibly recruited or denied access to education. This is a devastating strategy that aims to disrupt the flow of knowledge and extinguish the light of learning, leaving generations deprived of the opportunities that education can offer. But amidst these grim circumstances, educators continue to rise to the challenge. They transform abandoned buildings, makeshift tents, and even open spaces into makeshift classrooms, breathing life and resilience into the very spirit of learning. They defy threats to their lives and livelihoods, striving to provide a safe space for the pursuit of knowledge. In the face of hardship, they become beacons of hope, casting a light on the tenacious human spirit that can persevere even in the most hostile of conditions. This chapter offers a deep dive into the psychological and emotional toll that this educational battleground exacts on instructors and students alike. It exposes the incredible perseverance of individuals who fight to protect the sanctity of education in the face of relentless hardship, displaying the strength of the human spirit when faced with seemingly insurmountable problems (Khan, A. S. 2016). In conclusion, our voyage through the crucible of conflict zones reveals a terrain plagued with problems and complexities. Here, education emerges both as a victim and a lifeline. It remains a monument to the incredible strength of educators and students who persist in the pursuit of knowledge despite the barriers piled against them. Their stories serve as a tribute to the indomitable human spirit and a sharp reminder that, even in the worst of times, the flame of education may remain, offering a path toward a brighter and more stable future despite the chaos of conflict (Johnson, P. D. 2017).

## Educators as Beacons of Hope

This section of our journey delves deeper into the lives of extraordinary individuals who stand as beacons of hope amidst the tumultuous seas of conflict and despair. These are the educators who, against all odds, refuse to succumb to the pervasive despair that often engulfs conflict zones. Their stories are narratives of courage, resilience, and unwavering dedication, and they serve as a living testament to the indomitable human spirit (Wilson, E. K. 2019).

## Defying Despair

Here, we embark on a remarkable journey into the heart of human determination and tenacity. This is a realm where the very essence of bravery is redefined, as we encounter the inspirational stories of educators who choose to remain steadfast in conflict zones, despite the profound personal risks they confront. These individuals represent the unsung heroes of our time, whose unyielding dedication to nurturing the minds of the next generation, even in the face of overwhelming adversity (Smith, M. R. 2020), is nothing short of awe-inspiring. In the crucible of conflict, educators like these are faced with a choice: to either surrender to the pervasive despair that surrounds them or to defy it. They choose the latter. Their classrooms, often humble and makeshift, are transformed into sanctuaries of knowledge and hope, where children and young adults find refuge from the chaos and violence that loom outside. These educators are not just disseminators of information (Lee, J. Y. 2018); they are custodians of dreams. They understand that education is the cornerstone of a brighter future, and they refuse to let that cornerstone crumble. Their classrooms are beacons of normalcy in an otherwise chaotic world. In uncertainty, they provide structure. In the face of fear, they offer solace. Their stories are powerful reminders that, in the darkest of times, some refuse to let the flame of education be extinguished.

## Innovations in Teaching

The landscape of conflict zones is one marked by resource constraints and security concerns that are beyond the imagination of educators in more stable environments. In the face of these monumental challenges, educators in conflict zones demonstrate an extraordinary ability to adapt and innovate. This part of our journey explores the creative and innovative teaching methods they employ to ensure that knowledge continues to flow despite the odds stacked against them. Case studies and real-world examples reveal the remarkable resilience and adaptability of these unsung heroes. They've devised ingenious methods for teaching, often using minimal resources. Classes are held in the open air, in community spaces, or within the confines of makeshift shelters. Learning materials, scarce and valuable, are recycled, shared, and cherished. These educators understand that learning doesn't always require the latest technology or the most luxurious facilities; it requires passion, determination, and the belief that education is a lifeline. In conflict zones, the traditional boundaries of education are pushed to their limits. Teachers become counselors, mentors, and sometimes even surrogate parents to their students. They not only impart knowledge but also provide emotional support, helping children cope with the trauma and loss that conflict inflicts upon their young lives. It's a testament to the adaptability and dedication of these educators that they fill roles far beyond that of the traditional classroom. In our exploration of innovations in teaching, we witness the transformative power of education (Brown, S. L. 2017). It cannot only equip students with knowledge but also instill in them a sense of resilience, hope, and the belief that a better future is possible. The dedication of these educators serves as an example for all of us, a reminder that when passion meets purpose, even in the harshest of circumstances, the human spirit can triumph (Anderson, C. D. 2016).

## Empowering the Future: Recommendations for Addressing Educational Challenges

In the turbulent landscapes of refugee camps and conflict zones, where educational challenges loom as formidable obstacles, we stand at the crossroads of human potential and profound adversity. The need to chart a course that empowers individuals and communities, fostering an environment where quality education can thrive, is paramount. Education, as both a fundamental human right and an essential instrument for empowerment, resilience, and future prosperity, must transcend the boundaries of conflict and crisis. In this section, we embark on an exploration of concrete recommendations and strategies to address the educational challenges faced in these trying environments, with the goal of not just overcoming obstacles but also illuminating the path toward a brighter future (UNESCO, 2021). To meet the educational challenges posed by crises, a significant and concerted effort must be directed toward resource mobilization. Governments, international organizations, and non-governmental organizations (NGOs) should allocate a substantial portion of their financial and human resources to ensure that educational facilities are well-equipped, teachers are adequately trained, and students have access to quality learning materials. It is a commitment to investing in education that will considerably increase the delivery of educational services in these demanding environments. These resources should not be considered as expenses but rather as investments in human capital and, eventually, in the stability and development of countries emerging from violence (Save the Children, 2019).

Educators operating in crisis circumstances manage a complicated and frequently emotionally fraught terrain. Specialized training and professional development programs are required to equip them with the skills necessary to give psychological support, cope with trauma, and adjust to quickly changing situations. Moreover, systems for regular mentorship and emotional support for teachers should be implemented. Teachers are not merely sellers of knowledge but also pillars of support and direction for their students. Ensuring their well-being and efficiency is vital, for they carry the torch of learning in the darkest of times. In these complex situations, the conventional curriculum alone is unable to address the different demands of children. Flexible and inclusive curricula that expand beyond academic areas are crucial. They should incorporate life skills, peace education, social and emotional learning, and address the specific difficulties pupils confront, such as displacement and trauma. Tailoring education in this way is vital for the holistic development of individuals who are not only seeking information but also a sense of identity, belonging, and hope (ICRC, 2018).

Education does not operate in isolation; it thrives in the context of communities. Engaging local communities in the creation and management of educational initiatives is crucial. Communities should have a voice in creating the curriculum to ensure it matches their cultural values and ambitions. By integrating parents, community leaders, and local groups, a sense of ownership is generated. This engagement promotes the sustainability of educational efforts, as communities become not simply recipients but active contributors to the educational process. Innovation is the cornerstone of adaptability in crisis circumstances. Embracing novel delivery modalities, such as distant and online learning, can assist in overcoming logistical problems in providing education. Mobile learning, community-based schools, and digital resources have the potential to play a key role in reaching children who are geographically dispersed or unable to access regular classroom settings owing to insecurity. These creative ways represent the bridge between the isolation of conflict and the connectivity of learning (Smith, A. B, 2020). Education transcends academics; it covers the holistic well-being of students. Psychological and emotional support services must be an inherent element of the educational ecology. Schools should engage with mental health specialists and groups to provide counseling and support to students who have endured trauma and loss. These support systems are vital for providing a loving and sympathetic learning environment.

Raising worldwide awareness about the educational issues in refugee camps and conflict zones is crucial. Advocacy initiatives can assist in organizing resources, establishing political will, and creating a feeling of urgency to address these concerns. The international community, governments, and civil society organizations should work collaboratively to campaign for the right to education for all, regardless of their circumstances. The voices of advocacy should be unwavering, expressing the common understanding that education is a fundamental right that transcends conflict and catastrophe. To address the core causes of conflict and to promote peaceful coexistence, conflict resolution, and peace education should be introduced into the curriculum. Students should be provided with the skills and information necessary to contribute to conflict prevention, reconciliation, and peacebuilding in their communities. This integration will not only build a culture of peace but also assist in preventing future disputes from emerging (World Bank, 2017). In conclusion, solving the educational issues in refugee camps and conflict zones demands a holistic, forward-thinking strategy. We must move beyond short-term, emergency-focused efforts and undertake long-term planning to establish permanent solutions for education in these circumstances. This involves working towards the reintegration of displaced communities, reconstructing educational infrastructure, and investing in the reconstruction of societies impacted by conflict. Only by investing in the future can we encourage individuals to overcome misfortune, regain their ambitions, and contribute to the peace and prosperity of their societies.

In the end analysis, education, especially in the most adverse situations, is a potent weapon for transformation and a light of hope for the future. It is our collective responsibility to follow these guidelines and tactics, for in doing so, we enable individuals to not only survive but thrive, to not just endure but to blossom in the face of hardship. This is the path to empowering the future, one filled with opportunity, hope, and the enduring idea that education is the key to unlocking a brighter tomorrow.

## The Role of International Organizations and Aid

During the confusion and instability that define conflict zones, the role of international organizations and humanitarian agencies is nothing short of vital. This segment of our investigation digs into the invaluable contributions of these groups in promoting education in some of the most challenging and perilous environments on the globe. Their efforts are analogous to a beacon of hope that pierces through the darkness of conflict, delivering a glimmer of possibilities to those who need it most (UNHCR, 2019).

### Humanitarian Interventions: Nurturing Minds Amidst Adversity

The journey into the realm of humanitarian interventions in conflict zones reveals a story of dedication, resilience, and unwavering commitment. International organizations and relief agencies work tirelessly to provide access to education and maintain the fragile educational infrastructure in these areas, and their journey is fraught with challenges and achievements alike.

In conflict zones, educational infrastructure is often decimated, and even when physical structures still stand, they are frequently in disarray. Books are scarce, teachers are often in short supply, and the very act of attending school can be perilous. In such dire circumstances, international organizations and humanitarian agencies act as the lifeline, ensuring that the torch of education continues to burn (IOM, 2018), however dimly. These organizations, staffed with individuals of unwavering dedication, venture into dangerous situations. They navigate the treacherous paths of conflict zones, delivering textbooks, setting up temporary learning centers, and training local educators. They facilitate access to education for children who would otherwise be denied this fundamental right, and in doing so, they not only provide knowledge but also a sense of normalcy and stability. While their work is commendable, it is not without its perils. Humanitarian workers often put their lives at risk to ensure that education endures. They have faced violence, threats, and obstacles that would deter most, but their commitment to the cause remains resolute. Their achievements are measured not only in the number of classrooms reopened or textbooks distributed but also in the thousands of children whose lives are touched by the gift of learning during conflict (UNICEF, 2020).

## Empowering Local Initiatives: The Resilience of Community-Driven Education

In the crucible of conflict zones, foreign organizations and humanitarian agencies are not the only entities fighting relentlessly to sustain education. Local initiatives, generally grassroots efforts, play a key role in bridging the gaps left by larger groups. These programmes inspire and motivate local communities to take education into their own hands, building resilience, hope, and self-reliance. The value of empowering local efforts cannot be emphasized. While multinational organizations give vital help, they may not always be intimately familiar with the subtleties of local culture, customs, or the special needs of the community. Local efforts bridge this gap, delivering contextually relevant answers to the unique difficulties encountered in these locations. In conflict-affected regions, these grassroots activities are generally spearheaded by individuals who have an intimate grasp of the community's needs. They build informal classrooms, hold community workshops, and create a caring environment for learning, even under the most demanding conditions. Through these efforts, they convey a sense of agency to local communities, reminding them that they can be the architects of their own destinies. Empowering local initiatives is not just about delivering education; it's about restoring dignity and self-esteem to communities that have often been stripped of both by the ravages of violence (WFP, 2017). These activities are threads in the tapestry of resilience, and they serve a key role in repairing the social fabric that has been torn apart by violence. In conclusion, the function of international organizations and humanitarian groups in combat zones is a crucial one, delivering a ray of hope amidst despair. Their efforts, however tough and perilous, help keep the torch of education alive. At the same time, local initiatives underline the importance of community-driven efforts in bridging the gaps and developing resilience. Together, these institutions become the architects of transformation, reminding us that education can thrive even in the harshest of situations, and that hope can be reignited in the darkest of times (Save the Children, 2019).

## Navigating the Road to Recovery

As our profound trip through the furnace of conflict zones nears its conclusion, we change our emphasis to the road of rehabilitation that emerges in the wake of chaos. This section covers the various problems and unlimited opportunities that educators and students confront as they attempt to rebuild, heal, and reimagine their lives after enduring the ravages of violence. In this chapter, we dig into the stories of resilience, hope, and the incredible human power to recover and rebuild.

### Post-Conflict Rebuilding: The Phoenix Rising from the Ashes

In the aftermath of conflict, the process of healing is nothing short of a monumental task. Communities, once torn apart by turmoil and violence, are forced to struggle with the wounds and the various challenges that emerge. Here, we go into the complicated process of post-conflict recovery, reconciliation, and the arduous effort of reconstructing societal cohesiveness. In the center of this unstable landscape, education emerges as a critical catalyst, a force that helps these communities mend the cracks of conflict and opens the path for a brighter future. Education serves as a cornerstone in the process of regeneration (Mendenhall, M, 2018). Schools, sometimes decimated by the very battles that inflicted havoc, must be regenerated from the rubble. Educators, who themselves may be survivors of the conflict, act as pillars of hope and stability as they seek to restore a sense of normalcy. They help pupils through the road of recovery, addressing not merely the educational deficiencies but also the emotional wounds created by the tragedy of war. In doing so, they assist children in rediscovering their sense of identity, purpose, and self-worth. Education plays a dual role in post-conflict settings (International Alert, 2019). It delivers important knowledge and skills to individuals, helping them to rebuild their lives, their communities, and ultimately, their nations. But it also acts as a tremendous instrument for reconciliation. In classrooms where children from various sides of the dispute sit side by side, the seeds of understanding, tolerance, and unity are sown. The interchange of ideas, the formation of friendships, and the respect for varied perspectives are the first steps toward repairing trust in these broken communities. In this examination of post-conflict rebuilding, we find stories of individuals who have taken it upon themselves to be the architects of transformation. They have transformed war-ravaged landscapes into havens of study, setting the stage for peace to take root in the hearts and minds of the next generation. The journey to recovery is long and arduous, but with education as the cornerstone, there is hope that these cultures can emerge like phoenixes from the ashes, resilient and reborn (UNDP, 2020).

## Resilience and Trauma: Navigating the Emotional Landscape

The impact of conflict is not confined to the physical realm; it leaves profound emotional and psychological scars on educators and students alike. In this last chapter, we dive into the deep, often invisible scars that war inflicts, and we address the ways to cope with trauma and creating resilience. The scars of conflict run deep. Educators who have encountered the tragedies of war and loss are not immune to the emotional toll. They often shoulder the weight of not only their personal trauma but also that of the pupils under their care. The children they educate have been witnesses and occasionally victims of violence and instability. Navigating the emotional terrain becomes a challenge of utmost importance. The stories of tenacity that emerge from these conditions are awe-inspiring. Educators and kids alike display a remarkable capacity to heal, rebuild, and redefine themselves. They find solace in the power of education, which provides a sense of purpose and a route towards a better future. Community support, counseling services, and innovative mental health programs become lifelines on the route to recovery (Smith, L. H, 2017). In conclusion, the road to healing after the war is one filled with multiple challenges and opportunities. Education, as a strong tool for both reconstruction and healing, stands as a beacon of hope in these volatile times. It is through the dedication of educators and the tenacity of students that communities torn apart by violence may begin the process of healing, rebuilding, and ultimately, creating a future free from the shadows of conflict. The scars of the past may never totally vanish, but they serve as a witness to the indomitable human spirit, capable of enduring, thriving, and rising anew from the ashes of sorrow (UNESCO, 2016).

## CONCLUSION

As we draw the curtains on this profound voyage through the pages of "Educational Challenges in Refugee Camps and Conflict Zones," we stand in wonder before a magnificent tapestry woven with the strands of human perseverance, inventiveness, and an unyielding pursuit of knowledge. The educators who have heroically navigated the perilous seas of war, tragedy, and emergencies emerge as dazzling beacons of hope amidst the bleakest of settings, symbolizing the quintessential tenacity of the human spirit. Their experiences, etched across the canvas of adversity, constitute an elaborate mosaic that honours the transformational power of education. In times of disaster and conflict, education transcends its conventional role as a mere method of learning knowledge; it morphs into a profound source of solace, empowerment, and a luminous pathway leading to a brighter future. Amid apparently insurmountable problems, these educators exhibit the unshakable resolve to sustain the sacred flame of learning.

As we reflect on the narratives that have been recounted in this chapter, we are dragged into the circle of an obvious truth: education recognizes no bounds and refuses to submit to adversity. It continues resolutely amidst the worst of conditions, proving time and again that the human drive for learning and growth is invincible. In a world that seems to be perpetually teetering on the precipice of a crisis, this chapter stands as an enduring testament to the remarkable power of educators and students who, despite facing unparalleled challenges, demonstrate unequivocally that education can not only survive but thrive even in the most extraordinary of circumstances. Their stories etched over the backdrop of poverty, create a rich mosaic that respects the transformational potential of education. Education shifts from its traditional position as a simple technique of learning to become a meaningful source of comfort, empowerment, and an enlightening pathway leading to a brighter future during times of crisis and war. These educators serve as a bright example of the unrelenting determination to keep the sacred torch of learning in the face of seemingly insurmountable problems. We are drawn into the circle of an unavoidable reality as we study the experiences that have been presented in this chapter: education knows no borders and resists suffering. It continually proves that people's thirst for knowledge and improvement is insatiable despite the most challenging situations.

This chapter is a lasting testament to the incredible fortitude of instructors and students who, in the face of immeasurable hurdles, unambiguously demonstrate that education can not only endure but flourish even in the most unique settings. This chapter is a permanent tribute to the amazing power of teachers and pupils in a world that frequently seems to be on the approach of a tragedy. In a world beset by uncertainty and instability, their experiences serve as a siren call, encouraging us to deepen our commitment to learning and to promote education as a source of inspiration, courage, and change. The narratives we've read serve as the tangible symbol of humanity's indomitable spirit, a spirit that continues even in the most challenging circumstances. It is a spirit that attempts to illuminate the path ahead even though it is covered in darkness. It is an attitude that serves to remind us that learning is still a perpetual source of hope in the face of tragedy, a monument to the strength of the human mind and the unlimited potential that lives within each one of us. In our investigation of the obstacles faced by educators and children in crisis zones and refugee camps, we have witnessed the unyielding force of the human spirit, expressed by the unwavering dedication to education. We have witnessed how educators, often fighting against great difficulties, become beacons of hope, delivering not only knowledge but a lifeline to their students. We have marveled at the endurance of communities who refuse to let conflict extinguish the torch of learning and the transformative role that education plays in rebuilding lives and nations. As we take leave of this chapter, we carry with us the enduring stories of human resilience and the amazing impact of education. These stories encourage us to advocate the cause of education, seeing it as a force that transcends adversity and paves the path for a brighter, more hopeful future. They remind us that, even at the darkest of times, the human spirit endures, and the quest for knowledge continues to be a beacon of hope and a tribute to our unshakable resolve.

## REFERENCES

Anderson, C. D. (2016). The Role of Teachers in Restoring Normalcy in Conflict-Affected Communities: A Longitudinal Study. *The Journal of Educational Research, 109*(4), 425–437.

Anderson, R. E. (2019). The Psychological Impact of Conflict Zones on Children: Implications for Education. *Child Development Perspectives, 13*(3), 143–148.

Brown, S. L. (2017). Pedagogical Approaches in Conflict Zones: Strategies for Teacher Development. *International Journal of Educational Development, 60*, 101779.

International Alert. (2019). *Education and Peacebuilding: A Guide for Policymakers and Practitioners*. International Alert.

International Committee of the Red Cross (ICRC). (2018). *Education in Armed Conflict: Law, Policy, and Practice*. ICRC.

International Organization for Migration (IOM). (2018). *Education in Emergencies: Responding to the Needs of Displaced Populations*. IOM.

Johnson, P. D. (2017). The Role of Cultural and Ethnic Factors in Education in Conflict Zones. *Comparative Education Review, 61*(2), 280–302.

Khan, A. S. (2016). The Resilience of Education in Conflict Zones: A Comparative Analysis of Strategies. *Comparative Education, 52*(3), 334–352.

Lee, J. Y. (2018). The Impact of Teacher–Student Relationships on Resilience in Conflict-Affected Schools. *Journal of School Psychology, 56*, 1–15. doi:10.1016/j.jsp.2018.06.002 PMID:30340698

Mendenhall, M. (2018). Education for Peacebuilding in Post-Conflict Contexts: Recommendations for Policy and Practice. *Comparative Education, 54*(3), 349–368.

Save the Children. (2019a). *Ensuring Quality Education in Conflict-Affected Contexts: A Comprehensive Approach*. Save the Children.

Save the Children. (2019b). *Humanitarian Response Index: Education in Conflict and Crises*. Save the Children.

Smith, A. B. (2020). Transformative Education in Conflict Zones: Strategies for Sustainable Change. *Comparative Education*, *56*(1), 89–108.

Smith, J. A. (2018). Conflict Zones and Their Impact on Education: A Comprehensive Analysis. *Journal of Conflict Studies*, *38*(2), 45–68.

Smith, L. H. (2017). Post-Conflict Reconstruction and Education: Opportunities and Challenges. *Comparative Education Review*, *61*(4), 625–648.

Smith, M. R. (2020). Fostering Teacher Efficacy and Well-being in Conflict-Affected Schools. *Journal of Educational Psychology*, *112*(4), 683–698.

United Nations Children's Fund (UNICEF). (2020). *Education in Emergencies: A Handbook for the Field*. UNICEF.

United Nations Development Programme (UNDP). (2020). *Education for Sustainable Development in Post-Conflict and Post-Disaster Contexts: A Handbook*. UNDP.

United Nations Educational, Scientific and Cultural Organization (UNESCO). (2016). *Education for Peace, Reconciliation, and Conflict Prevention: Recommendations for a Holistic Approach*. UNESCO.

United Nations Educational, Scientific and Cultural Organization (UNESCO). (2021). *Education in Emergencies and Crises: Recommendations for Action*. UNESCO.

United Nations High Commissioner for Refugees (UNHCR). (2019). *Education for Displaced and Conflict-Affected Populations: Policy Recommendations*. UNHCR.

Williams, L. M. (2020). Social and Economic Forces in Conflict Zones: Effects on Education Access and Quality. *International Journal of Educational Development*, *82*, 102146.

Wilson, E. K. (2019). Teacher Resilience in Conflict Zones: A Qualitative Study. *Teachers and Teaching*, *25*(5), 567–584.

World Bank. (2017). *Learning in the Face of Adversity: The Role of Education for Conflict-Affected Communities*. World Bank.

World Food Programme (WFP). (2017). *Investing in the Future: A Comprehensive Approach to Education in Emergencies*. WFP.

# Chapter 7
# Pioneering Education for Afghan Girls in a Time of Crisis

**Ahmad Vali Behnood**
*Afghanistan Law and Political Science Association, Afghanistan*


## ABSTRACT

*This chapter explores ALPA's initiative to sustain education for Afghan girls, detailing its history, exile post-Taliban, and the growth of a robust online program for 500+ students. Despite the 70% exile of Afghan academics, ALPA engages educators to teach law and English, highlighting its vision for global collaboration. The concise narrative delves into challenges faced by Afghan girls in online education and ALPA's commitment to overcoming obstacles, concluding with the inspiring story of hope, resilience, and collective efforts in exile.*


## 1. INTRODUCTION

In the chaotic city of Kabul, Afghanistan, a transformative vision came to life in 2014, giving rise to the Afghanistan Law and Political Science Association (ALPA) in 2019. As one of the founding members, my journey with ALPA began when Professor Bashir Mobasher gathered an active group of secular law, political science, and Sharia law enthusiasts. We all met at Professor Mobashier's house, where our shared experiences from the Jessup competition and the diverse backgrounds of young graduates inspired the idea of creating an association for both secular and Sharia law students.

DOI: 10.4018/979-8-3693-1483-8.ch007

Through a series of meetings, we laid the groundwork for ALPA, drafting by-laws and navigating the bureaucratic landscape to register with the Ministry of Justice of Afghanistan. Our efforts gained momentum during the Jessup competitions, where we introduced ALPA to students and urged them to establish branches in their universities, triggering a wave of enthusiasm across the country.

ALPA's impact grew as it gathered academics from various corners of Afghanistan, earning compliments as a pioneering initiative welcomed by universities nationwide. In Afghanistan, there was an absence of an association to convene law students and academics. ALPA emerged as the pioneering initiative and acquired considerable approval upon its introduction, swiftly becoming the most accepted endeavor in the legal community. In 2020, ALPA reached a milestone by hosting its first Annual Conference at the American University of Afghanistan. This event provided a unique platform for academics to present their research papers, marking a significant jump forward for the academic society in Afghanistan. Some of the research papers presented in the conference had already yielded publications in prestigious international journals while several of them were poised for publication pending further improvement and scholarly development.

However, the narrative took an unexpected turn in August 2021, forcing ALPA to stop its activities as board members sought refuge in the US and Europe. Undiscouraged by challenges, ALPA embraced a new mission: addressing the educational needs of Afghan girls deprived of their right to education. The association offered courses in English, Dari, and Pashto languages, completing its winter semester in December 2023, and gearing up for an expanded offering in the next semesters of 2023 and 2024.

ALPA's ambitions extend beyond borders, seeking connections with universities in the US and Europe to pave the way for Afghan female students to pursue education abroad.

## 2. ALPA: A BRIEF HISTORY AND MISSION

In 2019, in Kabul, the Afghanistan Law and Political Science Association (ALPA) emerged as a transformative force. As one of the founding members, my journey with ALPA began among legal minds gathered by Professor Bashir Mobasher, uniting in our shared passion for law. From drafting by-laws to overcoming bureaucratic hurdles, our efforts gained momentum during the Jessup competitions, sparking enthusiasm nationwide. ALPA swiftly became a beacon of change, filling a void in the Afghan legal community.

ALPA's impact resonated as scholars from diverse corners of Afghanistan joined its ranks, earning acclaim as a pioneering initiative. The first Annual Conference in 2020 at the American University of Afghanistan marked a milestone, providing a platform for academics to present research and propelling ALPA into the heart of the legal discourse.

However, the narrative took an unexpected turn in August 2021, when the Taliban seized control, prompting ALPA's members to seek refuge abroad. Undeterred, ALPA embraced a new mission—providing education to Afghan girls deprived of their right to learn under a terrorist regime's rule. This transition showcased the association's resilience and commitment to a larger purpose.

As ALPA looks forward to the summer semester of 2024, with an expanded curriculum and the support of Darakht-e Danesh Academy, the organization stands poised to enhance its online education programs. The ultimate goal remains clear—to facilitate Afghan female students in continuing their education and finally pursuing higher education abroad, underscoring ALPA's dedication to empowering women globally.

From its inception to navigating the challenges of exile and innovating through online education, ALPA has proven its adaptability and resilience. The challenges were formidable, but ALPA, fueled by the aspirations of Afghan girls and supported by volunteer academics in exile, not only survived but thrived in the face of adversity.

## 3. THE TALIBAN TAKEOVER AND ALPA'S TRANSITION TO EXILE

In August 2021, Afghanistan experienced a dramatic political upheaval as the Taliban, an extremist insurgent group, swiftly seized control. This takeover, including the capital, Kabul, came in the wake of the withdrawal of US and NATO forces, ending their two-decade-long presence since the September 11 attacks. The Taliban's rapid advance led to the collapse of the Afghan government and the flight of President Ashraf Ghani, marking a critical shift in the nation's political and economic realms, and profoundly altering its educational landscape.

The Taliban's rise to power on August 15, 2021, brought about stringent restrictions on women's rights, especially in the realms of education and employment. This gradual intensification of constraints resulted in an outright ban on girls attending schools and universities, a move that has drawn widespread international condemnation and raised severe concerns about the future of women's education in Afghanistan. Now, over 2.5 million girls in Afghanistan are deprived of education.

Amidst this dire situation, the Afghanistan Law and Political Science Association (ALPA) stood out as a pillar of hope and resistance. Originating as the Afghanistan Law, Sharia and Political Science Association (ALSPA) in 2019, ALPA was founded by a group of enthusiastic academics in Afghanistan.

However, following the Taliban's takeover and the subsequent collapse of the Afghan government, the majority of ALPA members were compelled to flee Afghanistan, mainly relocating to the United States, due to fears of persecution. Despite being uprooted from their homeland and facing the challenges of displacement, ALPA members continued to underscore the importance of education, particularly for girls.

In their exile, ALPA effectively navigated these challenges and adapted to provide education to Afghan girls, now barred from traditional educational institutions under Taliban rule. Demonstrating resilience and commitment, ALPA conducted online classes to ensure the continuity of education for those most in need.

Recently, ALPA announced the successful completion of its fall semester, a significant achievement under the prevailing circumstances. This semester was marked by the involvement of over 20 dedicated instructors and the participation of hundreds of students across 15 varied courses. Characterized by vibrant discussions, innovative teaching methodologies, simulations, mock projects, and mental health exercises, the semester fostered a sense of unity and determination among both students and educators, united in their pursuit of knowledge and a brighter future.

This initiative is particularly crucial today, as an estimated 70 percent of Afghan law and political science professors live in exile. Through its efforts, ALPA aims to bridge the educational gap caused by the Taliban's restrictions, ensuring that Afghan women can continue to pursue their academic and professional dreams even in challenging circumstances.

## 4. ALPA'S ONLINE EDUCATION PROGRAM

In the aftermath of the Taliban takeover and recognizing the dispersed nature of its members, ALPA embarked on a groundbreaking initiative—the Online Education Program. In response to the comprehensive educational restrictions imposed by the Taliban, particularly affecting Afghan girls both in universities and schools, the Afghanistan Law & Political Science Association (ALPA) undertook a significant initiative to alleviate the deprivation in both higher education and secondary education spheres. The initiation of ALPA's efforts transpired during the summer semester of 2023, wherein the organization commenced its academic endeavors with a modest offering of three courses: Advanced English Reading, Introduction to Sociology: English Reading and Speaking, and Entrepreneurship & Law.

ALPA's online program was strategically conceived to cater to individuals facing bans from universities, with a particular emphasis on those expelled from Law faculties. To further enhance the clarity and disciplinary focus of its online courses, ALPA pursued an evolution in its approach. The subsequent fall 2023 semester witnessed the expansion of ALPA's curriculum, incorporating a diverse range of courses such as Mental Health: Emotional Wellness, Human Rights: English Reading & Speaking, Gender Studies: English Speaking, Research Method: English Reading, Discrimination and Society (Dari), Arts & English: English Listening, Academic Writing (Pashto), English for Beginners, Communication: Journalism (Dari) Academic Preparation: Advanced English, Advanced English Reading, Introduction to Politics: English Reading, Scholarship Preparations, Academic Writing (English), and International Humanitarian Law.

During the fall 2023 semester, ALPA effectively utilized Google Classroom as the primary platform for course delivery. This implementation facilitated students' anonymity options, allowing them to provide ALPA with aliases and alternative email addresses to ensure the security and privacy of students. Remarkably, ALPA undertook the significant endeavor of conducting its courses without any external financial support. All instructors volunteered their expertise, aligning with the commitment demonstrated by the board and administrative staff. Additionally, the expenses associated with the utilization of Google Classroom were covered by the dedicated members of ALPA, exemplifying the organization's resilience and commitment to providing educational opportunities in the face of challenging circumstances.

Following the completion of its Fall semester in 2023, the Afghanistan Law & Political Science Association (ALPA) embarked on the development of its programming for the upcoming summer semester of 2024. Scheduled to commence in January 2024, ALPA has structured an extensive curriculum, encompassing approximately 26 courses to cater to a diverse range of educational needs. The courses scheduled for the Summer semester are accurately designed to address a broad spectrum of subjects, including Legal Writing, Transnational Criminal Law, International Criminal Law, Global Citizenship, Sustainable Development and Charity Organizations, Political Systems and State Building, TOEFL, Introduction to Mobile Photography and Documenting Techniques, International Competitions, English for Beginners, Environmental Studies, Introduction to Gender Studies, Advanced Gender Studies, Human Rights: English Reading & Speaking, Mental Health: Emotional Wellness, Transitional Justice in Afghanistan, Research Lab, Scholarship Preparations (for Master's Degree), Constitution, Public Policy, Intermediate English, Intellectual Property Law, Afghanistan Documentaries: Listening & Speaking, Academic Writing II, Mental Health, and Entrepreneurship & Law.

Importantly, these courses are tailored not only to meet the educational needs of university students but also extend to some extent to accommodate secondary school students. ALPA's strategic planning involves a commitment to continually expand its course offerings in subsequent semesters. A notable development in the upcoming semester is the support received from Darakht-e Danesh Academy, enabling ALPA to utilize their Learning Management System (LMS) for course delivery. In addition to offering courses, ALPA has conducted workshops for both instructors and students, focusing on imparting crucial information about the courses and providing guidance on using teaching platforms such as Google Classroom.

ALPA is poised to enhance its online education programs further in subsequent semesters, with plans to establish partnerships with universities worldwide. The ultimate goal is to facilitate Afghan female students in pursuing master's degrees abroad, reflecting ALPA's unwavering commitment to realizing this significant objective. The organization remains dedicated to making education accessible and empowering Afghan women through academic opportunities on a global scale.

The Online Education Program served multiple purposes. It not only addressed immediate challenges posed by exile but also positioned ALPA as an innovative and adaptable organization. The virtual environment fostered a sense of community among members and students, bridging geographical distances and ensuring that the pursuit of academic excellence continued unabated. This initiative showcased ALPA's commitment to not just surviving but thriving in the face of adversity.

## 5. CHALLENGES AND OPPORTUNITIES

The challenges faced by ALPA were profound and multifaceted. The displacement of its members, the threat to academic freedom, and the uncertainty of the political landscape presented formidable obstacles. However, within these challenges lay unique opportunities for ALPA to redefine its role and impact. The very nature of exile prompted ALPA to innovate, leading to the creation of the Online Education Program. This not only addressed immediate challenges but also positioned ALPA as an organization capable of adapting to changing circumstances.

### Challenges

Education has always faced a myriad of challenges in Afghanistan. Irrespective of whether such challenges had to do with the culture or economic or many other factors, Afghan children, specifically girls, have for most of the part been subject to limitation when it comes to schooling and education. The past twenty years of international intervention brought with it a promising hope of an increase in school and university enrollments, and also increase in literacy rate in the country. Unfortunately, the Taliban takeover halted many of such inspirations and is pushing the country back into the dark ages of illiteracy for girls.

With relatively weaker fundamentals in more than half of the country, online education understandably faces some serious challenges. It this part, we will go over some of the main challenges that ALPA had to cope with regards to online education.

1. Non-familiarity with online education modalities: One of the main challenges of online education is the fact that not many students are well familiar with the online education system. Although Covid had already impaired face-to-face education, not many public and private institutions were able to adapt to the new reality—teaching online with some satisfactory extent. Except for a handful of higher education institutions and a very limited number of private schools, the educational system faced a total backdrop when covid hit. The backdrop even cost some students a year of their educational promotion. Although the Ministry of Higher Education, at least, tried to implement an online learning management system, it was not as successful as it should have been. The Ministry of Education unfortunately was not capable of making any meaningful attempt towards helping online education in school during the covid period.

This short coming unfortunately still impacts online education in Afghanistan. Although many institutions, including ALPA, have been actively providing online education in both school and university levels, not many students are familiar with how to learn online. This is particularly true in many rural areas. Advent and use of social media and smart phones have been particularly helpful with the generation living in the cities. Those in the rural areas have been crucially deprived of such luxuries of modern life.

2. Internet Connectivity: Internet connectivity remains one of the main challenges in the country. Even though many cell phone companies provide cellular data in many of the provinces, the quality of internet is still one of the basic issues students are facing. Online video conference and use of google meet or zoom have been very challenging. The bandwidth many students have access to does not support synchronous video conferencing, unfortunately. While students have technical switch of asynchronous models, playing and downloading a video of 30 minutes or more requires students to wait for hours before they are able to watch the recording. Such delays then impact students' completion of assignments and thus catching up with the course material.
3. Electricity: Electricity issues are one of the main hurdles to online education in Afghanistan. Even students with some level of satisfactory internet connectivity, lack of electricity bars them from substantial participation in class activities. A large number of provinces and rural areas do not have electricity or use locally built small water damps that provide a trivial amount of power. Such settings can only be used for lighting only and in limited instances charging smart phones. Using a PC or even a Desktop is not possible with the voltage provided by such settings. While this might not be a challenge with school education as course works can maintain a level of balance between content

delivery and students' homework and a small tablet may resolve the issue, university level education faces some serious problems. University level courses are understandably assignment intensive and require a fair amount of research and writing. Having access and being able to use a laptop, PC, or a similar device is key in online university level education which in turn requires access to electricity (and stable internet connectivity).

4. Economic hurdles: For many who are living in big cities such as Kabul, Herat, Mazar, or Qandahar, both internet and electricity might not be of primary concern, but the economy definitely is. With more than 95% of the population living below the line of poverty and more than 70% not able to resource three meals a day, the challenge should be more than apparent. An average use of social media for a person cost around a thousand Afghanis per month—that is five days wage for a daily wage worker in the Capital Kabul. Many of our students are not able to pay for the internet cost required for their classes. This is then a big challenge for families that have more than one child attending online classes.

## Opportunities

Even though students have been facing a fair number of challenges, it is inspiring to witness the number of students who are eager to study and are actively participating in their classes. Although the Taliban have banned in person school and university education, online education has provided blood to the veins of Afghan girls education. While starting classes through ALPA, the ALPA board and team of instructors who initiated this effort considered such an opportunity as a key benefit. The following provide an account the opportunities ALPA has been able to benefit from,

1. Girls' inspiration: one of the most important and key factors in online education success in Afghanistan post-Taliban takeover has to do with the new generations aspirations and their inspiration to learn. In spite of the challenges mentioned above, the number of students who want to learn and who are eagerly participating in their classes is a complete surprise. Different from the experience of the generation deprived of education in the first rule of Taliban who were initially highly impact by the civil war in the country right before Taliban control was established, the current generation have lived through two decades of educational opportunities and facilities open to them. Many of these girls witnessed their families, relatives, friends, or peers establish a successful life with the fortune of education. Such an experience has provided the current generation with a high motive and aspiration to learn and be a factor for positive change in their societies.

2. Families support: Connected to the first point raised above, it is similarly important to note that families have been supportive of their girl's continuation of education. With families experiencing the benefits of education in their own lives over the past two decades and understanding the risk of uneducated society, families have been very supportive of their girl's education. There are many families who are not able to maintain the very regular monthly expenses, but they make sure that their children are able to have access to sufficient facilities to participate in their online classes. This is particularly inspiring in families with more than one child studying online.
3. Online education: Many of us have memories or have heard stories of those girls between 1995-2001 who were completely deprived of their education and most of whom lost the opportunity to get education for the rest of their lives. This time, however, the conditions are apparently different with the advent of technology, specifically educational tech, and internet. With school being closed for upper secondary education right after Taliban takeover and universities closure at the end of 2022, a limited number of girls who had access to online facilities and could afford the costs felt the suffocation of education in the country. With some modes of online education being used after the spread of the Covid-19 virus, a large number of girls (and boys) who had the privilege of access to modern life facilities were able to switch to online education. This provided institution offering online education post-Taliban take over with precedent and basis. Although these students need a fair amount of training on using a so designed higher educational LMS, the process was much smoother than one could imagine.
4. International Support: another important factor in online education success in the Afghanistan is related to the support many institutions are getting from outside entities, which include universities, NGOs, and even governments. Some of these entities have provided a number of scholarships to Afghan girls, some have initiated their own online teaching modalities, and some have provided a limited amount of funding that has been crucial to the functioning of a number of institutions providing online education. Such supports have manifested as providing students with internet bundles, free access to online material, payment of instructors, or even distribution of tablets to those in need.
5. Afghan diaspora and philanthropists' individual support: Volunteer support of Afghan academics and scholars who are outside the country and also many other foreign academics and scholars has been one of the main drives of online education, at least in the case of ALPA. ALPA has conducted many individual classes prior to starting a full semester and a full semester in Fall of 2023 entirely based on volunteer support of such academics. Professors, PhD's holders, and

scholars have all come together and though importance courses within their area of expertise free of charge and without any expectation for Afghan girls. ALPA offered a total of 13 courses with 300+ enrollment in Fall of 2023. All of these courses which ranged from language skills or social sciences, mental health and legal studies were taught by Afghan or foreign scholars on volunteer basis. Many of these professors and scholars have pledged their commitment to volunteering for the coming semesters as well.

## 6. CONCLUSION

In the tumultuous landscape of Kabul, Afghanistan, the birth of the Afghanistan Law and Political Science Association (ALPA) in 2019 marked a transformative moment. As a founding member, my journey with ALPA began amidst a gathering of legal minds, driven by a shared passion for secular and Sharia law. From the early days of drafting by-laws to navigating bureaucratic challenges, our collective efforts gained momentum during the Jessup competitions, sparking enthusiasm across the nation. ALPA quickly became a beacon of change, filling a void in the Afghan legal community.

ALPA's impact reverberated as it gathered scholars from diverse corners of Afghanistan, earning acclaim as a pioneering initiative. The first Annual Conference in 2020 at the American University of Afghanistan marked a milestone, providing a platform for academics to present research and propelling ALPA into the heart of the legal discourse.

However, the narrative took an unexpected turn in August 2021, when the Taliban seized control, prompting ALPA's members to seek refuge abroad. Undeterred, ALPA embraced a new mission—providing education to Afghan girls deprived of their right to learn. This transition showcased the association's resilience and commitment to a larger purpose.

ALPA's response materialized through the Online Education Program, addressing the immediate challenges of displacement and educational restrictions. Despite facing profound challenges, including non-familiarity with online education, internet connectivity issues, electricity shortages, and economic hurdles, ALPA persevered. The commitment of instructors, the utilization of Google Classroom, and the support of members fueled the success of the fall 2023 semester.

The challenges were numerous, but within them lay opportunities. The inspiring eagerness of students to learn, coupled with the unwavering support of families, became crucial elements. Online education, though facing hurdles, presented a lifeline for Afghan girls, bridging the gap left by the Taliban's restrictions. International support, both from organizations and the Afghan diaspora, further bolstered the cause, reflecting a global commitment to education.

As ALPA looks forward to the summer semester of 2024, with an expanded curriculum and the support of Darakht-e Danesh Academy, the organization stands poised to enhance its online education programs. The ultimate goal remains clear—to facilitate Afghan female students in pursuing higher education abroad, underscoring ALPA's dedication to empowering women globally.

The journey of ALPA is a testament to the dynamism of academia intertwined with societal development in Afghanistan. From its inception to navigating the challenges of exile and innovating through online education, ALPA has proven its adaptability and resilience. The challenges were formidable, but ALPA, fueled by the aspirations of Afghan girls and supported by a global network, not only survived but thrived in the face of adversity.

In a world where education is often taken for granted, ALPA's story serves as a poignant reminder of the transformative power education holds, transcending borders and overcoming even the most daunting challenges.

# Chapter 8
# Planning the Big Escape:
## Perilous Journeys and Lessons Learnt

**Mimla Wardak**
*Levy Economics Institute, Bard College, USA*

## ABSTRACT

*This chapter recounts a poignant and inspiring narrative of resilience and personal growth in Afghanistan's political instability. The author's journey began at the American University of Afghanistan, where education became a symbol of defiance and optimism during the Taliban's rise. Education and empowerment remain the author's priorities despite the COVID-19 pandemic and political instability. The narrative vividly recounts the author's harrowing experiences of evacuation, the transition to life in Bishkek at the American University of Central Asia, and the author's eventual relocation to the United States to attend Bard College. The journey explores resilience, personal growth, and the relentless search for freedom and empowerment. This chapter celebrates family, self-discovery, and the resilience of optimism and persistence in the face of adversity, demonstrating the tenacity of individuals who pursue education and improvement.*

## PLANNING THE BIG ESCAPE: PERILOUS JOURNEY

In a sea of faces, I mastered to stand true. From my grandmother's words, wisdom anew. "Look beyond horizons," her voice pleaded, "Dream bigger, seek deeper, always aspire for more." As darkness fell, with the Taliban's sway. Their bans on learning, my hopes were nearly slain. Even her profound words seemed lost in the fray, But in silent defiance, I preferred to stay and pray. In her lessons, I discovered, amidst night's darkest pall, an unceasing spirit to rise after a fall.

DOI: 10.4018/979-8-3693-1483-8.ch008

In the frenetic tapestry of my life, there was always a prodding voice that echoed louder than the rest – my grandmother's. I spent most of my life with my grandmother, especially my early childhood years. She was not educated, yet she taught me the most profound and meaningful lessons of life, molding me into the person I am today. Her words became my compass. Every challenge, every obstacle, I met with the fervor she'd instilled in me.

However, as the clock's hands turned farther, a realization struck me: sometimes, no matter how much you desire or how hard you strive, obstacles beyond control can hinder your path, which can turn out to be the most heart-wrenching lesson of all. Still, amidst all the chaos of my life, my grandmother's words were like a guiding star, instilling in me courage and a resilient spirit. She taught me to see the world in new ways, to dream, and to fight for what I truly seek. This fighting spirit was my shield in hard times when everything seemed against my ambitions.

Let's begin from the start, "After graduating from Jamhuriat High School, I was eager to continue my education in an environment that valued diverse opinions and fostered personal growth. The American University of Afghanistan (AUAF) caught my attention as one of the premier institutions in the country, symbolizing hope and, for some, a sign of defiance. Whispers of caution from well-intentioned people warned me of security threats surrounding AUAF. Despite warnings about security concerns at AUAF, I was determined to follow my passion. Gaining admission to such a renowned university in the country was challenging, but my perseverance paid off, and I succeeded in embarking on this significant educational journey.

Initially, I joined AUAF in 2019 with Financial Aid as a Business Administration Student. However, in my second semester, due to my academic performance, I received a fully-funded scholarship from the US embassy. Embracing every opportunity at AUAF, my eyes were always set on a brighter sphere. AUAF has always been like a family to me, providing me with incredible friends and outstanding professors who supported me at every step. The university not only enhanced my academic skills but also fostered my social development. Being part of the vibrant AUAF family is a source of great pride for me. This chapter of my life embodies what I hold dear, echoing John Lennon's words, 'There is nowhere you can be that isn't where you're meant to be.' It's a testament to the belief that we are always exactly where we need to be in our journey. That is why even today I believe and support AUAF in its mission, "education must prevail."

Though everything was going so perfectly and well, but life has an odd way of deviating from what is intended. As I was finding my stride, the world was gripped by the pandemic. During the spring 2020 midterm exams, I remember walking with a friend in the AUAF yard, discussing post-break plans, unaware it would be our last day ever on campus, as the next day university closed due to COVID-19. This pandemic brought quarantine, isolation, loss, and fear, but our spirit remained unbroken. We shifted to online classes, continuing our education. AUAF successfully transitioned to this format. When we started online classes due to COVID-19, I was a sophomore, but by the time rumors circulated that in-person classes would resume in fall 2021, I was already a senior. Receiving an email from the AUAF president about the resumption of offline classes brought immense joy to every AUAF student, as everyone eagerly anticipated reuniting on campus. However, we were about to face challenges beyond our imagination.

The rumors began as whispers: one of the AUAF campuses had closed due to security threats, and our studies would remain online due to the deteriorating security situation in Afghanistan. Initially, I was skeptical, clinging to the hope of returning to offline classes. But reality proved otherwise. Yet, as fate would have it, another twist awaited, changing everything. At that moment, a saying resonated deeply with me: 'Death isn't the worst loss. It's when something dies inside you while you're still alive.' This profound realization struck me as a heavy cloud of concern and worry enveloped our nation. This growing anxiety stemmed not only from the immediate dangers and change but also from the realization that our normal way of life, the one we cherished and hoped for including our education at AUAF, was slipping away.

As for several months, reports emerged that the Taliban would ultimately take over Afghanistan, but I had confidence in my army and security forces that this would never happen. I believed that now we are stronger, better armed, and more educated and that my homeland will not succumb this time. They will fight for themselves since they now understand what is right and wrong. However, the speed at which several provinces fell astounded me. It was late at night, and the news of Jalalabad's falling without a shot being fired filled me with dread and disquiet. I remained up the whole night, afraid of what would occur ahead, what would happen to all of us? I kept calling my father, who was in the defense army and on duty at Jalalabad Airport that night. It was one of the most stressful nights I had ever experienced. My fears were not just for my father but for my country, my nation, and our future. In miserable and worrisome circumstances, the following day, I started working on my assignment, but I kept looking out from my home front window, through which a police roadblock could be seen. Seeing army troops and local police gave me the impression that everything was under control. It was the only silver line of hope that day.

Around 11:00 a.m. on August 15th, 2021, my mother who was a school teacher and was in school that day called and urged us to remain indoors and secure all windows and doors, as the Taliban had entered Pulcharkhe and were releasing prisoners. At the same time, my father returned home, I had never seen him that hopeless ever, he just said one thing burn all my documents and also your AUAF card and closed the door of his room. Burning those items — photographs of my father with his colleagues from his time as a pilot, his certificates, and cherished memories — was an excruciating ordeal. It symbolized not just the loss of documents, but a profound part of our lives and identities.

On that day, within just a few hours, my faith and optimism were completely shattered. It was startling and unbelievable to see how government troops melted away, and high-ranking officials fled in the twinkling of an eye. I cannot put into words how terrible and stressful that felt; all of my senses were numbed, and all I could see was a catastrophe and tragedy looming for all of us. In those tumultuous days, a whirlwind of emotions engulfed me: a visceral sense of betrayal, shock, and the oppressive feeling of impending doom. Days turned into nights, each progressively more frightening. Promises made were broken as quickly, adding to the sense of chaos. As I looked out at the world we once knew, now altered by factors beyond our control, I truly understood the weight of that old saying. Something had indeed died within me, even as I continued to breathe.

While writing this I went back to my diary which I wrote daily, Its pages are filled with daily contemplations and reflections from that period. Every word, every rushed scribble, was a reminder of a time when writing down all of my concerns and wishes on paper was mere solace to cope. Seeing those notes while writing this filled my eyes turbulent, tears flowed freely. It filled my heart with fear, and my mind raced, imagining the alternate reality where I remained trapped in that dire situation. A place where every day would be a stark reminder of lost dreams, stolen futures, and an oppressive existence. How terrible every human no matter what their gender is might feel knowing they have no future there but just to live as slaves in their own country, with no hopes, no dreams, no future. Recently, I saw a photograph of schoolgirls at their sixth-grade graduation, their faces streaked with tears. Their sorrow over lost opportunities and a future out of their hands is a harrowing sight. It's a painful reminder of the ongoing oppression, particularly for women, in my homeland.

Even as I penned this, my heart ached with empathy for those still trapped, yearning for a glimmer of hope in a landscape dominated by shadows. I still remember how terrified everyone was. When I went around, I saw sad, despairing, and scared faces. Though the sun shone as usual there was an atmosphere of grief and sorrow in my country. As a student at the American University of Afghanistan, I found myself in a similar unpleasant and dangerous position, concerned about what the future may hold. Every corridor echoed the same sentiments, "What Now." For me, August 15th, 2021, is a black day that I will never forget. Never had I skipped a class, never missed an assignment. Those days were different. The books lay unopened, the laptop untouched. How could I, when it felt like the walls around me were closing in? Nights were particularly agonizing. Every time I closed my eyes, the weight of reality pressed on my chest. It felt like a nightmare, I desperately wished to wake up from. But the dawn brought no reprieve, only a reminder of the Hideous truth.

I was concerned about a lot of things, particularly my halted education. I felt like I had come so far that there was no way I could quit or take a step back. The very core of my being revolted against the idea. But wasn't that the intent of the Taliban? To break us, to make us kneel, to force us into submission to their archaic ways? Yet, amidst the enveloping darkness, a flicker of determination ignited within me. Surrender was not an option. My education, and my dreams, represented a future brighter than any darkness they could cast. It was, and always would be, my beacon of hope.

As I grappled with these personal challenges, the wider situation in Afghanistan continued to evolve in ways that were both alarming and contradictory to the values I held dear. This discrepancy between my beliefs and the unfolding reality was not just a matter of divergent opinions; it struck at the very core of my understanding of Islam and governance. The Taliban's actions, so at odds with the peaceful teachings of Islam, highlighted a distressing paradox. The erosion of freedoms, especially for women and girls, became not just a symbol of lost rights but a glaring example of the misuse of religious and political power. It was in this environment of conflicting ideologies and practices that I found myself questioning the very nature of authority and faith in my country. How could the teachings of Islam, which emphasize peace and choice, be reconciled with the harsh realities imposed by those claiming to uphold these same principles? This struggle to understand and adapt to the changing landscape of my country was a journey of both internal conflict and external observation, a path fraught with complexity and contradiction.

I was also certain that things would worsen since there was a significant gap between what the Taliban was saying and what they were actually doing. I was unable to trust them, and I know they were capable of doing anything and still are. Their duplicity was evident to me, and while hope remained, trust was a luxury I couldn't afford. In the heart of the capital, where the walls of the presidential palace stood tall, murals of schoolgirls once painted a hopeful future were swiftly replaced, erasing all traces of dreams and aspirations for a brighter tomorrow for our girls. Their version of progress was marred with regress. This misuse of Islam, a faith rooted in peace, for authoritarian control confounds me. True Islam does not advocate for forcing others against their will. Living under such tyranny, where every aspect of life is governed by a rigid 'yes' or 'no,' goes beyond simple oppression. It's a suffocating existence, filled with endless rules and restrictions. It feels like being trapped in an invisible cage, a confinement that's arguably worse than physical imprisonment.

Among all the unknowns, news of evacuation on 31 August 2021 provided a window of hope and new beginnings, but waiting for several hours on buses for Airport gates to open was a nightmare. Hundreds of students went home with hearts full of sorrows and disappointments since the gates never opened, but our spirits still alight with the dream of escape. Coming back from that failed evacuation made me realize one thing it was not only me disparate enough to leave the oppression of my country but all, there were so many local families that day on the streets who just wanted to enter our buses, finding a way to escape the country.

Later on, students began receiving emails about evacuation via land, specifically through the Torkham border. I was part of a group scheduled to be evacuated on September 18, 2021. As I packed a small bag, memories of the last evacuation attempt made me doubtful about the success of this one. Despite my uncertainty, I followed the instructions: we left our homes around 11 pm, gathered at a friend's house, and departed in cars at 3 am. The farewells that day were particularly hard. My grandmother, along with my father and sister, dropped me off at the meeting point. My grandmother kissed my forehead, urging me to pray for Allah's protection on this uncertain journey. Her goodbye felt final as if she knew we might never see each other again. The sight of my father, with tears in both our eyes, was profoundly distressing and remains a painful memory.

The journey through the land route to the border was far from easy. We arrived at the border at 6 am, only to be denied crossing because we were unaccompanied by a 'mahram' (a male guardian). In the sweltering heat, we waited and persisted. After 10 hours, we were informed that the border would open for just 10 minutes, and it would be a matter of how swiftly we could run across. Driven by the hope of a better future, we all ran with such urgency that nothing else mattered. I vividly recall the exhausted and anxious faces of the students that day. Now, whenever we reunite, we reflect on that nearly impossible evacuation, acknowledging that it could not have been successful without our mutual patience, support, and communication. After staying for some days in Islamabad on October 3, 2021, we were evacuated through a charter flight to Bishkek Kyrgyzstan to pursue our education at the American University of Central Asia. The cool winds of Kyrgyzstan brought with them a whirlwind of emotions. As I stepped onto this new soil, nostalgia towed at my heartstrings. Thoughts of my family left behind clouded my vision, their safety, was an agonizing mystery.

Upon arriving in Bishkek, we were faced with new uncertainties, but this time, we felt safe. Settling down there was not easy. We cried, we fell, and sometimes even gave up, but we always stood back up, determined to pursue the goals for which we had left our country and beloved family. Life in this foreign land, with its unfamiliar culture and language, could have been extremely isolating, but thankfully, it wasn't, thanks to the supportive staff at AUCA. I always say I will be forever grateful to Bishkek and the American University of Central Asia for safeguarding our dreams at a time when they were most vulnerable. There, we built a community, creating our own enclave of comfort. Together, we persevered, learning and growing, and eventually graduating. The graduation I had always dreamt of was bittersweet. I had always wished for my grandmother, mother, father, and siblings to be there, watching me in my graduation attire. Seeing Kyrgyz families support their children during the ceremony brought tears to all of us, as we remembered our families and felt their absence. However, there were also moments of happiness and pride. Graduating as one of the top three students from AUAF and having the AUAF president attend the ceremony in support added to our joy.

During my two years in Bishkek, I developed a profound connection with that city. Yet, this bond was overshadowed by the heart-wrenching news of my beloved grandmother's passing. The anguish of attending her funeral virtually, through the small screen of a phone, and not being able to be there physically, has left an indelible mark on me. This period is etched in my memory, not just for the experiences in a new land, but for the pain of a significant loss felt from afar. Furthermore, life in Bishkek, while an improvement, remained shrouded in uncertainty. We often felt like temporary guests in a foreign land, and our stress levels increased as we awaited our interview calls for relocation, which arrived at different times for each of us. While we cherished our time and the memories we created in Bishkek, the prospect of moving to the US brought both hope and added anxiety. The process was fraught with challenges, including the slow pace of approvals and the looming uncertainty about our futures. Each passing moment felt heavier, as we clung to the dream of continuing our education in the United States.

As the time for relocation drew closer, I was filled with a sense of desperation and anticipation. But when the moment finally arrived and I received my flight ticket, I felt numb and unsure of how to react. The enormity of the change left me at a loss for words, a sentiment I expressed in my diary." 'Why is life so unpredictable that even I don't know how to react to its changes'? Saying goodbye once more, this time to people who had become like family to me in Bishkek, was difficult.

Looking back, I am filled with gratitude for the tireless efforts of organizations like the State Department, IOM, IIE, QSAP, and our host universities in the US. Their support was crucial in making this next chapter of our lives possible. Thus, this journey had its own share of bittersweet moments. While some of us expanded our wings and took flight, others stayed on the ground, bound by doubt and bureaucracy. The farewells at Manas International Airport weren't merely goodbyes; they were a wrenching separation from the family we had built. Tears shed there were for friendships, bonds that had become as pivotal as those with our own families. I distinctly remember the date: August 25, 2023. As the plane's wheels touched down on the tarmac in the United States, a wave of mixed emotions overwhelmed me. Relief, excitement, and anxiety intertwined, marking the beginning of the next phase of my journey. Although I had left behind the turmoil of my homeland and the uncertainties of Bishkek, I faced a new challenge: establishing myself in this vast and unfamiliar nation.

When I first arrived at JFK Airport and spent the following three days at Albany for legal formalities, everything seemed to pass in a blur. Although I was moving forward, I found it hard to keep up with the rapid changes around me. Settling into my new life in the US involved some unexpected adjustments, such as sharing rooms and living in mixed-gender dormitories, which were new experiences for me. Despite my long-held dream of coming to the US, these unexpected changes initially left me feeling a bit uneasy. I had eagerly awaited this opportunity, so feeling conflicted about these adjustments seemed ungrateful. However, I decided to focus on the positive aspects of my experience, trying not to dwell on the downsides. Still, there were real challenges to face, like the stress of job hunting and getting used to a new place. Being at Bard College, with its secluded location, added to these difficulties. But I saw this as my beginning, the starting point from where I would figure out my next steps. Instead of getting bogged down by complaints, I chose to embrace the changes, adapt, and learn from this new phase of my life.

Now, after four months in this new chapter of my life, I can affirm that the journey was indeed worth every challenge. I am filled with pride, not just for myself but also for my peers who have weathered numerous changes and difficulties alongside me. During this time, I've had the fortune of meeting some incredible people. Among them are the Levy Professors and the wonderfully kind ladies from Rhinebeck town. Their warmth and support have greatly eased my transition. Bard College, too, has played a pivotal role in making our lives here more comfortable. I am deeply grateful to all of them and harbor the hope of reciprocating their kindness in the best way possible when I am able.

Life, of course, is still filled with uncertainties, a truth I believe holds for everyone. There's much I have yet to figure out. I am looking forward to another graduation as an MSc student from the Levy Economics Institute of Bard College and pondering my next steps post-Bard. But amidst these unknowns, I remain thankful for everything I've experienced. Challenges may be constant in life, but I am confident that I will find my way, carving a path forward with the lessons I've learned and the resilience I've gained. Looking back, my journey resembles a complex maze, marked by constant movement, adaptation, and personal growth. Throughout it all, finding peace of mind has often seemed like a distant and elusive goal. Yet, I continue to harbor hope — hope for myself, for my peers, and for all those who dream of a better future.

As I pen down the final lines of my story, I see it not just as a narrative of my own journey but as a beacon of hope for anyone who reads it. 'Planning the Big Escape' was not just about geographic relocation; it was about the perilous journey of the spirit through adversity and change. From the depths of despair in my homeland to the challenges of adaptation to foreign lands, each step has been a lesson in resilience and determination. To those who find themselves amid their own escapes, whether literal or metaphorical, I want to say this: You possess an inner strength more formidable than you realize. Embrace it. Let my story be a reminder that even in the most harrowing times, we can find the courage to continue, to grow, and to aspire for more. No journey is without its trials, but it is in navigating these trials that we discover our true potential. So, as you tread your path, remember: the most significant escapes are those that lead us to discover our own resilience, redefine our dreams, and ultimately emerge stronger, ready to face whatever new adventures life holds. In this journey, every hurdle crossed and every barrier broken has been a chapter in my own book of growth and perseverance. The essence of 'Planning the Big Escape' extends beyond mere physical distances; it symbolizes the escape from fear, from doubt, and from the limitations I often placed on myself.

## LESSONS LEARNT

Thus, as I look back on the journey that has brought me to this point, I see a tapestry of learning, growth, and transformation, not just in an academic sense, but in the deepest realms of personal understanding and resilience. Today I understand that education is a beacon of hope, freedom – a precious commodity, and adversity helps you grow in ways, you never expected.

**Education as a Beacon of Hope**: Education, for me, has been more than just a pursuit of knowledge; it has been a beacon of hope and a tool of empowerment in the darkest of times. The challenges I faced have only deepened my appreciation for the power of education. It is not merely about acquiring information, but about shaping minds, fostering critical thinking, and nurturing the courage to question and challenge. My experiences have underscored the fact that education is a fundamental human right that should never be compromised, and it is the cornerstone for building a more just and enlightened society.

**Freedom – A Precious Commodity**: The value of freedom, both in thought and in life, has never been more apparent to me. Living under oppressive conditions where this freedom was severely curtailed has heightened my understanding of its importance. Freedom is not just the absence of physical constraints; it is the liberty to express, learn, and grow. It's about making choices for one's life and future. This newfound appreciation for freedom has become a guiding principle in my life, reminding me to cherish and advocate for these rights, not just for myself, but for all who are denied them.

**Personal Growth Amidst Adversity**: The adversity I faced was a crucible that forged my character and resolve. I learned that growth often comes dressed in the robes of challenges and that the human spirit is capable of remarkable resilience. The trials I endured taught me empathy, fortitude, and the importance of community and support. They also revealed to me the immense strength that lies in vulnerability and the courage to seek help when needed. This journey has been about discovering my capabilities and pushing beyond the boundaries I once set for myself.

**Looking to the Future**: These reflections form the bedrock of my aspirations for the future. I am committed to using my education and experiences to contribute meaningfully to society. My dream is to be an advocate for education and women's rights, especially in contexts where these fundamental rights are still a battleground. I aspire to be a voice for those who are silenced and a source of hope for those who are struggling.

# Chapter 9
# Planning the Big Escape:
## Lessons Learnt

**Hamidullah Ahmadi**
*American University of Afghanistan, Afghanistan*

**Gul Hasan Mohammadi**
*American University of Central Asia, Kyrgyzstan*

## ABSTRACT

*In the wake of the collapse of the Islamic Republic of Afghanistan by the Taliban in mid-August 2021, the lives of Hamidullah Ahmadi and Gul Hassan Mohammadi, two friends and classmates from the American University of Afghanistan (AUAF), took an unexpected turn. As the country fell into the hands of the Taliban, they faced many challenges that challenged their resilience and determination. Among the chaos and uncertainty, Hamidullah and Gul Hassan struggled to maintain their studies, jobs, and personal lives. The security situation deteriorated rapidly, making it increasingly difficult for them to pursue their dreams. However, undiscouraged by the hardships they faced, they embarked on a journey of determination and unwavering hope.*

## HAMIDULLAH AHMADI'S STORY

This is Hamidullah Ahmadi a student from Afghanistan currently living in Canada. Prior to the fall of Kabul, I have been studying law at the American University of Afghanistan with immense ambitions and hope to become a qualified commercial lawyer. I was also working as a legal advisor in a Kabul based organization to promote the health and well-being of smokers. I and my fellows from the American

DOI: 10.4018/979-8-3693-1483-8.ch009

University of Afghanistan's Vis Moot Court 2021-2022 were working tirelessly to prepare for the national round competition. Since I was in my summer break, I had some entertainment and fun activities like going to gym, swimming pool, and playing soccer with my school friends.

In late summer 2021 the political developments were alarming in Afghanistan as the Taliban occupied most major cities of the country. On August 8, 2021, I along with my university friends had a beautiful chat in a café in Kabul. Our discussion was about the reopening of the university campus and starting studies in person. Surprisingly, in August 15, 2021 everything has changed completely. It was unconceivable at the same terrifying that the Taliban regained power in Afghanistan and the recent developments and progress of the country were in true danger. After the fall of Kabul, I thought everything is over and I would not be able to continue my education anymore. I imagined myself inside a broken ship without a clear destination. I strongly can claim that I spent one of the most chaotic and painful eras of my life which I have ever experienced before. My mind was frizzed with no ability to think and decide. I was experiencing high level of stress and anxiety struggling that what should I do cope the challenges.

In the next pages I would like to briefly narrate my story that how I dealt with the emergency I had faced after the Taliban's return which challenged my resilience and determination regarding my education and safety. I would like to present my narrative in a chronological order with important details so to inspire my other fellows who are living under the Taliban-ruled Afghanistan. Especially, I hope my story motivate girls and women who are despaired and see bright future in Afghanistan that if fingers crossed then I am sure everything works out. The key is to have a positive attitude towards the emergency situations that happen in our life by thinking that what we should do to make it right, and find solutions instead of focusing on barriers and obstacles.

Few days prior to the fall of Kabul, I received a confidential email from the administration of the American University of Afghanistan notifying me that the recent political developments are alarming for AUAF students, saying that we must provide the university with our information and should be prepared when called for leaving. After the fall of Kabul, I received another email from my university that they are working closely with their partners to evacuate AUAF students. In the meantime, the university emphasized us to take necessary security precautionary measures until we get to safety. One day prior to US withdrawal from Kabul Airport, I received a mobilization email from my university that I have to arrive at a specific location near to Kabul Airport to get out of Afghanistan. I got ready at night in a matter of a few hours. In the early morning, I left my family behind, and decided to leave the country with a small bag containing my laptop, educational materials, food stuff and a few clothes. When I got at the specific location, unfortunately, I received an email instructing me to leave the area and return back home, acknowledging that the operation cancelled. I was very disappointed, feeling that I would not be able to get out of the country and find safety.

After the complete US withdrawal, my university could succeed to evacuate a considerable number of the students via Torkham border to Pakistan and then to American University of Central Asia (AUCA) in Bishkek, Kyrgyzstan in September 2021, and another group to American University of Iraq-Sulaimani (AUIS) in Kurdistan region of Iraq through Kabul Airport in October 2021. Unfortunately, I was not among those lucky students since the selection was random.

The situation seemed immensely hopeless. I was depressed due to the recent incidents in the country. However, I didn't quit. I started my new semester at AUAF which was delivered online. Beside that I was working on a number of options to get out of Afghanistan and continue my education. Between the time period of October 2021 to September 2022 I tried a considerable number of options to get out of the chaotic situation that was happening in Afghanistan. Some of the options were through immigrant pathway to European countries, the US and Canada. There were also some non-immigrant options on the table to get to the above countries. I tried a significant number of those options, however, I could not succeed. At one stage, I became hopeless, deciding to give up. However, my brother advised me that since I have a satisfying educational background, as well as good command of English proficiency so it would be feasible for me to focus on education pathway. Furthermore, he advised me that educational pathway is more realistic and requires less time to be process comparing to immigrant pathway.

After considering this distinction, I started working on a number of educational options. At the outset, I considered the US, UK, and Canada as my priority since they were English spoken countries, and I felt that the ground would be wider for me in those countries. Beside that I considered Germany as a potential option since it was a prominent destination for international students. Although, I was continuing my L.L.B at AUAF online, in the meanwhile I started self-studying of German language at home via various platforms, including YouTube, DW Learn German and Duolingo App. The requirement to be admitted in German taught programs in a university in Germany was B2 German language proficiency level. Hence, I started to fulfill this requirement.

In March, 2022 I was notified by one of my friend who was in the process of getting a German student visa, that the processing time for German student visa in German Embassy-Islamabad takes about 18 months for Afghans after the fall of Kabul. This news shocked me and I understood that it would be very time consuming option. Therefore, I drew my focus on the US universities since the processing time for the US student visa (F1) was very less time consuming for Afghan nationals in Pakistan. I prepared for admission to US universities. I took Duolingo English Test with the financial assistance of the Yalda Hakim Foundation. I scored a substantial number in the test that met the English requirements of considerable number of US universities. I started applying to a number of universities in the US.

While in Afghanistan between August 2021 and July 2023, I mostly just stayed home and focused on my online education. I worked to find ways out of Afghanistan during this time. I also supported myself by working as a Legal Advisor and Social Media Officer for a project called Tobacco Harm Reduction Afghanistan (THR-Afghanistan) to promote the health and well-being of smokers in Afghanistan. I worked remotely from home for this project from June 2021 to June 2022, until the completion of the project I was hired for. On the other hand, during this time, I applied to several schools in the USA from Afghanistan, in the hopes of obtaining a study visa to the USA so that I could leave Afghanistan. I knew that universities in the USA would recognize my studies at the American University of Afghanistan and that my grades at the American University of Afghanistan would allow me to enroll in a U.S. school. I applied to the University of Toledo in Ohio, and to The New School in New York City. I applied to the University of Toledo because it had relatively low tuition, and to The New School because they agreed to recognize studies and credentials from the American University of Afghanistan.

I was accepted to the program of Law and Social Thought at the University of Toledo, and I received a partial scholarship to attend. I did not have money to cover the rest of my tuition, so after I was accepted, I contacted my former employer and asked them if they would be willing to help with my tuition. I explained that I was in danger in Afghanistan and that this would be a way out of Afghanistan for me. They agreed to pay my tuition for the University of Toledo if I was able to get to the United States. I searched online for ways to reach the United States and learned about an educational organization called the New University in Exile Consortium. I emailed them to ask for assistance in getting to the United States, and they agreed to pay for my plane tickets and other travel expenses. I had to obtain a visa from the Consulate General of the USA in Karachi, Pakistan, and then I travelled to Pakistan for the first time in December 16, 2022 for my visa interview. Subsequently, I traveled to Pakistan via Spin Boldak Border putting my life in a great danger to follow up and provide additional information on my student visa. Fortunately, I obtained my US F1 visa and I flew from Pakistan to the USA on July 20, 2023.

When I landed in the US, at the second week of my arrival in US things changed dramatically. I faced to another crucial emergency situation. The sponsoring organization for my tuition could not afford to cover the whole duration of academic program which was for two years. I was very anxious about struggling to secure funds to start my program. However, with immense resilience I could found a safe way to immigrate to Canada through a legal pathway which I was eligible for. Then I immigrated to Canada in August 4, 2023. In the meantime, I continue my online studies at the American University of Afghanistan. Finally, with all pass and downs, I could safely get out of Afghanistan and find a safe refuge in Canada. Meanwhile, I have the opportunity to continue my education in peace, without any trouble and barriers. I see a bright future in Canada, and I am optimistic to become a qualified licensed attorney.

After experiencing and passing all these ways and confronting many challenges and barriers, I found that in times of adversary, stress and anxiety will only exacerbate the problems and challenge human judgements. Instead, I found that maintaining resilience is the key to emerging from challenging and tough situations. Moreover, I perceived the power of education in strengthening my determination and resilience in the times of chaos.

## GUL HASSAN'S STORY: SUCCESS LIES BEYOND THE SWAMP

I am Gul Hassan Mohammadi, a student from Afghanistan. In late 2018, upon completing high school, it was usual to pass the university entrance exam called Kankor in order to pursue higher education in Afghanistan. However, considering the low quality of education offered by most, if not all, universities in Afghanistan, I did not set my sights on this exam. Instead, having a good command of the English language, I sought out scholarship opportunities, with my ultimate goal being to pursue higher education abroad, particularly in the United States. Unfortunately, due to financial constraints stemming from my low-income background, traveling to the United States seemed unattainable. Nevertheless, attending the American University of Afghanistan (AUAF), the most esteemed institution in the country, remained a feasible yet valuable option.

To get admitted to AUAF, I had to pass an admission process. I prepared diligently for each stage, knowing that this was my best chance to achieve my academic dreams. I was overjoyed when I received the acceptance letter from AUAF, along with a generous scholarship that covered most of my tuition fees. I felt proud of myself for overcoming the challenges and obstacles that I faced in my journey. I was eager to start my studies at the law department at AUAF, where I hoped to learn from the best professors and peers and explore new horizons of knowledge and opportunity.

After completing the second semester at the university, I was offered a position as a translator at a renowned company in Kabul. Balancing work and studies proved to be a challenge, but it marked a turning point in my life as I began to acquire invaluable knowledge through both endeavors. Subsequently, I earned a promotion within my organization and, in July 2021, signed a new contract as a permanent employee with a substantial salary. Filled with hope for the future, I wholeheartedly pursued my professional and academic aspirations. However, everything changed on August 15, 2021.

On that fateful day, the Taliban captured Afghanistan's capital city of Kabul, seemingly marking the end of the war in Afghanistan and a total victory for the terrorist group. The president of Afghanistan, Ashraf Ghani Ahmadzai, fled the country, leaving behind a power vacuum and a state of chaos. I watched in horror as the Taliban entered the presidential palace and declared their rule over the country. I realized that not only my life, like the lives of millions of other citizens, was in grave danger, but my educational journey and achievements were also in great danger.

With the fall of the republic government to the Taliban, my dreams seemed shattered. It was inconceivable that Afghanistan had regressed to the state it was in back in 1996 when the Taliban first seized power. Temporarily, I lost my job, which was the initial effect of the government's collapse. Given the tough circumstances, I believed that the university would not resume its activities, as many students did not possess the courage or motivation to continue their studies under a terrorist-led regime. In addition to the political and security concerns, I grappled with personal family issues. My older brother had been diagnosed with schizophrenia a year before, which has been a great source of sorrow for me and my family. I also had lost my father, who supported me constantly, the previous year. The Taliban's takeover further complicated my already challenging life. Everywhere I turned, I encountered nothing but despair. However, as the summer break drew to a close, the university resumed its activities, albeit with a slight delay due to the issues mentioned above that had affected nearly everyone in Afghanistan.

I was relieved and grateful that AUAF had not given up on its mission to provide quality online education to students in Afghanistan despite the risks and uncertainties; I decided to resume my studies as well, hoping to find some solace and purpose in learning. I was inspired by the resilience and courage of my professors and classmates, who continued to teach and learn under the shadow of the Taliban. I realized that education was not only a means to achieve my personal goals but also a way to contribute to the development and progress of my country. I resolved to make the most of the opportunity that AUAF had given me and to use my knowledge and skills to serve my community and my nation.

As I commenced my studies, I gradually arrived at a crucial realization: the importance of resilience in the face of adversity. I convinced myself that even in the midst of complete chaos and a lack of employment opportunities, succumbing to disappointment would only exacerbate the situation. Undoubtedly, losing hope is the true calamity that befalls us all. I have learned that hopelessness itself does not kill dreams, but rather, succumbing to hopelessness can extinguish them. Consequently, I endeavored to focus on my studies and immerse myself in books to maintain my focus and normalize the situation for myself.

### *Planning the Big Escape*

Upon the conclusion of my first semester after the country fell to the Taliban, I achieved one of the highest term GPAs, which made me eligible for the prestigious "President's High Honor Award" for the third time since joining AUAF. This award is bestowed upon students who exhibit exceptional academic performance each semester. A few months after the Taliban's takeover, I commenced my job, albeit remotely, as a Reviewer and Technical Translator with a consultant contract. Though the organization did not offer me a permanent contract then, which I had before the takeover of the country by the Taliban, it was still a tremendous opportunity that I cherished deeply because I needed to support myself and my family financially besides my studies. Despite the hardships and uncertainties that I faced, I never gave up on my dreams and ambitions. I continued to work hard and excel in both my academic and professional endeavors, proving to myself and others that I was capable of overcoming any challenge.

In the wake of the anarchy two years ago, life in Afghanistan became increasingly challenging as a multitude of issues plagued the country day by day. Access to electricity and a reliable internet connection were scarce. With power outages occurring throughout the day in Kabul, especially during the harsh winter months, I found myself visiting different restaurants and cafes to charge my laptop. This allowed me to stay on top of my assignments and prevent falling behind in my studies. I even would join some classes when I was sitting inside a loud restaurant when my laptop was plugged in.

I believe such a situation tested my patience and perseverance, as I had to cope with the noise and distractions around me while trying to focus on my classes and assignments. I also had to deal with the security risks and the transportation costs of moving from one place to another. Sometimes, I would not find any place with electricity or internet, and I would have to miss my classes or submit my assignments late. I felt frustrated and helpless at times, but I never gave up on my education. I knew that it was the only thing that could give me hope and empowerment in such a bleak and uncertain environment. I also appreciated the support and understanding of my professors and classmates, who helped me whenever I faced any difficulties or challenges.

Furthermore, the security situation worsened following the Taliban's takeover, as sporadic suicide and explosive attacks became commonplace in Kabul, particularly in the western part of Kabul, where my ethnic fellows, Hazaras, were targeted particularly. Like thousands of others, I lived in constant fear, unsure if I would return home alive. The threat of attacks loomed over the streets, educational institutions, shrines, and even public transportation in my neighborhood leaving no place safe. To be honest, I felt angry and helpless at the injustice and oppression that the Taliban and other extremist groups inflicted on the innocent people of Afghanistan, especially the Hazaras, who have faced centuries of discrimination and persecution. However,

I witnessed the courage and the compassion of the people who risked their lives to help the victims and the survivors of the attacks.

Adding to the peril, Taliban militants began conducting house-to-house searches, putting my life and the lives of my family members at significant risk if any affiliation with an American university was discovered. Two days before our house was about to be searched, I had to burn my university student ID card and all the books I had borrowed from the AUAF library before the close of the university. These books bore the logos of AUAF, serving as evidence of my association with the Americans, whom the Taliban considered their sworn enemies. Like many of my classmates who burned their books, I hardly convinced myself to destroy them because they held too much value for me.

Despite these overwhelming challenges, I resisted, unwavering in my commitment to continue my studies online and graduate at any cost. For nearly two years, I pursued my education under constant fear and anxiety. Finally, my determination paid off, and I graduated with outstanding performance. Shortly after completing my undergraduate studies, my high GPA earned me a place in the fully-funded scholarship for the master's degree program at the American University of Central Asia (AUCA), where I am pursuing my LL.M. degree.

I was overjoyed and grateful for this opportunity, as it meant that I could continue my education in a safe and supportive environment. I was also excited to explore a new country and culture, meet new people, and make new friends. I moved to Bishkek, Kyrgyzstan, where AUCA is located, and started my master's program in September 2023. I was impressed by the quality and diversity of the faculty and the students and by the facilities and the resources that AUCA offered. I felt welcomed and accepted by the AUCA community, and I quickly adapted to the new setting.

As I reflect on my past, I recall a profound quote from one of my teachers at school. He said, "Success lies beyond the swamp." I think that enduring relentless challenges, threats, and scarcity was hard, but it was all worth the effort. At least, it was for me. I realize that I endured countless hardships but remained steadfast in my pursuit of education. If I had given up when the anarchy erupted in mid-August 2021, I would not have completed my undergraduate studies or obtained my diploma. Indeed, my unwavering dedication during the darkest days in Afghanistan was rewarded with the opportunity to pursue higher education at AUCA. I believe my determination and resilience, even in the face of difficulty, pave the way for success. I also realize that I am not alone in this journey and that I have been blessed with the support and guidance of many people who have helped me along the way. I am grateful to my family, who have always encouraged me and loved me unconditionally. I am grateful to my friends, who have always been there for me and shared my joys and sorrows.

As Hamidullah and Gul Hassan share their remarkable journeys, they hope to inspire others to never lose sight of their dreams, no matter how impossible the obstacles may seem. Their stories, generally narrated in the following pages, are a testament to the power of education, resilience, and unwavering hope in the face of hardships. Through their experiences and success stories, they aim to ignite a flame of hope in the hearts of those affected by the tumultuous events in Afghanistan, reminding them that even in the darkest times, there is always a light of resilience.

# Chapter 10
# Teachers' Professional Development in the Emergency Contexts:
## Resilience of Teachers in Teaching the Rohingya Children in Bangladesh

**M. Mahruf C. Shohel**
https://orcid.org/0000-0002-4048-0577
*University of Roehampton, UK*

**S. M. Rashedujjaman**
https://orcid.org/0009-0004-0879-5570
*International Rescue Committee, Bangladesh*

**Mahbuba Rahman Sofin**
https://orcid.org/0009-0003-2678-5109
*International University of Business Agriculture and Technology, Bangladesh*

**Kaysar Saurav**
*Save the Children, Bangladesh*

**Mohammad Abu Bakar Siddik**
https://orcid.org/0000-0003-3998-1005
*Primary Teachers' Training Institute, Dhaka, Bangladesh*

**Tahmina Akter**
https://orcid.org/0000-0001-6762-9688
*Norwegian Refugee Council, Bangladesh*

## ABSTRACT

*This book chapter explores the critical role of professional development in equipping teachers to navigate the challenges and barriers of emergency contexts. It focuses specifically on the resilience exhibited by teachers in teaching the Rohingya refugees in Bangladesh. The Rohingya refugee crisis, characterised by forced displacement*

DOI: 10.4018/979-8-3693-1483-8.ch010

*and limited access to education, presents unique challenges and barriers for both teachers and students. Drawing upon secondary sources and theoretical frameworks, this chapter unpacks the various dimensions of teachers' professional development for both the Rohingya and the host community teachers in such adverse circumstances in refugee camps in Bangladesh. It highlights the adaptive strategies employed by the Cox's Bazar Education Sector and the teachers themselves to enhance their pedagogical skills, foster socio-emotional well-being, and address the diverse needs of the Rohingya learners. Furthermore, the chapter analyses the organisational support systems, policies, and interventions aimed at promoting teacher resilience and effectiveness in emergency settings.*

## 1. INTRODUCTION

In the refugee context, faced with challenges and barriers like language barriers, geo-politics and limited resources, paramount importance must be placed on professional development (PD) for teachers. This equips them to support refugees through quality education and fosters their resilience (Pherali, 2020). The Rohingya crisis, characterised by forced displacement and limited access to education (Shohel, 2022 & 2023), presents significant challenges for both teachers and students living in refugee camps in Bangladesh. This chapter explores the critical importance of PD for teachers working in emergency contexts, with a specific focus on the resilience exhibited by teachers in teaching the refugee children in Cox's Bazar Rohingya camps in Bangladesh.

The latest Rohingya crisis, unfolding since August 2017, has led to the displacement of hundreds of thousands of the Rohingya people, particularly from Myanmar to neighbouring Bangladesh (Shohel et al., 2022, 2023a, 2024a). This mass movement has resulted in overcrowded Rohingya refugee camps and strained resources, exacerbating the challenges faced by the Rohingya refugees in accessing education (Suma, 2022). Despite the adversities, teachers in the refugee camps have demonstrated remarkable resilience and commitment to providing education to the Rohingya children even under difficult circumstances (Nilsen et al., 2023).

The concept of Teachers' Professional Development (TPD) in emergencies goes beyond simply equipping teachers with new teaching techniques (Crompton et al., 2021). In these situations, teachers themselves are often refugees who have experienced trauma and displacement. Their own well-being directly impacts their capacity to effectively support their students (Zee & Koomen, 2016). Based on existing literature and derived theoretical frameworks, this chapter aims to explore the various ways TPD in emergency settings can pave a crucial path for supporting the teacher to be resilient in complicated and challenging situations. It highlights the adaptive strategies employed by teachers to enhance their pedagogical skills, foster socio-emotional well-being, and address the diverse needs of the Rohingya learners. Additionally, the chapter analyses the organisational support systems, policies, and interventions aimed at promoting teacher resilience and effectiveness in emergency settings. This chapter seeks to provide a comprehensive understanding of the resilience exhibited by teachers and the factors influencing their PD in emergency contexts. Thus, by examining the complexities surrounding TPD in the Rohingya emergency contexts, this chapter seeks to provide insights for policymakers, practitioners, and researchers striving to support teachers in similar crisis-affected settings.

## 2. BACKGROUND AND THE CONTEXT

### 2.1 The Rohingya Crisis in Bangladesh

Since August 2017, Bangladesh has faced one of the most severe humanitarian crises in recent history (Faisal & Ahmed, 2022). More than 742,000 Rohingyas have fled violence and persecution in Myanmar's Rakhine State (UNHCR, 2023; Ullah, 2016), adding to approximately 203,431 Rohingyas already living in Cox's Bazar district who entered Bangladesh between July 2005 and August 24, 2017 (Sohel et al. 2024b; Khatun, 2017). The influx of the Rohingya refugees into Bangladesh has led to the establishment of overcrowded refugee camps in Cox's Bazar, stretching resources and infrastructure to their limits (Alam, 2022; Mohiuddin & Molderez, 2023), resulting in multifaceted challenges, including those related to education (Prodip, 2023; Shohel, 2022 & 2023). According to the UNHCR population factsheet (UNHCR, 2024), about 942,619 Rohingyas are currently living in 33 camp locations in Cox's Bazar district excluding Bhasan Char relocation site, a remote island where currently about 35,000 Rohingyas are residing. Within these 33 camp locations 43,018 registered Rohingyas from previous influxes are living in two registered camps: Kutupalong Registered Camp in Ukhiya subdistrict and Nayapara Registered Camp in Teknaf subdistrict.

The Rohingya crisis in Bangladesh has led to significant challenges for both registered and non-registered refugees. However, the Rohingyas, registered or non-registered, are not acknowledged as refugees by the Government of Bangladesh (GoB). The host-country, Bangladesh, being a non-signatory country to the 1951 United Nations Refugee Convention or its 1967 Protocol, does not recognise the Rohingyas as refugees but rather identifies them as Forcibly Displaced Myanmar Nationals (FDMN). The Rohingyas from the registered camps face difficulties in economic, health, food, residence, social, and security issues, while the Rohingyas from other non-registered camps prioritise economic and social issues (Sultana et al., 2023).

The fact regarding the remarkable resilience often displayed by teachers in the face of emergencies like refugee crises often goes unnoticed and even ignored, although it is crucial for ensuring the continuation of education for displaced children. This resilience can be attributed to several key factors, like strong social support networks encompassing colleagues, students, and the broader community, who are often credited with buffering the negative impacts of stress in such situations. Additionally, the development of positive coping mechanisms, such as humour or reframing challenges, by teachers has been linked to a greater capacity to overcome adversity (Agyekum & Otoo, 2020).

## 2.2 Challenges in Access to Education

Access to education has been severely limited for the Rohingya refugees in Bangladesh due to various factors, including the lack of adequate educational facilities, language barriers, and socio-economic constraints (Shohel, 2022). Many Rohingya refugees, mostly young children, have experienced disruptions in their schooling or have never had access to formal education, further exacerbating their vulnerability (Shohel, 2023; Ahmed, 2021). Moreover, the rapid influx of refugees has strained the capacity of local educational institutions and personnel, leading to overcrowded classrooms and insufficient resources. According to the UNHCR data, 52% of the total Rohingya population in Cox's Bazar are children, and about 17% of the total population aged between 5-17 years (UNHCR, 2024). These large numbers of children need educational intervention for their future growth, wellbeing and protection from further harm of the crisis. According to the 'Joint Multi Sectoral Needs Assessment (J-MSNA): Camp-level findings' (Rohingya Response Bangladesh, 2023), 61% of school-aged children (referred to as 5–18-year-old) were enrolled and regularly attended learning facilities (with 80% attendance rate) during the 2022-2023 school year. However, the Government of Bangladesh has persistently excluded refugees from formal education. Thus, the state policy of exclusion has created local misunderstandings about refugees' right to education (Karim, 2023).

In the Rohingya refugee camps in Bangladesh, people continue to face inequity regarding their entitlement to education, despite all the existing challenges (Karim & Hussain, 2019). The lack of access to a formal and accredited education system has extremely concerning implications for the educational prospects and well-being of children and adolescents in these difficult circumstances. Therefore, it hinders the access to the host country's education system or the establishment of any proper educational infrastructure for children in the refugee camps. Despite alternative approaches to providing education by the Cox's Bazar Education Sector, there are other challenges due to the lack of adequate resources including classrooms, qualified teachers, textbooks, and educational materials. A study reveals that inadequate funding, time management, teacher training, and other logistical support systems make it difficult to provide basic education to the Rohingya refugee children (Rahman et al., 2022). The refugee community already faces challenging circumstances in the adverse living conditions in the refugee camps, and the lack of access to high-quality education for refugee children makes the situation more vulnerable by impeding their long-term growth and future prospects.

The main challenges to provide education for the Rohingya refugees include the ongoing problems with language and curriculum. It is indispensable to understand the language that is delivered for educational purposes for this refugee community. According to a report (Cox's Bazar Education Sector, 2018), 15% of the refugee population who prefer Burmese and English, 50% of the refugee population prefers Rohingya as their language of instruction, despite the fact that it is a verbal language without a standardised and internationally recognised written script. Moreover, effective learning is tremendously limited due to the language barriers that exist between the host community and the refugees, as well as the mismatch between the existing curriculum and the educational requirements of the displaced population (Mahmud & Nalifa, 2020). Due to language issues, it may be difficult for many Rohingya refugees to fully participate in the educational system because they are not fluent in the language of their new curriculum in Burmese language. Communication difficulties between teachers and students can also get affected from such language related issues.

According to a report from the Refugee Relief and Repatriation Commissioner (RRRC) office (2023), around 30,000 children are born in refugee camps every year. In the absence of schooling, these youngsters are more vulnerable to exploitation and have fewer opportunities for personal growth and development. Besides, environmental problems increase socioeconomic inequality, with particular challenges in access to education, skill development, and limited opportunities for young people (Mitu et al., 2022). Despite the government's restrictions on formal education, certification requirements, access to technology, language barriers, unfavourable learning environments, gender disparities, a shortage of qualified teachers, the COVID-19 pandemic, and numerous other issues, the Learning Competency Framework and Approach (LCFA) curriculum was used in Cox's Bazar education sector to deliver informal education (Shohel et al., 2023b; Sadia, 2021) until July 2022, before the Myanmar Curriculum (MC) Scale Up was initiated (Akter, 2023). Dropouts due to financial difficulties are not uncommon, particularly among older learners who might have to work to support their families and the girls who are married off during adolescence.

Events like the January 2023 fire in Camp 5 displaced families and damaged learning facilities, disrupting the education of thousands of the Rohingya children (Hossain et al., 2023). The challenges to accessing education even increased when the Myanmar Curriculum first started to scale up in July, 2022 due to its unfamiliarity to students what they have been taught in LCFA. On the other hand, Myanmar Curriculum scale-up also has its own challenges as all the key subjects are to be taught in Burmese language, except English as a core subject (Akter, 2023).

## 2.3 Impact on Teachers

Teachers working in the Rohingya refugee camps face unique challenges and barriers in delivering education to the Rohingya children. Limited training and resources often leave them struggling to create effective learning environments. Language barriers can further complicate communication and comprehension in the teaching and learning process. Trauma from violence and displacement can manifest in behavioural issues among students, requiring additional support from teachers (Shohel et al., 2022). The absence of a formal curriculum specifically designed for the Rohingya children necessitates adaptation on the part of teachers.

Despite these significant challenges, teachers in the camps have consistently displayed remarkable resilience and dedication. They frequently improvise teaching methods and materials, tailoring them to the diverse needs of their students (UNHCR, 2022). This fosters an environment where learning can still flourish even in the face of adversity. Despite political restrictions on education, community-based education networks and teachers employ strategies to secure basic education for Rohingya youths (Nilsen et al., 2023). The government of Bangladesh, in collaboration with UNICEF, has launched initiatives to improve educational outcomes for Rohingya refugee children. The implementation of the Myanmar Curriculum Pilot is a step towards providing a more structured learning experience (Farzana, 2023). It was scaled up from July 2022 on a rolling basis, starting from Kindergarten (KG), grades 1 and 2 in 2022, and grades 3, 4 and 5 from the 2023-2024 academic year (Haque et al., 2023). However, ongoing challenges remain. The camps face various vulnerabilities, such as overcrowding, limited access to clean water and sanitation, and poor health infrastructure, which further complicate the mental health of the Rohingya population, impacting both students and teachers (Shohel et al., 2024a; Islam et al., 2022; O'Connor & Seager, 2021). It is not hard to understand that the teachers in the Rohingya camps have their own unique challenges and barriers. The Rohingya teachers are also part of the refugee community and can have traumatic experiences back in Myanmar. On the other hand, the host community teachers may have challenges related to the drastic socio-economic changes, employability and inflation.

## 3. THEORETICAL FRAMEWORK

The theoretical framework guiding this chapter encompasses various perspectives on education in emergency contexts, teacher resilience, professional development, and intersectionality and contextual factors. Drawing upon established theories and conceptual frameworks, this section provides a theoretical framework through which to analyse the resilience exhibited by teachers in teaching the Rohingya children in Bangladesh and the role of teachers' professional development in supporting their effectiveness.

### 3.1 Education in the Emergency Contexts

The right to education is universally recognised, even during the crisis. Education in emergency contexts refers to the provision of learning opportunities for individuals affected by crises, including natural disasters, armed conflict, displacement, and other emergencies (Shohel, 2022). These emergencies disrupt normal schooling and can have a negative impact on children's development and well-being. The right to education is recognised as a fundamental human right and a critical component of humanitarian response and recovery efforts in emergency settings. Education in emergencies (EiE) aims to mitigate the negative impact of crises on individuals' well-being, resilience, and future prospects by providing safe spaces, psychosocial support, inclusive, and quality learning environments (Shohel, 2023). Research suggests they can play a critical role in promoting long-term recovery for children and communities affected by emergencies (Heltne et. al., 2019). These programmes can reduce the risk of child exploitation by providing structured learning environments and keeping children occupied, EiE programmes can help mitigate the risk of child labour, recruitment into armed groups, or other forms of exploitation. Besides, the safe spaces and psychosocial support offered through EiE programmes can contribute to reducing anxiety, depression, and other mental health challenges faced by children in emergencies (Shohel et al., 2022). In addition, schools can act as focal points for communities during emergencies, fostering a sense of normalcy and promoting social interaction and collaboration.

### 3.2 Teacher Resilience

Resilience theory posits that individuals can adapt positively to adversity and overcome significant challenges through internal and external resources (Southwick et al., 2014). The concept of teacher resilience is crucial within the framework of EiE. In the context of education, teacher resilience refers to the capacity of teachers to withstand and bounce back from stress, adversity, and professional challenges while maintaining their commitment to teaching and learning (Mansfield et al., 2016; Gu & Day, 2007). In order to sustain the dedication, enthusiasm, and efficacy of promoting students' learning, teachers must possess the resilience to effectively manage and recover from the stresses, setbacks, and difficulties that come with being in the teaching profession (Mansfield et al., 2016). Key components of teacher resilience include self-efficacy, optimism, problem-solving skills, social support, and a sense of purpose and meaning in teaching (Hascher, Beltman, & Mansfield, 2021).

Understanding and fostering teacher resilience is essential for ensuring educational continuity and quality in disrupted contexts. Teachers who exhibit resilience can create safe and supportive learning environments for learners who themselves may be experiencing trauma and displacement (Howard & Johnson, 2004). This focus on teacher resilience complements other aspects of the EiE framework, such as curriculum adaptation and community engagement, by highlighting the importance of investing in teachers' well-being and professional development (Barbora, 2021). In the context of the personal focus on teacher resilience, researchers have concerns about the negative impact on teachers of stress and psychological burden, which can lead to a loss of satisfaction with their chosen career or a reduced sense of wellbeing (Beltman, 2020).

## 3.3 Teachers' Professional Development in EiE

Professional development encompasses systematic efforts to enhance teachers' knowledge, skills, and dispositions to improve teaching and learning outcomes. With the ultimate goal of raising achievement among learners, TPD includes deliberate and ongoing attempts to improve teachers' knowledge, abilities, and instructional methods through a range of professional learning opportunities (Darling-Hammond et al., 2017; Rahman & Siddik, 2022). Effective TPD initiatives are characterised by their alignment with teachers' needs, goals, and contexts, as well as their relevance to student learning and achievement. TPD can take various forms, including workshops, seminars, mentoring, coaching, collaborative learning communities and cycles, and action research, among others (Admiraal et al., 2021). Both training and on-the-job professional support for teachers and youth workers should be at the heart of any education programme for crisis-affected populations. These populations have to face a difficult future and need the best possible education services (Sinclair, 2002).

The principal difference between education for refugees and other crisis-affected people is that refugees are separated from the education system of their own country, which has its own curriculum, its own assessment and examination procedures, its own system of progressing from one level of education to the next, and its own approaches to teacher training and certification. When refugees are first displaced, the curriculum in refugee schools is normally the one which teachers and students are familiar with. Providing a familiar curriculum is both practical and psychologically reassuring. The use of the curriculum of the country of origin also helps children re-enter their own education system after repatriation and is termed as 'education for repatriation'. If possible, the refugees continue to use home country textbooks and sit home country examinations (Sinclair, 2002).

When there is only intermittent contact with field staff, the quality of education depends on their motivation and competence. It is crucial to maximise in-service training of teachers so that they can continue to work effectively when cut off from normal support. The training of headteachers and senior teachers as 'mentors' who can provide hands-on support to newly trained or untrained teachers is also especially important in these circumstances. Another measure which can help teachers is to be sure that all schools or teachers have access to complete sets of textbooks and teacher guides (replaced when necessary) and to provide handbooks and motivational materials for field trainers, supervisors, headteachers, and teachers (Sinclair, 2002).

In the emergency contexts, establishment of a task force or entity for the renewal of teacher training is essential. This should include accelerated pre-service training and in-service training structured to lead to a recognised teaching qualification. Preparation of project proposals for training of trainers is needed to develop and deliver effective TDP programmes. At the same time, rehabilitation and modernisation of local teacher-training institutions. This ensures, if appropriate, collaboration to promote literacy, language, and life skills studies by technical and vocational students. Additionally, skills training linked to literacy programmes should be part of the proposal (Sinclair, 2002).

Some of the international discourse on EiE addresses practical challenges like logistics and resource limitations. These discussions also emphasise the importance of helping students and teachers heal from trauma, as well as equipping them with valuable life skills (Sinclair, 2002). Within teacher-training programmes, it may be useful to combine studies of education for populations affected by conflict and natural disasters with studies relating to 'silent emergencies' prevalent in the country concerned. By preparing and encouraging teachers to deal with crises in their neighbourhood and school, they may be more able to respond if there is a large-scale catastrophe. Academic researchers and teacher trainers can develop their knowledge of EiE by working intensively with teachers during local emergencies, and can subsequently share their experience and research findings with the wider international community (Sinclair, 2002).

## 3.4 Intersectionality and Contextual Factors

Intersectionality theory provides a valuable lens for understanding the experiences of teachers in emergency contexts. This theory highlights the interconnected nature of social identities and experiences, such as gender, ethnicity, class, religion, and displacement status (Cho et al., 2013). These identities intersect and influence on shaping individuals' experiences of adversity and building resilience (Yosso, 2002). For example, a female Rohingya teacher in a Bangladeshi refugee camp might navigate challenges related to her gender, ethnicity, and displacement status simultaneously. This complex interplay of identities will likely shape her ability to access PD opportunities, her experiences within the school system, and her capacity to cope with emergencies.

Contextual factors further shape the experiences of teachers in emergency settings. These factors could include the following:

*Cultural Norms:* Cultural expectations surrounding gender roles, teaching styles, and student-teacher relationships can influence how teachers approach their work (Madrid Akpovo, Nganga & Acharya, 2018).

*Institutional Structures:* The policies and practices of education ministries, NGOs, and other organisations involved in education provision can either support or hinder teachers' professional development (UNESCO, 2018).

*Policy Frameworks:* National and international education policies can shape the curriculum, resources, and professional development opportunities available to teachers working in emergencies (Nicolai, 2003).

*Community Dynamics:* Relationships between schools, families, and community leaders can influence teacher morale, engagement, and sense of support (Baquedano-López, Alexander & Hernandez, 2013).

By considering the interplay between individuals (through intersectionality), organisational (institutional structures and policies), and contextual factors (cultural norms and community dynamics), this theoretical framework provides a multifaceted lens for analysing and interpreting data on teacher resilience and professional development in emergency contexts. This comprehensive approach allows for an in-depth understanding of the challenges and opportunities faced by teachers and informs the development of more effective policies and practices to support them.

## 4. EXISTING LITERATURE

The literature on teacher resilience, professional development, educational technology and teacher training in emergency contexts provides valuable insights into the challenges, strategies, and interventions relevant to teaching the Rohingya refugees in Bangladesh. This section reviews key findings and theoretical perspectives from existing research, policy documents, and academic and grey literature, focusing on topics such as teacher training, psychosocial support, community engagement, and educational interventions targeting displaced populations.

## 4.1 Education in the Emergency Contexts

Education in emergency contexts is recognised as a fundamental human right and a critical component of humanitarian response and recovery efforts in crisis-affected settings. Research highlights the unique challenges faced by teachers and students in emergency settings, including displacement, trauma, loss of infrastructure, and disruption of schooling. Studies emphasise the importance of providing safe, inclusive, and quality learning environments for displaced populations, as well as addressing the psychosocial and educational needs of affected children and youth.

Based on the field of educational psychology, the study addresses the necessity of efficient, teacher-centred, trauma-informed training to improve teachers' self-efficacy in meeting the educational requirements of refugee children (Castellanos, 2018). Trauma-informed practice objectives typically concentrate on students' challenging behaviours and the situations that they are dealing with outside of the classroom. (Venet, 2023). To educate vulnerable adolescents who struggle in school because of traumatic histories from abuse, neglect, or violence, trauma-informed positive education blends teaching practices from two practice paradigms: trauma-informed education and positive education (Brunzell et al., 2019). It sets emphasis on establishing a secure and encouraging learning environment.

The majority of refugee children are living in low-resource environments and endure hardships caused by past traumatic events, post-migration poverty, and challenges. Through the establishment of the first locally-led collaborative public art education programme in refugee camps, the international non-governmental organisation Artolution (www.artolution.org) has pioneered the emerging field of community-based public art education in emergencies, exploring art instruction and emergency preparedness (Frieder, 2020). Moreover, adolescent Rohingyas in Bangladesh refugee camps employ pedagogical approaches, frequently facilitated by community-based education networks, to guarantee their access to elementary education (Nilsen et al., 2023). It appears that humanitarian education providers should actively involve the refugee community, which holds significant insights into the worries and aspirations of the young Rohingya people.

Recently, researchers have emphasised the need to prioritise delivering early childhood education and care (ECEC) services, which offer indispensable physical, emotional, and cognitive support to vulnerable refugee and migrant populations in resource-constrained situations (Stevens et al., 2023). A study argues that community development strategies during a humanitarian crisis require some outside engagement and bureaucratic oversight because of the lack of assets and resources (Islam, 2019). After analysing the issues of the Rohingya crisis, the humanitarian community has placed a strong emphasis on sustainable learning strategies, including the opening of new learning centres, the strengthening of weather and disaster management-related issues in educational materials, and the use of alternative learning methods (such as home-based learning, mobile learning, and radio-based teaching) (Zaman et al., 2020).

According to the authors' connection to the research evidence, money transfer programmes can boost student attendance, and physical education, early childhood development initiatives, or internet game-based solutions can improve learning outcomes like second language acquisition (Palik & Østby, 2023). It indicates that other interventions, including theatre workshops, have no impact on learning a second language. Researchers from 30 countries worldwide came to the conclusion that, despite students growing up surrounded by technology in the 21st century, many learners had limited prior engagement with online learning for educational purposes before the pandemic (Bozkurt et al., 2020), which indicates that engagement with e-learning substantially rose following the pandemic emergency learning context.

## 4.2 Teacher Resilience

Teacher resilience has emerged as a central focus of research in the field of education, particularly in contexts characterised by adversity, stress, and uncertainty. Studies highlight the importance of factors such as self-efficacy, optimism, coping strategies, social support, and professional development in promoting teacher resilience and well-being.

Research suggests that resilient teachers are better able to adapt to challenging circumstances, maintain positive relationships with students and colleagues, and sustain their commitment to teaching and learning despite external pressures. Study conducted by Habib (2020) shows that female teachers are more positive than male teachers when it comes to attitudes towards the vulnerable Rohingya refugees, suggesting that female teachers' behaviour and communication in the classroom are more interactive than male teachers due to the female teachers' language ability. It appears that teachers are more sympathetic and less discriminatory towards the Rohingya children. Also, the 'International Child Development Program' intercession did not altogether decrease the negative state of mind of the teachers towards their Rohingya displaced children which implies teachers were not completely empathised towards these powerless traumatised children. Their conception of the Rohingya children was somewhat stigmatised and they see understudies? as a question, not completely as individuals. Teachers' state of mind towards the refugee learners tended to avoid those learners from their zone of compassion (Habib, 2020)

The study found that in Myanmar and Bangladesh, refugee children are deprived of basic education due to discrimination and severe factors, including age, gender, religious background, ethnic identification, and economic position (Rahman et al., 2022). Even while speaking with teachers, refugees' insecurity is fuelled by cultural and psychological biases, and as a consequence, they have no faith in anyone at the school (McBrien, 2005). In addition, it is obvious that refugees may have encountered prejudice, stigma, and discrimination in their new areas, which can cause them to lose faith in establishments such as educational institutions. Refugees can view schools as hostile or unwelcoming due to past unpleasant encounters with teachers or authorities.

*Figure 1. The characteristics of a resilience teacher*

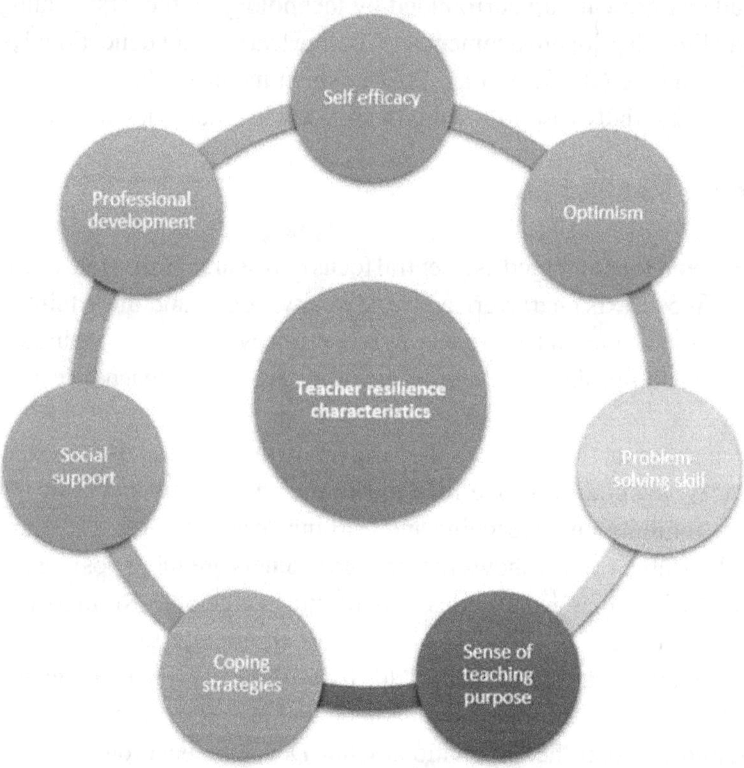


Teachers at the Temporary Learning Centres (TLC) who do not speak Burmese have difficulty teaching students because all of the learning resources are in Burmese except English as a subject (Folven, 2022). Experienced teachers with strong language and teaching skills could have a significant impact on developing education in the camps, such as by assisting in Burmese language and translation (Olney et al., 2019). In the Rohingya refugee camps, teachers face various difficulties, which have been mentioned previously, but they can be more resilient following different aspects (see Figure 1) due to their dedication to providing refugee children with education. Teachers expect that all people can learn, and knowledge can improve a generation.

Considering the global context of emergencies, teachers may face different kinds of challenges during their professional lives compared to the Rohingya crisis. Teachers encounter a variety of emergencies, including those related to the environment, natural disasters, and armed conflict. Following a natural disaster such as a Hurricane, schools frequently concentrate on developing a recovery plan that addresses the learners' urgent needs, but teachers' needs are not given priority, so they are left to deal with interruptions in their personal and professional lives on their own (Cannon et al., 2020). In order to make communities aware of natural disasters, teachers must have access to efficient and geographically restricted teaching materials, which will also improve community resilience (Boyd et al., 2021). The educational system can be disrupted due to armed conflict. Millions of children suffer from armed conflict areas, with terrible incidents interrupting their everyday lives and leaving an indelible scar. Children's education is a fundamental right that is impacted by this fact (Cervantes-Duarte & Fernández-Cano, 2016). Besides, teachers who work in places affected by war can face challenges to their safety and security, such as acts of violence, attacks on schools, forced relocation, and disruptions in educational services as a result of political turbulence or armed conflict. Coping strategies offer a chance to assist teachers' well-being by using an asset-based strategy; thus, in low-resource, crisis, and conflict-affected contexts, teachers face tremendous stress and challenges in their roles, yet they also exhibit resilience, persistence, and dedication (Falk et al., 2019).

## 4.3 Teacher's Professional Development in EiE

Professional development plays a critical role in enhancing teachers' knowledge, skills, and efficacy, enabling them to meet the diverse needs of students and navigate complex educational environments. Effective PD initiatives are characterised by their alignment with teachers' needs, goals, and contexts, as well as their focus on job-embedded learning, collaboration, and reflective practice. Studies emphasise the importance of ongoing, sustained PD in fostering teacher growth, resilience, and effectiveness, particularly in emergency settings where they face heightened challenges and constraints.

TPD in emergencies is crucial for addressing the needs of providing education during emergencies and ensuring effective pedagogical adaptation. However, a pedagogical approach is called Essence of Learning (EoL), developed by Caritas Switzerland with the goal of providing emotional and educational support to children in crisis situations; the technique allows instructors to teach a traditional curriculum using relaxation and play to help children heal from traumatic circumstances and regain their ability to study (Coombes & Ponta, 2022). Training teachers on innovative and culturally appropriate teaching strategies can help them better engage the Rohingya children in meaningful learning experiences. This may involve interactive and hands-on activities, the use of technology, and individualised instruction to address a variety of learning requirements.

According to the study, a psychosocial curriculum and assistance for children who are exposed to violence, crime, and conflict are necessary; therefore, it is clear that all instructors require PD and assistance while working with refugee children (Aydin et al., 2019). This involves training to deal with children who have encountered trauma and violence, as well as training in peace education. Teachers who receive trauma-informed training can better recognize and respond to the trauma that many Rohingya children have endured. This includes tactics for building safe and supportive school environments as well as promoting kids' social-emotional well-being. Meanwhile, teachers not only advance their PD but also assist students who are facing challenging situations.

## 4.4 Educational Technology in the Emergency Contexts

Educational technology presents an opportunity to improve teachers' professional development. Although technology plays an essential role in enhancing TPD, it is also being used more and more to help address the issues by providing flexible, affordable, and extensive options on a wide range of educational topics (Powell & Bodur, 2019). The integration of technology and pedagogical implementation in teacher training and development may face challenges in the digital era due to proper investment that falls behind the effective preparation of teaching practices (Yurtseven Avci et al., 2019), especially for the Rohingya refugee teachers and students. A study focused on the use of digital teaching competency in training curriculum materials, with an emphasis on three knowledge domains, including social awareness, pedagogical compatibility, and technical competence (Instefjord & Munthe, 2017). To support teachers in further PD and provide students with successful learning experiences, many digital tools are being utilised in education. Thus, due to technology integration, teachers need to develop their technological skills.

An empirical study analysis revealed four categories of Information and Communications Technology (ICT) applications in teacher development, including Web conferencing, Web 2.0 that supported teacher development and community, video-based training, and mobile technology for teacher learning (Hu et al., 2021). On the other hand, students most frequently use applications including Moodle (70%) and Google Meet (94%) as their distant learning platforms. In addition, they utilise Zoom, Skype, Viber, and Telegram for academic purposes and to stay in contact with their lecturers (Dushkevych et al., 2020). Furthermore, online learning platforms provide a virtual space to continue education during emergency situations, as well as offer indispensable resources for teachers and students.

The engagement of students in emergency online learning systems has multiple characteristics, all of which were assessed as having an exceptionally high engagement, such as positive conduct (79%), work engagement (78%), and attendance (75%) (Kurniawati & Fauziyah, 2022). In the context of technology and learning development in the Rohingyas, BRAC, a national non-governmental organisation (NGO), set up a comprehensive plan and introduced a technology-based, home-based teaching and learning mechanism called "Home School" to ensure the continuation of learning for the refugee students (Prodip & Roy, 2023). Teachers and students can foster resilience and flexibility in difficult circumstances by utilising online learning tools to continue their educational journeys undisturbed in emergency situations. In order to facilitate PD and engage with students, teachers can use online platforms to conduct teaching, evaluate student progress, and give feedback on time.

## 4.5 Teacher Training and Support at the Rohingya Camps

Teacher training and support are essential components of efforts to enhance the quality of education and support the resilience of teachers working with displaced populations. Research suggests that targeted professional development initiatives, including training workshops, mentoring programmes, and peer learning communities, can improve teachers' pedagogical skills, cultural competence, and socio-emotional well-being. Studies emphasise the importance of contextualised, culturally responsive approaches to teacher training and support, as well as the need for ongoing monitoring and evaluation to ensure the effectiveness and sustainability of interventions.

Although there are curriculum discrepancies between Burmese and Bangladeshi educational establishments for the Rohingya children, refugees also face challenges in obtaining basic education due to inadequate international funding, poor teaching quality, scheduling conflicts, and other logistical support (Rahman et al., 2022). The evidence gathered indicated that there is a significant gap and that teachers require additional training in basic pedagogy, subject matter, and Myanmar language, as well as capacity development (Chowdhury Mohammed, 2023). Furthermore, while host community teachers are qualified to teach the Bangla curriculum, they have to actively engage in teaching the Myanmar curriculum to the Rohingya children staying in Bangladeshi refugee camps.

Research demonstrates that teachers with extensive training are more effective in their work with refugee children (Richardson et al., 2019). Evidence from studies has revealed that it is not possible to include both theoretical and practical knowledge in a lecture in such a short time (Rahman et al., 2022). Besides, it is highlighted that pedagogy training is only provided for two days, which means that good teachers cannot be produced in that duration of time. With the aim of teachers' effective training, the International Child Development Programme (ICDP) prepares teachers to better understand and support children in vulnerable situations, successfully prevent child abuse and neglect, and change their views toward students who are living in refugee camps (Habib, 2020).

Over the past few decades, experiential learning pedagogy has become a transformational research paradigm that increases social equality by bridging the gap between knowledge and practice through community participation and social action (Wallerstein & Duran, 2010). Using an experiential learning approach ensures more engagement with the learning process and accelerates retention through access to practical teachings. However, maintaining the relationship is essential for generating socio-culturally relevant materials. Mapped Educational Specialist Knowhow (MESH) and Voluntary Service Overseas (VSO) offer great potential to enhance the training of volunteers and learning provision, although elements might be altered to boost impact while training empowers a marginalised group to enhance children's learning (Laxton et al., 2020).

Direct observation revealed the existing difficulties in treating the Rohingya refugees in Bangladesh, and the concerns include the practitioners' lack of clinical expertise, moral judgement, language and cultural competency, clinical supervision, and self-care (Islam & Mozumder, 2021). Apart from the challenges, following the study advises expanding access to materials for developing linguistic and cultural competency, making use of technology platforms for supervision, training, and service delivery, offering instruction on ethical and evidence-based practice and intervention models, implementing a peer supervision system, and encouraging more cooperation and coordination between the organisations.

## 4.6 Teacher Training and Support in Similar Global Crises

In the EiE context, practical aspects of rapid response include the deployment of teachers – often requiring training for volunteers, shelter arrangements, education supplies, and materials. It has been advocated that education programmes should include a major component of training for teachers as well as youth and adult teachers, and provide incentives to avoid teacher turnover (Sinclair, 2002). Teachers and youth leaders can identify those who need help because of problems related to post-traumatic stress, sickness, malnutrition, disability, abuse, and so on (Sinclair, 2002). The 'teacher emergency package' proposed by Sinclair (2002) could help as there are examples of the distribution of educational materials and the development of teacher education. Examples of successful implementations include the distribution of educational materials and the design of open-learning approaches for teacher training in Somalia; and support for temporary schools in Angola and cross-border support for schooling in war-torn Southern Sudan.

Teachers in crisis situations are often under-resourced, and require additional support through training and materials. In emergency situations, teachers may also face unfamiliar challenges and require training regarding the psychosocial effects of trauma. All teachers, or at least a core group per school, along with youth workers, should receive urgent training in this area (Sinclair, 2002). Teachers should also be equipped with resources to manage their well-being and adapt their teaching methods to meet the needs of traumatised students. Governments and teacher-training institutions working with crisis-affected populations should incorporate the psychosocial dimension into all teacher-training programmes (Sinclair, 2002).

In many crisis-affected communities, including the Rohingya community, teachers are underqualified and untrained. Even trained teachers may have had limited exposure to modern teaching methods. The crisis may provide an opportunity to introduce a new approach to teacher training that can help even inexperienced teachers to develop the skills required for effective teaching, as well as taking into account the special needs of children exposed to some form of crisis (Sinclair, 2002). A key challenge is ensuring the recognition of training provided in emergencies. For example, the International Rescue Committee (IRC) offered extensive training to refugee teachers from Liberia and Sierra Leone who had taken refuge in Guinea. The Ministry of Education in Liberia subsequently recognised the good performance of returnee teachers but had difficulty in awarding qualified teacher status, which required completion of a specified training curriculum. A compromise was reached where teachers received training while in exile, and received a basic-level teacher qualification. This case highlights the importance of ensuring that emergency training programmes incorporate elements required for qualified teacher status in the home country, with proper documentation (Sinclair, 2002).

The Jesuit Refugee Service (JRS) in Northern Uganda funded Sudanese teachers' participation in the national distance education teacher training program allowing them to obtain qualified teacher status (Sinclair, 2002). Similarly, following the Kosovo crisis, the new government recognised this teaching experience of Kosovar Albanians in the 'parallel' (unrecognised) school system by giving two years' credit towards a new teacher qualification. These teachers then completed one more academic year of studies through summer courses, at the end of which they obtained a 'Bachelor of Teaching' degree (Sinclair, 2002).

To increase engagement among Lebanese teachers working with refugee students, several initiatives were undertaken focusing on cost-effective, technology-enhanced, problem-solving learning approaches. These initiatives also emphasise the importance of technology in both TPD and student engagement (Metni, 2022). The intervention utilised low-cost technologies similar to those found in maker-spaces, promoting knowledge-building in the real world, highlighting the role of technology in enhancing learning experiences.

## 4.7 Community Engagement and Collaboration

Community engagement and collaboration are critical for promoting the holistic development and empowerment of displaced populations, including the Rohingya children and families. Research highlights the importance of involving local communities, parents, caregivers, and stakeholders in decision-making processes, programme implementation, and advocacy efforts. Studies demonstrate the positive impact of community-based approaches to education, including parent involvement, volunteerism, and partnerships with local organisations and NGOs, in supporting teachers and improving educational outcomes for displaced children.

By synthesising findings from the literature on teacher resilience, professional development, and education in emergency contexts, this review provides a theoretical foundation and conceptual framework for understanding the challenges and opportunities teachers are facing in teaching the Rohingya children in Bangladesh. It also informs the design, implementation, and evaluation of interventions aimed at supporting teachers and improving educational outcomes for displaced populations, while highlighting gaps in knowledge and areas for future research and action.

In order to effectively address the needs of the Rohingyas and provide life-sustaining support in the refugee camps, several non-governmental organisations (NGOs) have built supportive relationships with the Bangladeshi government, UN agencies, and other funding bodies over the past three years (Ahmed & Das, 2021). By establishing temporary schools and hiring teachers, NGOs are offering informal education, but these efforts are tragically insufficient. According to a pilot initiative launched by the Government of Bangladesh (GoB) and UNICEF, 10,000 children living in the camps got access to formal education from sixth to ninth grade, based on the Myanmar Curriculum (Reidy, 2020), however it had to be postponed for more than a year due to COVID-19 pandemic. It appears that the host government did not genuinely want to integrate refugees into local services and facilities, especially access to education in public institutions, to prevent geopolitical and socio-cultural problems locally (Habib et al., 2023). Despite challenges, the Rohingya community and civil society organisations participated in negotiations with the government and were able to secure, in part, an inclusive space for the growth of educational opportunities through conversations, discussions, and resilience.

Decision-making power in Cox's Bazar is divided across sectors, most of which are run by UN agencies and a small number of international and national NGO co-representatives. Undertaking significant choices concerning humanitarian programming, camp-in-charges (Bangladeshi government officials) frequently work informally with 'majhis' or block-level community leaders (Islam & Naing, 2023). An annual Joint Response Plan (JRP), which is the result of bilateral discussions between the Bangladeshi government and the United Nations, defines the year's strategic goals and financial requirements and shapes the refugee response in Cox's Bazar (Kurtzer et al., 2021). The JRP has continuously prioritised protection and assistance for refugees as well as host communities since 2018. A review of the plan's goals from 2018 to the present reveals a progressive transition away from programming that is solely humanitarian in nature and towards one that is more development-focused. Beyond these conflicting points of view and the drawbacks of formal and informal contradiction, the term generally contributes to a broader understanding of the aims, technologies, and participants of humanitarian action and draws attention to the sometimes-overlooked experiences of local volunteers who try to navigate various assistance scales and resolve the conflicts that go along with them (Thomas et al., 2018). Enhancing collaboration may help the area's comprehensive education programme, which would eventually benefit the refugee children there as well. Moreover, education is a valuable asset that follows them everywhere they go, and the international community must offer refugee children chances to broaden their horizons, improve their minds, and support their communities' growth and development.

## 5. METHODOLOGY

Based on secondary sources, this chapter uses both qualitative and quantitative data to explore the resilience of teachers and the effectiveness of professional development initiatives in the context of teaching the Rohingya children in Bangladesh. A comprehensive review of existing literature on education in emergencies, teacher resilience, professional development, and education in emergency contexts has been carried out. The literature review encompasses peer-reviewed articles, reports, policy documents, and academic publications, focusing on relevant topics such as teacher training, psychosocial support, community engagement, and educational interventions targeting displaced populations. Document analysis involved a synthesis of relevant documents, including programme reports, training materials, policy documents, and monitoring and evaluation reports. Documents provide valuable insights into the design, implementation, and outcomes of professional development initiatives, as well as organisational support systems and interventions implemented by the Education Sector and humanitarian agencies. This document analysis helped to contextualise both qualitative and quantitative findings within broader policy and programmatic frameworks, enabling a more comprehensive understanding of the factors shaping teachers' professional development and resilience in emergency contexts.

## 6. FINDINGS

### 6.1 Resilience of Teachers in the Rohingya Refugee Camps

Teachers in the Rohingya camps have exhibited remarkable resilience in the face of numerous challenges. Despite a confluence of obstacles, including limited resources, overcrowded learning environments, communication challenges, and student experiences with trauma, teachers have exhibited a steadfast commitment to providing displaced children with educational opportunities. Even under severe limitations in material support, such as a lack of printed learning materials, instructional tools, and appropriate classroom space, teachers in the Rohingya refugee camps have displayed remarkable resilience through improvisation and adaptation of their pedagogical approaches.

For instance, some teachers utilised locally available materials, such as recycled paper and handmade educational aids, to create interactive learning experiences for their students. In addition to that, teachers provide emotional support to students in times of disaster, fire incidents and in other psychological breakdown although not in a specialised way but moreover as a ground-level support.

### 6.2 Improvisation and Innovation in Teaching Methods

In response to the unique challenges posed by the Rohingya crisis, teachers have showcased improvisation and innovation in their teaching methods. Drawing upon their resourcefulness and ingenuity, teachers have developed creative approaches to engage learners and facilitate learning, often incorporating elements of experiential and participatory pedagogy to foster active student involvement in the learning process (Kong, 2021; Mebert et al., 2020). For instance, they may use recycled materials like cardboard, plastic bottles, and fabric scraps to make flashcards, charts, and other visual aids. By leveraging these locally available materials, teachers enhance the effectiveness of their lessons and promote interactive and participatory learning experiences for students.

Teachers within the Rohingya camps prioritise the establishment of safe learning environments where students feel comfortable sharing their experiences through storytelling activities. This can involve utilising role-playing exercises, artistic expression through drawing, or facilitating group discussions as tools for exploring emotional responses and developing coping mechanisms. Additionally, teachers actively collaborate with members of the Rohingya community to weave traditional narratives, songs, and artistic expressions into the curriculum. This fosters a sense of cultural kinship and fosters active participation among students in the learning process.

## 6.3 Collaboration and Support Networks

Community-based education networks and collaboration with humanitarian organisations have played a crucial role in supporting teachers in the refugee camps. Despite the lack of formal training, teachers have received support and guidance through various channels, enabling them to navigate the challenges of teaching in emergency contexts effectively. By fostering peer learning groups and teacher learning cycles (TLC) among teachers, these networks have facilitated the exchange of best practices, resources, and support systems, thereby enhancing teachers' resilience and effectiveness in delivering education in emergency settings.

The peer groups and TLCs open the scope for teachers in the Rohingya refugee camps to provide emotional support and solidarity to one another recognizing the emotional toll of working in an emergency context. Teachers training also involves sessions on psychosocial support (PSS) and wellbeing to enhance informal support networks where teachers can share their experiences, express their concerns, and seek advice from colleagues. Additionally, some organisations and NGOs also offer psychosocial support programmes and counselling services to help teachers cope with stress, trauma, and burnout.

## 6.4 Role of Teachers' Professional Development

Teachers' professional development initiatives have emerged as essential mechanisms for enhancing teachers' capacity and resilience in the Rohingya refugee camps. Targeted training programmes, informed by evidence-based practices and contextual considerations, have equipped teachers with the necessary skills and competencies to address the diverse needs of displaced learners effectively. These initiatives have focused on areas such as trauma-informed pedagogy, culturally responsive teaching, and language proficiency, aligning with the specific challenges and priorities of teaching in emergency contexts. Various organisations and NGOs operating in the Rohingya refugee camps conduct in-service training programmes to enhance the PD of their teachers. These programmes cover a wide range of topics, including pedagogy, classroom management, assessment. trauma-informed teaching practices, and psychosocial support.

Capacity-building workshops and trainings under TPD programme under the Cox's Bazar Education Sector, are designed to equip teachers with the knowledge, skills, and resources necessary to address the unique challenges of teaching in an emergency context i.e., child protection, refugee rights, conflict resolution, and cultural sensitivity etc. By participating in capacity-building activities, teachers also enhance their understanding of the socio-political context and develop culturally responsive teaching practices that promote equity, diversity, and inclusion.

In the Rohingya refugee camps, some organisations offer online learning platforms and digital resources endorsed by the Education Sector to facilitate TPD in the Rohingya refugee camps such as- Learning Management System (LMS) app developed by Save the Children for host community teachers and Education Sector Master Trainer. These platforms provide access to self-paced courses, webinars, instructional videos, and educational materials on topics relevant to refugee education. For example, teachers may access modules on language instruction, literacy development, numeracy skills, or social-emotional learning. By leveraging online learning platforms, teachers can engage in continuous professional development (CPD) and self-learning regardless of geographical constraints or resource limitations.

## 6.5 Importance of Mentoring in the Emergency Contexts

The professional development of teachers in emergency contexts encompasses multiple dimensions, each of which plays a critical role in enhancing educators' effectiveness, resilience, and capacity to address the diverse needs of the Rohingya refugees in Bangladesh. Mentoring is important for PD programmes in high-risk contexts (Sabieh, 2023). In the Rohingya refugee camps, experienced teachers often take on mentoring roles to support novice or less experienced teachers. These mentorship relationships provide opportunities for new teachers to learn from the expertise and experience of their more seasoned colleagues. Mentors offer guidance, feedback, and practical advice on various aspects of teaching, including lesson planning, classroom management, and student engagement.

In an emergency context, cross-cultural mentoring initiatives for teachers play an important role in facilitating mutual learning and understanding. In the Rohingya refugee camps, the Rohingya teachers learn from qualified or experienced host teachers about local context, culture, norms and social differences. On the other hand, host teachers can learn Rohingya culture, and their unique needs from the Rohingya teachers. This opportunity helps symphony of the teaching-learning processes in the learning facilities within the Rohingya camps. Besides, Cox's Bazar Education Sector has been continuously developing a master trainers pool from different humanitarian organisations that also increases the scope of learning based on the needs of both teachers and students.

Given the traumatic experiences that many refugee children have endured, trauma-informed mentoring provides crucial support for teachers in addressing the emotional and psychological needs of their learners. Mentors trained in trauma-informed approaches offer guidance on creating safe and supportive learning environments, recognising signs of distress, and implementing trauma-sensitive teaching practices. For example, a mentor teacher may help another teacher develop strategies for building trust with students, managing challenging behaviours, and promoting resilience and coping skills.

## 6.6 Role of Technology in Teacher's Professional Development

Technology can significantly increase access and effectiveness of professional learning communities in emergency settings (El-Serafy, 2022). However, there are significant barriers to implementing technology-supported professional learning communities, which need to be addressed to fully leverage the potential benefits of technology in TPD in an emergency context. For example, online learning platforms offer access to a variety of courses and resources related to pedagogy, classroom management, and subject-specific instruction. Educational apps and digital resources are used to support teachers' capacity building in the Rohingya refugee camps.

Mobile learning technologies, such as smartphones and tablets, are used to deliver PD content to teachers in the Rohingya refugee camps. Teachers may access training materials, videos, and educational resources directly on their mobile devices, allowing for flexibility and convenience in learning.

## 6.7 Initiatives for Teacher's Professional Development

Initiatives like Teacher's Professional Development projects under Education Sector, such as the one Save the Children implemented under LCFA education phase, have provided valuable guidance for designing and implementing effective training programmes for teachers and facilitators. By emphasising adaptable, contextualised, and practice-based learning approaches, these initiatives have informed the development of tailored PD initiatives that meet the unique needs of teachers working in the Rohingya refugee camps. However, Just like effective emergency response requires coordinated action from various agencies, the TPD framework's development or adaptation needs inclusive participation from all stakeholders, especially those operating in emergency settings. Their voices and experiences must be central to shaping the framework.

The Rohingya crisis serves as a powerful case study for the importance of effective TPD programmes in emergency settings. Initiatives like the TPD by Save the Children, in collaboration with the Cox's Bazar Education Sector, offer valuable lessons in crafting training that empowers teachers . This project, and others like it, highlight the key role of adaptable, contextualised, and practice-based learning approaches.

### 6.7.1 Adapting to the Chaos

Imagine a classroom crammed with students, a stark reality for many Rohingya teachers. Effective TPD programmes address these challenges. Training for overcrowded classrooms might focus on cooperative learning techniques, where students work together in small groups, maximising engagement in a limited space. Conversely, teachers in more open-air settings might explore project-based learning activities, allowing students to move around and collaborate. Save the Children's project reportedly provided training modules with diverse approaches, allowing teachers, especially young Rohingya teachers, to choose methods that best fit her specific classroom dynamics.

### 6.7.2 Weaving Culture into Learning

Culturally sensitive training ensures teachers can effectively connect with their students. Many TPD programmes integrate the Rohingya language elements and storytelling techniques into training modules. This builds on familiar cultural practices for both teachers and students. Additionally, collaborating with the Rohingya community leaders allows for the development of culturally relevant learning materials that resonate with learners'' backgrounds. Imagine the impact on the young Rohingya children when they see stories and characters from their own heritage reflected in their lessons.

### 6.7.3 From Theory to Practice

Effective TPD goes beyond dry lectures. Save the Children's project, along with many others, reportedly utilised a hands-on approach. Workshops, peer learning activities, and opportunities for teachers to experiment with new techniques in a supportive environment are key. This allows teachers like Mohammed, a seasoned teacher, to refine his skills and gain confidence in implementing the learned strategies in his classroom.

The Rohingya crisis underscores the need for inclusive collaboration in TPD development. The voices and experiences of all stakeholders, especially those directly involved, are crucial. This includes teachers like Aisha and Mohammed, community leaders, other NGOs, and government education officials. By working together, they can inform the design and delivery of culturally relevant training, maximise impact, and avoid duplication of efforts. Ultimately, this creates a stronger support system for teachers navigating the complexities of the Rohingya education response, fostering a more hopeful future for displaced children.

## 6.8 Impact of Training on Teacher Competence

Training programmes have contributed to enhancing teacher competence and confidence in delivering quality education in the Rohingya camp settings. By providing practical guidance on adapting teaching methods to challenging environments and promoting student well-being, and Social and Emotional Learning (SEL), professional development initiatives have empowered teachers to overcome obstacles and foster positive learning outcomes. In the face of the immense challenges posed by the Rohingya refugee crisis, well-designed training programmes have emerged as a powerful tool for bolstering teacher competence and confidence. These initiatives go beyond simply delivering traditional teaching methods, instead addressing the unique needs of both teachers and displaced learners within the camps.

A critical focus of these training programmes is equipping teachers with strategies to adapt their pedagogy to overcome the challenges they face. Workshops might delve into active learning techniques, like cooperative learning or gamification, to keep students engaged despite limitations on space and resources. These approaches foster student participation and collaboration, maximising learning opportunities even in overcrowded classrooms.Furthermore, the training programmes prioritise student well-being and social-emotional learning (SEL). Techniques like mindfulness exercises, relaxation techniques, and peer support activities are introduced, empowering students to build resilience, manage emotions, and create a more supportive classroom environment.

The focus doesn't stop there. Training programmes often emphasise fostering culturally relevant teaching practices. This might involve collaborating with the Rohingya community members to develop learning materials that resonate with the learners'' cultural background, or integrating traditional games and songs into lessons. This approach fosters a sense of normalcy and strengthens the connection between children, their culture, and the learning process. Finally, training programmes recognise the importance of teachers' well-being. By establishing support groups and mentorship opportunities, they create safe spaces for teachers to share experiences, learn from each other, and develop coping mechanisms for the demanding realities of teaching in emergency contexts.

By promoting these various aspects, training programmes play a vital role in strengthening teacher competence within the Rohingya refugee camps. Equipped with the skills and knowledge to navigate complex challenges, teachers are empowered to deliver quality education and create a more positive learning environment for displaced children. This fosters a sense of normalcy and hope for the future, a crucial element in the Rohingya refugee response.

These are just a few examples of how well-designed training programmes contribute to teacher competence. By providing practical skills and fostering resilience, these initiatives equip teachers to deliver quality education and create a more positive learning environment for displaced children in the Rohingya refugee camps.

## 6.9 Challenges and Opportunities in Teacher's Professional Development

Teacher's professional development has the potential to improve the quality of teaching, thus improving education for children living in crisis (Darling-Hammond, Hyler, et al., 2017). However, in EiE, teachers face unique challenges and may be the only resource children have for learning (Richardson et al., 2018). Humanitarian actors and researchers are gathering knowledge on how to provide TPD that will meet the needs of EiE teachers, as well as positively impact the quality of teaching (Tubbs Dolan, 2018).

There is little research regarding teacher training in EiE, especially in school systems affected by prolonged violent conflict (Burde et al., 2017). Meanwhile, leaders in countries receiving large numbers of refugees or dealing with displaced populations struggle to find qualified teachers (Mendenhall et al., 2018). In light of the current global teacher shortage, children affected by crises are often taught by underqualified teachers, which negatively impacts the quality of their education (Mendenhall et al., 2018).

The Rohingya crisis in Cox's Bazar, Bangladesh, highlights a crucial challenge faced by TPD initiatives in EIE settings is the language barrier. The Rohingya community primarily uses the Rohingya dialect in daily life, which lacks a widely recognised written script. This creates a significant obstacle and frustration in delivering effective training programmes for teachers. For example, if a trainer presents complex teaching methods in English or Bangla (the local language of the host community) language, many Rohingya teachers struggle to grasp the finer points of the training. Thus, language barrier hinders their ability to implement new strategies in their classrooms.

Here is how dedicated teachers and TPD initiatives are working to bridge this gap:

*Building Bridges with Translators and Interpreters:* Skilled translators, often resourced from the Rohingya community and interpreters act as vital bridges during workshops and training sessions. They ensure clear communication and allow teachers to fully grasp the intricacies of the training content.

*Creating Resources in Multiple Language:* Developing training materials in Rohingya, Bangla, and English (where applicable) empowers teachers to access information in a language they understand. This might involve translated handouts, visual aids with key concepts presented in multiple languages, and glossaries with simplified terms.

*Learning from Each Other:* Peer Support and Buddy Systems: Facilitating peer learning opportunities and establishing buddy systems leverage the language skills of some teachers to support those facing a language barrier. Teachers with stronger language skills can translate key points or explain concepts in simpler terms for their peers, fostering a more collaborative learning environment.

By addressing the language barrier through these approaches, TPD initiatives are trying to ensure equitable access to training for all Rohingya teachers. Equipped with the skills and knowledge assembled from training programmes, teachers are empowered to deliver quality education to displaced children, ultimately fostering a more supportive and positive learning experience in the Rohingya camps.

## 6.10 Interplay of Social Identities and Experiences

Intersectionality theory underscores the interconnected nature of social identities and experiences, highlighting the influence of factors such as gender, ethnicity, class, and displacement status on individuals' resilience and coping mechanisms. In the context of the Rohingya refugee camps, teachers' experiences of adversity and resilience are shaped by a complex interplay of intersecting identities and contextual factors, necessitating tailored approaches to PD and support.

The Rohingya teachers in camps are themselves refugees who have been displaced from their homes due to conflict, persecution, or violence. The experience of displacement and forced migration can have profound impacts on teachers' well-being, resilience, and sense of identity. For instance, the Rohingya teachers may bring a deep understanding of the trauma, and resilience of their students, having experienced similar challenges in their own lives.

Gender dynamics play a significant role in shaping the experiences and opportunities of teachers in the Rohingya refugee camps. Women may face unique challenges and barriers in accessing education, PD, and leadership roles compared to their male counterparts. For example, female teachers may experience restrictions on mobility, limited access to training opportunities, or societal expectations regarding caregiving responsibilities. However, the experiences of the Rohingya teachers extend beyond just gender and class. Intersectionality theory emphasises the interplay of various social identities. For instance, teachers with disabilities may face challenges accessing TPD due to physical limitations or non-inclusive training materials. TPD programmes can address this by providing accessible training formats and fostering collaboration among teachers with and without disabilities.

Similarly, ethnicity and pre-displacement experience shape teacher needs. Non-Rohingya teachers might struggle with communication and cultural differences. TPD programmes can equip them with culturally sensitive teaching methods and promote collaboration with the Rohingya colleagues. Additionally, leveraging on the prior experience of teachers who worked in formal education systems can enhance TPD by building upon existing skills while offering necessary adaptations for the refugee camp context. This can be achieved through targeted training modules and mentorship programmes connecting experienced teachers with those new to the camp setting. By acknowledging these diverse social identities, TPD and well-being initiatives can create a more supportive learning environment where all Rohingya teachers can thrive and contribute their unique talents to educating displaced children.

## 6.11 Cultural Norms and Contextual Factors

Contextual factors, including cultural norms, institutional structures, and policy frameworks, significantly impact TPD and resilience in emergency settings. By acknowledging the socio-cultural context of education in the Rohingya refugee camps and fostering culturally responsive pedagogy and support systems, stakeholders can enhance the effectiveness and sustainability of PD initiatives.

Cultural values and beliefs about education shape the attitudes, priorities, and practices of teachers in the Rohingya refugee camps. For example, teachers may prioritise certain subjects or teaching methods based on cultural perceptions of their importance or relevance to learners' futures. Additionally, cultural attitudes towards authority, hierarchy, and decision-making may influence the dynamics of professional relationships and collaboration among teachers. Cultural norms and limited resources in the Rohingya refugee camps significantly impact teacher training and development initiatives. Understanding the socio-cultural context of education is crucial. For instance, teachers might prioritise religious studies or practical skills based on community values. Additionally, cultural norms around hierarchy might influence teacher-student relationships.

Effective TPD programmes acknowledge these factors. Culturally responsive pedagogy equips teachers to integrate the Rohingya language and practices into lessons, fostering a more meaningful learning experience for students. Limited resources necessitate flexibility. TPD programmes can utilise low-tech resources or peer coaching models, leveraging existing expertise within the camp community. Collaboration with the Rohingya community leaders, parents, and religious figures further enhances TPD. This fosters a more supportive learning environment and integrates community values into the education system. Furthermore, TPD programmes can explore ways to incorporate valuable indigenous knowledge about traditional medicine, agriculture, or environmental practices into the curriculum. This approach fosters cultural pride and connects learning to practical skills relevant to the camp environment. By acknowledging these contextual factors, TPD initiatives can be designed to be more effective, sustainable, and culturally relevant for both teachers and students in the Rohingya refugee camps.

# 7. RECOMMENDATIONS

## 7.1 Dimensions of Teachers' Professional Development

Research suggests that resilient teachers participated in both official TPD programmes, which are organised by the government or the school, and informal TPD, which they initiated on their own. Teachers expressed agreement that needs-based TPD programmes made them more resilient by giving them the tools they need to stay in the teaching profession. However, the teacher recommended that TPD programmes include materials that address teachers as individuals who require attention in addition to their teaching practices and as such, their well-being. As a result, resilient teachers recommended that TPD design include contents related to mental health and spirituality (Larasati & Kuswandono, 2023*)*.

### 7.1.1 Identity, Motivation and Wellbeing

Identity, motivation and wellbeing play vital roles in providing education in emergencies (Kirk & Winthrop, 2013). In the context of emergency settings such as the Rohingya crisis in Bangladesh, addressing the identity, motivation, and wellbeing of teachers is paramount to ensuring the effectiveness and resilience of the education system. Teachers working in these challenging environments often face immense pressure, trauma, and uncertainty, which can significantly impact their professional identity, motivation levels, and overall well-being.While student well-being and academic achievement are undoubtedly crucial aspects of education, neglecting teacher well-being can have a detrimental impact on overall educational outcomes. This can manifest in a variety of ways, including increased teacher burnout, decreased job satisfaction, and ultimately, a higher teacher turnover rate. A demoralised and unsupported teaching force can struggle to create a positive and stimulating learning environment for students. A recent analysis of teacher training materials reveals a concerning trend: teacher well-being is positioned as a means to an end, namely, enhancing student success. These resources, though acknowledging the importance of teacher well-being for positive student outcomes, often undermine this very goal. This is achieved by framing teachers solely as resources for student learning, and by failing to equip them with adequate tools to address their own well-being needs.

Recognizing and celebrating the unique skills and experiences that Rohingya teachers bring to the classroom can further empower and motivate them. TPD programmes can achieve this by incorporating opportunities for teachers to share their own cultural knowledge and pedagogical approaches. This fosters a sense of professional identity and validates the valuable contributions that the Rohingya teachers make to the education of displaced children. Furthermore, fostering a sense of agency and ownership over their professional practice can be a powerful motivator for the Rohingya teachers. This can be achieved by involving them in curriculum development, decision-making processes related to classroom management and instructional strategies, and ongoing programme evaluation. By valuing their input and expertise, TPD programmes can create a more collaborative and empowering learning environment for both teachers and trainers.

However, a more holistic approach is required. TPD programmes should explicitly address the multifaceted nature of teacher identity, fostering a sense of professional purpose and empowerment alongside strategies for self-care and stress management. Despite this shortcoming, the analysis also identifies a glimmer of hope. The very materials that potentially undermine teacher well-being also offer opportunities for teachers to cultivate self-care strategies and develop coping mechanisms, ultimately fostering their own resilience (*Teacher Wellbeing Resources Mapping & Gap Analysis | INEE*, 2021).

## 7.1.2 Pedagogical Skills Enhancement

Given the diverse learning needs and backgrounds of the Rohingya learners, teachers require specialised pedagogical skills to facilitate effective teaching and learning experiences. Professional development initiatives focus on enhancing teachers' abilities to design culturally responsive and inclusive instructional strategies, differentiate instruction to meet individual learning needs, and integrate technology and multimedia resources into their teaching practices. Adaptive strategies include peer learning, lesson study groups, and collaborative planning sessions, where teachers share best practices, develop innovative teaching approaches, and reflect on their pedagogical practices in light of student outcomes.Additionally, effective TPD programmes for the Rohingya teachers go beyond generic skills. They address the unique needs of students who may have experienced trauma. This means equipping teachers with trauma-informed practices to create safe, supportive classrooms. Mindfulness techniques can also be valuable tools for students managing anxiety. Additionally, culturally appropriate approaches ensure teachers can address the social and emotional needs of the Rohingya children. It is also important to acknowledge that language can be a hurdle. TPD programmes can equip teachers with effective multilingual instruction strategies. This might involve incorporating the Rohingya language alongside the primary language of instruction. Visuals and graphic organisers further support comprehension for students developing fluency. Finally, providing differentiated learning activities ensures all students can access the curriculum, regardless of language proficiency.

### 7.1.3 Socio-Emotional Well-being

Working in emergency settings can take a toll on teachers' mental, emotional, and physical well-being, as they navigate challenging circumstances, witness trauma, and experience stress and burnout. Professional development programmes prioritise teachers' socio-emotional well-being by providing psychosocial support, stress management techniques, self-care strategies, and opportunities for personal and professional reflection. Initiatives such as mindfulness training, counselling services, peer support groups, and wellness workshops aim to build teachers' resilience, coping skills, and sense of self-efficacy in managing the emotional demands of their work.

### 7.1.4 Cultural Competence and Sensitivity

Teachers working with the Rohingya learners must demonstrate cultural competence and sensitivity to effectively engage with students. Professional development efforts focus on promoting intercultural understanding, empathy, and respect for diversity among teachers, enabling them to create inclusive and welcoming learning environments. Training programmes provide opportunities for teachers to deepen their understanding of the Rohingya culture, language, history, and traditions, as well as develop strategies for building positive relationships with students and families.

### 7.1.5 Contextual Adaptation and Flexibility

Teachers in emergency contexts must demonstrate adaptability, flexibility, and resourcefulness in responding to dynamic and unpredictable circumstances, such as changes in student enrolment, curriculum requirements, and available resources. Professional development initiatives emphasise the importance of adaptive teaching practices, problem-solving skills, and improvisation techniques, enabling teachers to navigate challenges and seize opportunities for innovation. Strategies include participatory needs assessments, action planning exercises, and peer learning communities, where teachers collaborate to develop contextually relevant solutions to educational challenges.

### 7.1.6 Addressing Diverse Learner Needs

The Rohingya learners often come from diverse linguistic, cultural, and educational backgrounds, requiring teachers to employ inclusive and flexible teaching strategies. Professional development focuses on equipping teachers with the skills to address the diverse needs of their students. Training may include sessions on differentiated instruction, inclusive education practices, and understanding the socio-cultural context of the Rohingya community to create a more inclusive and responsive learning environment.

### 7.1.7 Collaborative Learning Communities

Recognising the importance of collaboration in fostering teacher development, professional development initiatives encourage the formation of collaborative learning communities. These communities provide a platform for teachers to share experiences, insights, and best practices. Collaborative learning may take the form of peer mentoring, joint lesson planning, and participation in professional learning networks, creating a supportive ecosystem that enhances the collective resilience of teachers.

### 7.1.8 Use of Adaptive Technologies

Technology can play a vital role in overcoming challenges associated with the lack of resources and infrastructure in emergency settings. Professional development programmes may focus on training teachers to effectively utilise adaptive technologies for remote learning, virtual collaboration, and accessing educational resources. Technology can increase access to professional learning communities and enhance the effectiveness of professional learning communities.

By addressing these dimensions, PD initiatives aim to empower teachers with the skills and support needed to navigate the complexities of teaching the Rohingya children in emergency contexts. This multifaceted approach not only enhances individual teacher resilience but also contributes to the overall quality and inclusivity of education for displaced populations. Through targeted training, support, and collaboration, teachers can play a crucial role in ensuring access to quality education and promoting the holistic development of the Rohingya learners in emergency settings.

## 7.2 Organisational Support Systems and Interventions

### 7.2.1 Training and Professional Development Programmes

The Education Sector collaborates with international organisations, NGOs, and local partners to develop and implement training and PD programmes tailored to the needs of teachers working with the Rohingya learners. These programmes focus on enhancing teachers' pedagogical skills, cultural competence, trauma-informed practices, and socio-emotional well-being, enabling teachers to provide quality education and support to displaced children. Training workshops, seminars, and webinars cover topics such as inclusive education practices, child protection, psychosocial support, and curriculum adaptation, equipping teachers with the knowledge and skills needed to address the diverse needs of the Rohingya refugees.

### 7.2.2 Strengthen Teacher Support Networks

Organisations operating in the Rohingya refugee camps can strengthen and enhance teacher support networks to provide teachers with access to resources, mentorship, and emotional support. These networks may take the form of non-formal structures within learning facilities or informal communities of practice among teachers. For example, organisations may facilitate peer support groups, mentorship programmes, or counselling services to help teachers navigate the challenges of working in an emergency context and maintain their well-being.

### 7.2.3 Integrating Trauma-Informed Practices

Integrating trauma-informed practices into teacher training curricula is critical to supporting the mental health and well-being of both teachers and students. By equipping teachers with the knowledge and skills to recognise and respond to trauma-related issues effectively, PD initiatives can foster a supportive learning environment conducive to academic success and resilience.

### 7.2.4 Mentorship and Coaching

Mentorship and coaching programmes provide ongoing support and guidance to teachers, allowing them to receive personalised feedback, mentorship, and PD opportunities. Experienced teachers, educational specialists, and humanitarian workers serve as mentors and coaches, offering advice, encouragement, and practical strategies to enhance teachers' effectiveness and resilience. Mentorship relationships foster a sense of belonging, professional growth, and continuous learning among teachers, empowering them to overcome challenges and thrive in their roles.

### 7.2.5 Resource Provision and Infrastructure Support

The Education Sector works to provide teachers with access to essential resources, materials, and infrastructure needed to deliver quality education in emergency settings. Efforts include the provision of teaching materials, textbooks, school supplies, and classroom furniture, as well as the construction and maintenance of school facilities and learning spaces. By addressing resource gaps and infrastructure deficiencies, organisations enable teachers to create safe, conducive, and inclusive learning environments for the Rohingya learners, enhancing their educational experiences and outcomes.

### 7.2.6 Policy Development and Advocacy

The Education Sector advocates for policy development and implementation to support teachers and improve the quality of education for the Rohingya children in Bangladesh. Advocacy efforts focus on issues such as teacher recruitment and training, curriculum development, accreditation and certification, teacher incentives and retention, and collaboration with relevant government agencies and stakeholders. By influencing policy decisions and mobilising resources, organisations can create an enabling environment for teachers to thrive and contribute to the long-term educational development of the Rohingya communities.

### 7.2.7 Community Engagement and Partnership:

Organisations actively engage with local communities, parents, caregivers, and stakeholders to build partnerships and mobilise support for education initiatives targeting the Rohingya learners. Community-based approaches involve the participation of community leaders, volunteers, and parents in decision-making processes, programme implementation, and monitoring and evaluation of activities. Through community engagement and partnership-building efforts, organisations strengthen the social fabric, resilience, and sustainability of education programmes, while fostering trust, ownership, and collaboration among all stakeholders.

By implementing these organisational support systems and interventions, the Education Sector and humanitarian agencies can effectively support teachers in providing quality education and promoting the resilience and well-being of the Rohingya children in Bangladesh. By addressing the multifaceted needs of teachers and creating a supportive and enabling environment, national and international organisations contribute to the overall educational development and empowerment of displaced populations in emergency settings.

## 7.3 Community Engagement and Collaboration

### 7.3.1 Participatory Needs Assessment:

Community engagement begins with a participatory needs assessment process, whereby organisations consult with community members, including teachers, parents, and local leaders, to identify priorities, challenges, and opportunities related to education. Needs assessment activities may include surveys, focus group discussions, interviews, and community meetings, allowing stakeholders to voice their concerns, share insights, and contribute to decision-making processes. By involving communities in the identification of needs and priorities, organisations ensure that education initiatives are responsive to local contexts, cultural norms, and aspirations, fostering a sense of ownership and empowerment among community members.

## 7.3.2 Capacity Building and Training

Community engagement efforts include capacity-building and training programmes targeting parents, caregivers, and community leaders to support teachers and enhance educational outcomes for the Rohingya children. Training sessions cover topics such as child development, positive parenting practices, school-home partnerships, and community mobilisation strategies, equipping stakeholders with the knowledge and skills needed to actively support children's learning and well-being. By empowering parents and caregivers to play an active role in their children's education, organisations strengthen the home-school connection and create a supportive learning environment that extends beyond the classroom.

## 7.3.3 Volunteerism and Parent Involvement

Organisations promote volunteerism and parent involvement in school activities and decision-making processes, encouraging community members to contribute their time, resources, and expertise to support teachers and enrich the educational experience of the Rohingya learners. Volunteer initiatives may include mentoring programmes, tutoring sessions, extracurricular activities, and parent-teacher associations, fostering positive relationships and collaboration between schools and communities. By harnessing the collective efforts and talents of community members, organisations enhance the educational opportunities and outcomes for the Rohingya children, while promoting social cohesion and solidarity within displaced communities.

## 7.3.4 Advocacy and Social Mobilisation

Community engagement efforts include advocacy and social mobilisation campaigns aimed at raising awareness, mobilising support, and advocating for the rights and needs of the Rohingya children and teachers. Advocacy activities may involve community rallies, media campaigns, storytelling sessions, and community dialogues, empowering stakeholders to voice their concerns, demand accountability, and drive positive change. By amplifying the voices and agency of community members, organisations strengthen advocacy efforts, mobilise resources, and galvanise collective actions to address systemic barriers and challenges affecting education in emergency settings.

### 7.3.5 Partnership and Collaboration

Collaboration with local organisations, NGOs, government agencies, and other stakeholders is essential for leveraging resources, expertise, and networks to support teachers and improve educational outcomes for the Rohingya learners. Partnerships may involve joint planning, resource-sharing, capacity-building initiatives, and coordination mechanisms, enabling stakeholders to work together towards common goals and objectives. By fostering collaboration and synergy among diverse actors, organisations maximise their impact and effectiveness in addressing the complex and interconnected challenges facing education in emergency contexts.

### 7.3.6 Promoting Community-Driven Solutions

Sustaining long-term PD initiatives requires a multifaceted approach that addresses systemic challenges and promotes community ownership and capacity building. Investing in local leadership development, mentorship programmes, and resource mobilisation efforts can help foster sustainable solutions that empower teachers to thrive in emergency settings and ensure continuity of education for displaced children.

Through community engagement and collaboration, organisations can harness the collective wisdom, resources, and resilience of communities to support teachers and enhance the quality of education for the Rohingya children in Bangladesh. By fostering partnerships, empowering stakeholders, and mobilising collective action, organisations contribute to the holistic development and empowerment of displaced populations, while promoting social justice, equity, and inclusion in education.

## 7.4 Promoting Sustainable Solutions

Sustaining professional development efforts requires a comprehensive approach that addresses both immediate training needs and long-term capacity-building goals. By investing in local leadership development, mentorship programmes, and community-driven initiatives, stakeholders can promote sustainable solutions that empower teachers to thrive in emergency settings and ensure continuity of education for displaced children.

### 7.4.1 Long-Term Capacity Building

Initiatives focused on promoting sustainable solutions prioritise long-term capacity building and PD for teachers in the Rohingya refugee camps. Rather than offering short-term interventions, these initiatives aim to build the skills, knowledge, and resilience of teachers over time. For example, organisations may implement multi-year training programmes that provide ongoing support and mentorship to teachers, allowing them to deepen their expertise and adapt to evolving educational needs over an extended period of time.

### 7.4.2 Training for Trainers (ToT) Programmes

Training of Trainers (ToT) programmes are designed to build a pool of master trainers who can then cascade down PD activities to a wider group of teachers. In the Rohingya refugee camps, organisations may implement ToT programmes to develop local teachers as facilitators and mentors for their peers. For instance, experienced teachers may undergo intensive training on instructional techniques, leadership skills, and humanitarian learning principles, and then serve as trainers for subsequent teacher training workshops and initiatives.

### 7.4.3 Local Ownership and Leadership

Promoting sustainable solutions in TPD involves fostering local ownership and leadership within the refugee community. Organisations may collaborate with camp education authorities, community leaders, and teacher associations to co-design and implement PD initiatives that reflect the needs and priorities of the Rohingya teachers. By empowering local stakeholders to take ownership of PD efforts, initiatives are more likely to be contextually relevant, culturally sensitive, and sustainable in the long term.

### 7.4.4 Integration with Formal Education Systems

Sustainable solutions in TPD can aim to integrate with formal education systems and structures within the Rohingya refugee camps. Rather than operating in isolation, PD initiatives align with national and international education standards, frameworks, and curricula. For example, organisations may work closely with government ministries of education or refugee education agencies to ensure that teacher training programmes are coherent with broader educational policies and goals.

### 7.4.5 Monitoring and Evaluation

Promoting sustainable solutions requires ongoing monitoring and evaluation to assess the effectiveness and impact of TPD initiatives. Organisations may establish monitoring and evaluation frameworks to track key indicators such as teacher competencies, student learning outcomes, and programme sustainability over time. For example, regular assessments, surveys, and feedback mechanisms may be implemented to gather data on the quality and relevance of PD activities and inform continuous improvement efforts.

## 7.5 Building Resilience through TPD

Teacher resilience is important for navigating the complex and challenging environments of emergency contexts, such as the Rohingya crisis in Bangladesh. By investing in Teacher Professional Development (TPD) initiatives that explicitly target the development of resilience skills and capacities, education stakeholders can empower teachers to effectively cope with adversity, adapt to changing circumstances, and sustain their commitment to providing quality education to displaced populations. Effective TPD programmes to increase the resilience for teachers in the Rohingya camps should focus on several key areas:

### 7.5.1 Content Knowledge and Pedagogy

Training should address the specific needs of the context of curriculum, including language support and culturally responsive teaching methods (Save the Children, 2018).

In the challenging context of the Rohingya refugee camps, TPD programmes must go beyond generic curriculum delivery. Training should address the specific needs of the Rohingya learner population.The educational landscape in the Rohingya refugee camps poses a significant hurdle: providing quality education to a diverse student population amidst resource constraints. However, Teacher Professional Development (TPD) programmes step up to the plate, empowering teachers with the tools to develop a comprehensive curriculum adaptation strategy.

Addressing language barriers is paramount. Bilingual resources and integrating the Rohingya language into daily activities pave the way for language fluency. Technology, including translation apps and mobile learning platforms, ensures access to education regardless of connectivity. Additionally, hands-on learning, such as incorporating the traditional Rohingya crafts into maths lessons, enhances comprehension and engagement.

Beyond textbooks, education embraces local wisdom. Lessons incorporate agricultural practices and traditional medicine, while the Rohingya folklore enriches historical understanding. Art forms like music and dance enliven subjects like maths, fostering cultural pride and providing avenues for student expression.

Emotional well-being is prioritised through trauma-informed practices and mindfulness techniques tailored to the Rohingya cultural norms. Teachers are trained to recognize and support students who may require additional mental health assistance.

Finally, community involvement is key. Parents, leaders, and cultural experts contribute to the curriculum, ensuring it reflects the community's values and experiences. This collaborative effort ensures a holistic and culturally sensitive educational experience for the Rohingya children, nurturing hope and knowledge alike.

### 7.5.2 Psychosocial Support

Teachers can benefit from training on self-care and psychosocial support techniques to manage their own well-being and support their students (Teach for All, 2019). Providing trauma-informed training and psychosocial support as part of TPD initiatives can help teachers develop the knowledge, skills, and strategies needed to recognize and respond to the diverse emotional and psychological needs of displaced learners and of the teachers themselves. By fostering a deeper understanding of psychosocial issues in adverse camp settings and its impact on learning and behaviour, teachers can create safe and supportive learning environments that promote healing, resilience, and academic success.

Incorporating stress management and self-care strategies into TPD programmes can be helpful for teachers to develop resilience by equipping them with practical tools and techniques for managing their own wellbeing in the face of challenging circumstances. By prioritising self-care practices such as mindfulness, relaxation techniques, and boundary setting, teachers can better cope with stress, prevent burnout, and maintain their physical, mental, and emotional well-being over the long term.

### 7.5.3 Community Collaboration

Building collaborative relationships with parents, community leaders, and other stakeholders is paramount for creating a supportive and inclusive learning environment. TPD initiatives should recognise the transformative potential of community collaboration and leverage it to enhance the quality and relevance of education for displaced learners. Effective TDP programmes go beyond simply informing parents and community leaders about school activities. They recognize the power of collaboration and its ability to strengthen the learning environment (UNESCO, 2018). Engaging Parents as Partners in Education: Incorporating parents as active partners in the education process fosters a sense of ownership and investment in their children's learning journey. Effective TPD programmes provide opportunities for parents to participate in decision-making processes, contribute their perspectives and insights, and collaborate with teachers to support their children's academic and socio-emotional development. By nurturing positive and collaborative relationships between teachers and parents, TPD initiatives can bridge the gap between home and school, promote parental involvement in learning activities, and reinforce the importance of education within the community.

#### 7.5.3.1 Empowering Community Leaders as Advocates for Education

Engaging community leaders as advocates for education amplifies the impact of TPD programmes and mobilises support for educational initiatives within the broader community. Effective TPD initiatives recognize the influence and authority of community leaders and actively involve them in shaping educational policies, advocating for resource allocation, and promoting the value of education as a catalyst for social change and community development. By empowering community leaders as champions for education, TPD programmes can harness local expertise, cultural insights, and social networks to address the unique needs and challenges faced by displaced learners and their families.

#### 7.5.3.2 Facilitating Collaborative Partnerships and Networks

Facilitating collaborative partnerships and networks among schools, NGOs, governmental agencies, and community-based organisations strengthens the collective capacity to address the multifaceted needs of displaced learners and promote holistic development. Effective TPD programmes provide platforms for collaboration, knowledge sharing, and joint problem-solving, enabling stakeholders to leverage their collective expertise, resources, and influence to overcome barriers and drive positive change. By fostering a culture of collaboration and cooperation, TPD initiatives can foster innovation, resilience, and sustainability within the education ecosystem, leading to improved outcomes for displaced learners and their communities.

#### 7.5.3.3 Promoting Cultural Responsiveness and Inclusivity

Promoting cultural responsiveness and inclusivity within TPD programmes ensures that educational initiatives are responsive to the diverse cultural, linguistic, and socio-economic backgrounds of displaced learners and their communities. Effective TPD initiatives prioritise culturally relevant pedagogies, practices, and materials that honour and respect the cultural heritage, identities, and experiences of learners, thereby fostering a sense of belonging, dignity, and empowerment. By embracing diversity and inclusivity as core principles, TPD programmes can create learning environments that celebrate differences, promote mutual respect, and cultivate empathy and understanding among students, teachers, and community members.

#### 7.5.3.4 Fostering Sustainable Community-Led Solutions

Fostering sustainable community-led solutions within TPD programmes empowers local stakeholders to take ownership of educational initiatives and drive meaningful change from within. Effective TPD initiatives prioritise capacity-building, leadership development, and participatory decision-making processes that empower community members to identify their own needs, assets, and priorities, and develop contextually relevant solutions that are sustainable and scalable. By investing in the agency and autonomy of local communities, TPD programmes can catalyse grassroots innovation, resilience, and self-reliance, leading to lasting improvements in educational access, quality, and equity for displaced learners and their communities.

By fostering strong partnerships with the broader community, TDP programmes can tap into a wealth of untapped potential. This collaboration creates a more supportive and enriching learning environment for the Rohingya children .

## 7.6 Recommendations for Policy and Practice

Based on the exploration of teacher resilience, professional development, and the complexities of EiE contexts, a number of initiatives need to be undertaken to implement the TPD programme, firstly integrating teacher resilience as a core component of EiE frameworks. Allocate resources for programmes that promote teacher well-being, such as self-care, stress management workshops and social support networks. Additionally, invest in context-specific PD activities that equip teachers with the skills needed to address the unique challenges of their emergency setting. This might involve training on topics like trauma-informed pedagogy, inclusive education practices, and culturally responsive teaching. Furthermore, learning facilities (nonformal or informal learning spaces or centres) need to provide a congenial learning environment. Community leadership can play a crucial role in fostering teacher resilience. It is also important to encourage collaborative learning environments, positive feedback, and provide opportunities for professional growth within schools operating in emergency settings. Additionally, EiE programme designers should actively engage teachers in the development and implementation of EiE interventions. This participatory approach can empower teachers and ensure programmes are responsive to their needs and realities.

## 7.7 Recommendations for Future Research

The field of EiE can benefit from further research on longitudinal studies that track the impact of TPD programmes on teacher resilience and student learning outcomes in emergency contexts. Moreover, research on the effectiveness of existing strategies could be done to promote teacher resilience, considering factors like cultural background, prior teaching experience, and specific emergency context. Another grey area is the integration of technology in EiE, to get a comprehensive understanding of provisions and the uses of technology, investigations into the role of technology need to be conducted in supporting TPD and fostering collaboration within EiE programmes, particularly in remote or resource-limited settings.

By implementing these recommendations and prioritising further research, it could create a more supportive and empowering environment for teachers working in emergency contexts. This will ultimately contribute to delivering high-quality education for displaced children and youth, even in the most challenging circumstances.

# 8. IMPLICATIONS

Education in emergencies (EiE) plays a critical role in supporting children and youth affected by crises. It goes beyond simply providing basic literacy and numeracy skills. Research by Aguilar & Retamal (2009) highlights how EiE offers a sense of normalcy, fosters resilience, and equips children with the tools they need to rebuild their lives. This is particularly important as children in emergencies are often exposed to trauma and disruption. Denying children access to quality education has significant negative consequences. Without a strong educational foundation, they may struggle to acquire the basic skills or knowledge needed to contribute positively to society as adults. Furthermore, they are also at greater risk of exploitation, child labour, or recruitment into violent groups (UNHCR, 2016).

EiE has emerged in response to these challenges, which are shaped by humanitarian aid and developmental organisations, international politics, and comparative and international education frameworks (Lerch & Buckner, 2018; Pherali, 2019). Halman et al. (2018) summarised that the humanitarian paradigm presents a coherent framework as to why education should be given a prominent role in humanitarian response due to its lifesaving and life sustaining potentials.

Sustainable Development Goal 4: "Ensure inclusive and quality education for all and promote lifelong learning" (UNHCR, 2016, p. 6). The global mandate of education for all facilitated the shift of education from primarily a development goal to being also included in humanitarian work (Lerch, 2017). Education in Emergencies is a priority as a response to children's right to learn and grow. It is also a priority because it is considered an enabling right, contributing to accessing other basic rights, such as healthcare and social services (European Commission, 2019) and to supporting communities' psychosocial wellbeing (Burde et al., 2017). By providing safe spaces, fostering social interaction, and promoting a sense of hope, EiE programmes can help communities heal and rebuild. Moreover, EiE includes the span from humanitarian aid to development work and also draws on educational theory and practice, it is a significant emerging global organisational and professional field (Lerch, 2017).

Not only is access to education necessary for children living in crisis contexts (Halman et al., 2018), but the quality of the education they receive is also significant (Piper et al., 2020). Education can be lifesaving and protective, especially if it is of high quality (Monaghan, 2019). Furthermore, researchers have linked the quality of teaching to improved student outcomes. Darling-Hammond, Campbell, et al. (2017) have analysed systems worldwide and have found that teacher quality is vital in the most successful systems. Therefore, investing in TPD is essential for ensuring quality EiE programmes.

TPD programmes equip teachers with the knowledge, skills, and strategies needed to navigate the challenges of emergency contexts. Darling-Hammond et al. (2017) highlighted the importance of teacher quality in successful education systems and also focused on areas such as trauma-informed pedagogy, inclusive education practices, culturally responsive teaching, and self-care strategies. By fostering teacher resilience, TPD programmes can empower teachers to better support their students and contribute to the overall success of EiE interventions.

## 9. CONCLUSION

This study explored the resilience of teachers working in the Rohingya refugee camps in Bangladesh. The resilience exhibited by teachers in the face of adversity is very important and plays a pivotal role in professional development by facilitating its effectiveness in emergency contexts. Understanding the specific challenges faced by teachers in the Rohingya camps, such as limited resources, cultural differences, and trauma, highlights the need for comprehensive TPD interventions. These interventions should equip teachers with the knowledge, skills, and strategies needed to support the Rohingya learners effectively. The background and context of the Rohingya crisis in Bangladesh highlight the urgent need for comprehensive interventions to address the educational challenges faced by the Rohingya refugees and support the resilience of teachers working in emergency settings. The Rohingya refugee crisis in Bangladesh serves as a powerful case study, setting the stage for a comprehensive examination of the challenges, strategies, and support systems involved in ensuring access to quality education for the Rohingya refugees in Bangladesh. By understanding the unique socio-economic, cultural, and political dynamics at play, policymakers, practitioners, and researchers can develop targeted strategies and initiatives to improve access to quality education for the Rohingya learners. This includes not only supporting teachers' resilience but also strengthening teacher development programmes to ensure quality education for all Rohingya children in the refugee camps in Bangladesh.

## REFERENCES

Admiraal, W., Schenke, W., De Jong, L., Emmelot, Y., & Sligte, H. (2021). Schools as professional learning communities: What can schools do to support professional development of their teachers? *Professional Development in Education*, *47*(4), 684–698. doi:10.1080/19415257.2019.1665573

Aguilar, P., & Retamal, G. (2009). Protective environments and quality education in humanitarian contexts. *International Journal of Educational Development*, *29*(1), 3–16. doi:10.1016/j.ijedudev.2008.02.002

Ahearn, M. C. (2005). Professional development for teachers in crisis and post-crisis contexts: A review of the literature. *Journal of Refugee Education*, *12*(2), 189–210. http://file.snnu.net/res/20126/18/018526a6-3cbf-4c9d-ac0f-a0740094aa75.pdf

Ahmad, M. M., & Nusrat, R. (2022). Human Trafficking and Crimes in the Rohingya Refugee Camps in Bangladesh. In E. D. Essien (Ed.), *Handbook of Research on Present and Future Paradigms in Human Trafficking* (pp. 26–42). doi:10.4018/978-1-7998-9282-3.ch003

Ahmed, K. (2021). Educating Rohingya Children and Youth in Emergencies: A Reflection From the Field. In R. Gordon, K. Ahmed, & M. Hosoda (Eds.), *Evolving Multicultural Education for Global Classrooms* (pp. 1–20). IGI Global., doi:10.4018/978-1-7998-7649-6.ch001

Akpovo, S. M., Nganga, L., & Acharya, D. (2018). Minority-World Preservice Teachers' Understanding of Contextually Appropriate Practice While Working in Majority-World Early Childhood Contexts. *Journal of Research in Childhood Education*, *32*(2), 202–218. https://doi.org/10.1080/02568543.2017.1419321

Akter, T. 2023. Implementation of Myanmar Curriculum (MC) in the Rohingya Refugee Camps of Bangladesh: A Case Study on Implementation Scope and Challenges. [unpublished master's dissertation]. Nottingham, University of Nottingham.

Al-Naabi, I., Kelder, J., & Carr, A. (2021). Preparing teachers for emergency remote teaching: A professional development framework for teachers in higher education. *Journal of University Teaching & Learning Practice*, *18*(5), 52–72. Advance online publication. doi:10.53761/1.18.5.4

Alam, M. S. (2022). Rohingya Refugees in Bangladesh: Exploitations and Challenges for Host Community. *Global Journal of Human Social Science*, *22*(H2), 67–72. doi:10.34257/GJHSSHVOL22IS2PG67

Aydin, H., Gundogdu, M., & Akgul, A. (2019). Integration of Syrian refugees in Turkey: Understanding the educators' perception. *Journal of International Migration and Integration*, *20*(4), 1029–1040. doi:10.1007/s12134-018-0638-1

Baquedano-López, P., Alexander, R. A., & Hernandez, S. J. (2013). Equity Issues in Parental and Community Involvement in Schools: What Teacher Educators Need to Know. *Review of Research in Education*, *37*(1), 149–182. https://doi.org/10.3102/0091732X12459718

Barbora P. (2021). The Teacher's Wellbeing in Teacher-Student Interactions. Lifelong Education: the XXI century, 33, 103-115. Doi.org/ doi:10.15393/j5.art.2021.6691

Beltman, S. (2020). Understanding and examining teacher resilience from multiple perspectives. Cultivating Teacher Resilience, 11-26. https://doi.org/ doi:10.1007/978-981-15-5963-1_2

Boyd, K. J., Gold, A. U., & Littrell, M. K. (2021). Teaching practices around natural hazards and community resilience in Colorado. *Journal of Geoscience Education*, *70*(4), 460–473. https://doi.org/10.1080/10899995.2021.1964319

Bozkurt, A., Jung, I., Xiao, J., Vladimirschi, V., Schuwer, R., Egorov, G., ... Paskevicius, M. (2020). A global outlook to the interruption of education due to COVID-19 pandemic: Navigating in a time of uncertainty and crisis. *Asian Journal of Distance Education*, *15*(1), 1–126. doi:10.5281/zenodo.3778083

Brunzell, T., Stokes, H., & Waters, L. (2019). Shifting teacher practice in trauma-affected classrooms: Practice pedagogy strategies within a trauma-informed positive education model. *School Mental Health*, *11*(3), 600–614. doi:10.1007/s12310-018-09308-8

Cannon, S. R., Davis, C. R., & Fuller, S. C. (2020). Preparing for the Next Natural Disaster: Understanding How Hurricanes Affect Educators and Schooling. AASA Journal of Scholarship & Practice, 17(2), 6-16. https://eric.ed.gov/?id=EJ1259714

Caritas Switzerland. (2019). *Teachers in Crisis Contexts: Cox's Bazar*. Retrieved from https://inee.org/resources/promising-practices-teacher-well-being-management-and-school-leadership https://inee.org/resources/promising-practices-teacher-well-being-management-and-school-leadership.

Castellanos, F. (2018). Teaching refugee children: increasing teacher self-efficacy through trauma-informed trainings *(Doctoral dissertation)*. http://hdl.handle.net/2152/65961

Cervantes-Duarte, L., & Fernández-Cano, A. (2016). Impact of armed conflicts on education and educational agents: A multivocal review. Revista Electrónica Educare, 20(3), 1. https://doi.org/10.15359/ree.20-3.12

Cho, S., Crenshaw, K. W., & McCall, L. (2013). Toward a Field of Intersectionality Studies: Theory, Applications, and Praxis. *Signs (Chicago, Ill.)*, *38*(4), https://doi.org/0097-9740/2013/3804-0001$10.00

Chowdhury, M. J. I. (2023). Teachers' preparedness to implement Myanmar curriculum for the Rohingya children in the Rohingya camps of Cox's Bazar (Doctoral dissertation, Brac University). http://hdl.handle.net/10361/20106

Coombes, A., & Ponta, O. (2022). Building community engagement and teacher support in education: Qualitative findings from process evaluations in two exceptional settings. *Journal of Development Effectiveness*, *15*(3), 302–319. doi:10.1080/19439342.2022.2067889

Cox's Bazar Education Sector. (2018). Joint Education Needs Assessment: Rohingya Refugee in Cox's Bazar. https://reliefweb.int/report/bangladesh/joint-education-needs-assessment-rohingya-refugee-cox-s-bazar-june-2018

Crompton, H., Burke, D., Jordan, K., & Wilson, S. W. (2021). Support provided for K-12 teachers teaching remotely with technology during emergencies: A systematic review. *Journal of Research on Technology in Education*, *54*(3), 473–489. doi:10.1080/15391523.2021.1899877

Crompton, H., Burke, D., Jordan, K., & Wilson, S. W. (2021). Support provided for K-12 teachers teaching remotely with technology during emergencies: A systematic review. *Journal of Research on Technology in Education*, *54*, 473–489.

Darling-Hammond & Gardner. (2017). *Effective Teacher Professional Development*. Retrieved March 15, 2024, from https://learningpolicyinstitute.org/sites/default/files/product-files/Effective_Teacher_Professional_Development_REPORT.pdf

Dushkevych, M., Barabashchuk, H., & Hutsuliak, N. (2020). Peculiarities of student distance learning in emergency situation condition. Revista Romaneasca pentru Educatie Multidimensionala, 12, 71-77. https://doi.org/10.18662/rrem/12.1sup2/248

El-Serafy, Y., Adam, T., & Hassler, B. (2022). The effectiveness of technology-supported teacher professional learning communities in emergency settings. In S. Gravett & N. Petersen (Eds.), *Future-Proofing Teacher Education: Voices from South Africa and Beyond*. Routledge. doi:10.4324/9781003185499-13

Faisal, M. M., & Ahmed, Z. (2022). Stateless Rohingya Refugees in Bangladesh: Life and Livelihood Challenges. *Journal of Social and Development Sciences*, *13*(4), 11–23. doi:10.22610/jsds.v13i4(S).3326

Falk, D., Varni, E., Johna, J. F., & Frisoli, P. (2019). Landscape review: Teacher well-being in low resource, crisis, and conflict-affected settings. Education Equity Research Initiative. https://www.edu-links.org/sites/default/files/media/file/TWB%20 Landscape%20Review_June%202019.pdf

Farzana, T. (2023). Humanism and Skill Development among the Rohingya Refugees through Content-based Instruction. *International Journal For Multidisciplinary Research*, *5*(1), 1–18. doi:10.36948/ijfmr.2023.v05i01.1388

Folven, I. (2022). Education in Crises. Education for What?-The Understood Purpose of Education for Rohingya Refugees in Cox's Bazar, Bangladesh, and Implications on Their Education in Times of Crisis.

Frieder, M. L. (2020). The Rohingya Artolution: Teaching Locally Led Community-Based Public Art Educators in the Largest Refugee Camp in History. Doctoral dissertation, Teachers College, Columbia University. doi:10.7916/d8-n1dz-c651

Gu, Q., & Day, C. (2007). Teachers resilience: A necessary condition for effectiveness. *Teaching and Teacher Education*, *23*(8), 1302–1316. https://doi.org/10.1016/j.tate.2006.06.006

Habib, I. N. (2020). *The Effects of International Child Development Program (ICDP) to Improve Teachers Attitude Towards the Rohingya Refugee*. Retrieved from https://www.duo.uio.no/handle/10852/80864

Habib, M. R., Chowdhury, A. R., & Uldanov, A. (2023). Creating pathways to opportunity: Non-formal educational 'Inclusion' for Rohingya refugee children in Bangladesh. *Journal of South Asian Development*. Advance online publication. doi:10.1177/09731741231202872

Haque, S., Dow, A., & Hasan, M. (2023, September). ERICC Research Agenda for Cox's Bazar, Bangladesh. https://inee.org/sites/default/files/resources

Hascher, T., Beltman, S., & Mansfield, C. (2021). Teacher wellbeing and resilience: Towards an integrative model. *Educational Research*, *63*(4), 416–439. doi:10.1080/00131881.2021.1980416

Heltne, U. M., Dybdahl, R., Elkhalifa, S., & Breidlid, A. (2019). Psychosocial Support and Emergency Education: An Explorative Study of Perceptions among Adult Stakeholders in Sudan and South Sudan. *Sustainability*, *12*(4), 1410.https://doi.org/10.3390/su12041410

Howard, P. L., Auwarter, A., & Gallimore, R. (2016). Teacher professional development in complex emergencies: Building capacity for social-emotional learning. *Peabody Journal of Education*, *91*(2), 226–243. https://journals.sagepub.com/doi/10.1177/00224871231153084

Howard, S., & Johnson, B. (2004). Resilient teachers: Resisting stress and burnout. *Social Psychology of Education*, *7*(4), 399–420. doi:10.1007/s11218-004-0975-0

Hu, D., Yuan, B., Luo, J., & Wang, M. (2021). A review of empirical research on ICT applications in teacher professional development and teaching practice. Knowledge Management & E-Learning. *International Journal (Toronto, Ont.)*, *13*(1), 1–20. https://doi.org/10.34105/j.kmel.2021.13.001

Instefjord, E. J., & Munthe, E. (2017). Educating digitally competent teachers: A study of integration of professional digital competence in teacher education. *Teaching and Teacher Education*, *67*, 37–45.https://doi.org/10.1016/j.tate.2017.05.016

Inter-agency Network for Education in Emergencies. (2022) A Guidance Note for Teacher Wellbeing in Emergency Settings. https://inee.org/resources/guidance-note-teacher-wellbeing-emergency-settings

Inter-agency Network for Education in Emergencies (INEE). (2021). *Teacher Wellbeing Resources Mapping & Gap Analysis | INEE*. (2021, August 19). https://inee.org/resources/teacher-wellbeing-resources-mapping-gap-analysis

IslamA.MozumderT. A.RahmanT.ShatilT.SiddiqueA. (2022, March 15). Forced Displacement, Mental Health, and Child Development: Evidence from the Rohingya Refugees. https://doi.org/ doi:10.31235/osf.io/b4fc7

Islam, I., & Naing, Z. (2023). Five years lost: Youth inclusion in the Rohingya response. *Journal on Migration and Human Security*, *11*(1), 99–108. doi:10.1177/23315024231160765

Islam, M. S. (2019). Communal peace in Bangladesh after Rohingya crisis: An assessment of different approaches to community development. *Journal of Prevention & Intervention in the Community*, *48*(3), 256–271. doi:10.1080/10852352.2019.1625604 PMID:31244395

Islam, M. S., & Kaiy, M. (2022). Provision of education to Rohingya refugee children in Bangladesh: Exploring the forms of discrimination and intersectionality. *Journal of Refugee Studies*, 1–22. doi:10.1093/jrs/feac022

Islam, M. S., & Şahin, A. (2023). Impact of Forcibly Displaced People's Identity Crisis on Host State's Security: The Case of Rohingya Refugees in Bangladesh. Saudi Journal of Humanities and Social Sciences. doi:10.36348/sjhss.2023.v08i04.002

Islam, S., & Mozumder, M. K. (2021). Challenges to providing quality mental health services for forcibly displaced populations: The case of Rohingya refugees in Bangladesh. *Asian American Journal of Psychology*, *12*(3), 234–240. doi:10.1037/aap0000195

Joint education needs assessment: Rohingya refugee in Cox's Bazar. (2018, June 1). ReliefWeb. https://reliefweb.int/report/bangladesh/joint-education-needs-assessment-rohingya-refugee-cox-s-bazar-june-2018

Karim, A., & Hussain, F. (2019). 'When Will You Start Teaching the REAL Curriculum?': Challenges and Innovations in Education for Rohingya Refugees. In P. Nielsen & H. C. Kimaro (Eds.), Information and Communication Technologies for Development. Strengthening Southern-Driven Cooperation as a Catalyst for ICT4D. ICT4D 2019. IFIP Advances in Information and Communication Technology (Vol. 551). Springer. https://doi.org/10.1007/978-3-030-18400-1_55

Karim, M. U. H. F. (2023). Refugee Education in Bangladesh: A Nation-State Paradox. *Journal of Population and Social Studies*, *31*, 762–782. http://doi.org/10.25133/JPSSv312023.042

Kirk, J., & Winthrop, R. (2013). Teaching in contexts of emergency and state fragility. In J. Kirk, M. Dembélé, & S. Baxter (Eds.), *More and better teachers for quality for all: Identity and motivation, systems and support* (pp. 221–139).

Kong, Y. (2021). The Role of Experiential Learning on Students' Motivation and Classroom Engagement. *Frontiers in Psychology*, *771272*(12), doi:10.3389/fpsyg.2021.771272

Kurniawati, N., & Fauziyah, F. (2022). Fostering students' engagement in synchronous learning using interactive web-based media. Indonesian EFL Journal, 8(1), 13-22. https://garuda.kemdikbud.go.id/documents/detail/2570935

Kurtzer, J., Vaidya, S., & Islam, I. (2022, September 21). *Four Years On: An Update on the Rohingya Crisis*. https://www.csis.org/analysis/four-years-update-rohingya-crisis

Kurtzer, J., Vaidya, S., & Islam, I. (2022, September 21). *Four Years On: An Update on the Rohingya Crisis*. https://www.csis.org/analysis/four-years-update-rohingya-crisis

Larasati, R., & Kuswandono, P. (2023, April 2). Enhancing Teachers' Resilience through Teacher Professional Development. *Language Circle: Journal of Language and Literature*, *17*(2), 391–402. doi:10.15294/lc.v17i2.42673

Laxton, D., Cooper, L., Shrestha, P., & Younie, S. (2020). Translational research to support early childhood education in crisis settings: A case study of collaborative working with Rohingya refugees in cox's Bazar. Education 3-13, 49(8), 901-919. doi:10.1080/03004279.2020.1813186

Mahmud, E., & Nalifa, M. (2020). Barriers to Institutional Education for Refugees: A Case Study of Rohingya Community in Bangladesh. Business, economics and sustainable development.

Mansfield, C. F., Beltman, S., Broadley, T., & Weatherby-Fell, N. (2016, February). Building resilience in teacher education: An evidenced informed framework. *Teaching and Teacher Education*, *54*, 77–87. doi:10.1016/j.tate.2015.11.016

McBrien, J. L. (2005). Educational needs and barriers for refugee students in the United States: A review of the literature. *Review of Educational Research*, *75*(3), 329–364. doi:10.3102/00346543075003329

Mebert, L., Barnes, R., Dalley, J., Gawarecki, L., Ghazi-Nezami, F., Shafer, G., Slater, J., & Yezbick, E. (2020). Fostering student engagement through a real-world, collaborative project across disciplines and institutions. *Higher Education Pedagogies*, *5*(1), 30–51. doi:10.1080/23752696.2020.1750306

Metni, E. (2022). *Exploring Lebanese Teachers' Engagement in a Low-Cost, Technology-Enhanced, Problem-Solving, Orientated Learning Intervention with Refugee Children* (Doctoral dissertation, UCL (University College London)). Retrieved from https://discovery.ucl.ac.uk/id/eprint/10151196/

Mitu, K., Jones, N., Vintges, J., & Devonald, M. (2022). Climate risks and truncated opportunities: How do environmental challenges intersect with economic and social disadvantages for Rohingya adolescents in Bangladesh? *Sustainability (Basel)*, *14*(8), 4466. doi:10.3390/su14084466

Mohiuddin, M., & Molderez, I. (2023). *Rohingya influx and host community: A reflection on culture for leading socioeconomic and environmental changes in Bangladesh.* European Journal of Cultural Management and Policy. doi:10.3389/ejcmp.2023.11559

Molnar, B. E., Bradshaw, C. P., & Buka, S. L. (2015). Building resilience in Teachers of Traumatised Students in the Classroom. *School Psychology Quarterly*, *30*(4), 381–394. https://www.researchgate.net/publication/304010563_Classroom_Ready_Building_Resilience_in_Teacher_Education

Nicolai, S. (2003). *Education in Emergencies: A tool kit for starting and managing education in emergencies, Save the Children.* Retrieved from https://inee.org/sites/default/files/Staffing_Building_Team_Supporting_Staff.PDF.

Nilsen, M. J., Olney, J., Maung, K., Karim, L., Ahmad, S., Haque, N., & Mubarak, H. R. (2023). Community-Led Education among Rohingya Refugees and the Politics of Refugee Education in Bangladesh. *Journal of Refugee Studies*, *36*(4), 712–735. doi:10.1093/jrs/fead037

O'Connor, K., & Seager, J. (2021). Displacement, Violence, and Mental Health: Evidence from Rohingya Adolescents in Cox's Bazar, Bangladesh. *International Journal of Environmental Research and Public Health*, *18*(10), 5318. doi:10.3390/ijerph18105318 PMID:34067724

Palik, J., & Østby, G. (2023). Interventions to improve refugee children's access to education and quality learning: A scoping review of existing impact evaluations. *International Review of Education*, *69*(1-2), 227–247. doi:10.1007/s11159-023-10004-2 PMID:37313292

Pherali, T., Abu Moghli, M., & Chase, E. (2020). Educators for Change: Supporting the Transformative Role of Teachers in Contexts of Mass Displacement. *Journal on Education in Emergencies*, *5*(2), 147. doi:10.33682/gtx6-u5q8

Powell, C. G., & Bodur, Y. (2019). Teachers' perceptions of an online professional development experience: Implications for a design and implementation framework. *Teaching and Teacher Education*, *77*, 19–30.https://doi.org/10.1016/j.tate.2018.09.004

Prodip, M. A. (2023). Education for what? The politics of education for Rohingya refugees in Bangladesh. *Politics & Policy*, *51*(6), 1028–1045. Advance online publication. doi:10.1111/polp.12541

Rahman, M. M., Shindaini, A. J., & Abdullah, A. B. (2022). undefined. *Asia Pacific Education Review*, *24*(3), 433–445. doi:10.1007/s12564-022-09770-9

Rahman, M. M., Shindaini, A. J., & Husain, T. (2022). Structural barriers to providing basic education to Rohingya children in the Kutupalong refugee camp, Cox's Bazar, Bangladesh. *International Journal of Educational Research Open*, *3*, 100159. doi:10.1016/j.ijedro.2022.100159

Rahman, M. S., & Siddik, M. A. (2022). Need for Teachers' Professional Development in a Low-Resource Context During and After COVID-19: A Bangladesh Perspective. *International Journal of Teacher Education and Professional Development*, *5*(1), 1–14. doi:10.4018/IJTEPD.295547

Reidy, K. (2020, February 10). Expanding education for Rohingya refugee children in Bangladesh. UNICEF. https://www.unicef.org/bangladesh/en/stories/expanding-education-rohingya-refugee-children-bangladesh

Richardson, J., Karam, C. M., & Afiouni, F. (2019). The global refugee crisis and the career ecosystem. *Career Development International*, *25*(1), 1–13. doi:10.1108/CDI-04-2019-0104

Rohingya Response, Bangladesh (2023). Joint Multi-Sectoral Needs Assessment (J-MSNA): Camp-level Findings. https://rohingyaresponse.org/wp-content/uploads/2024/04/BGD_CXB_J-MSNA-2023_Factsheets_Camps_Feb-2024.pdf

RRRC. (2023). Report: At a glance. Office of The Refugee Relief and Repatriation Commissioner, Cox's Bazar. Retrieved from: https://rrrc.portal.gov.bd

Avci, Z. Y., O'Dwyer, L. M., & Lawson, J. (2019). Designing effective professional development for technology integration in schools. *Journal of Computer Assisted Learning*, *36*(2), 160–177.https://doi.org/10.1111/jcal.12394

Sabieh, C. (2023). Teacher Education and Professional Development: Prepping Educators for Emergency Education Through Project-Based Learning and Technology. In D. Tafazoli & M. Picard (Eds.), *Handbook of CALL Teacher Education and Professional Development*. Springer. doi:10.1007/978-981-99-0514-0_24

Sadia, S. (2021). Experiences and challenges of Non-Governmental Organizations (NGO)s professionals in ensuring the right to education for the Rohingya refugee children in Cox's Bazar. *Instituto Universitário de Lisboa*. http://hdl.handle.net/10071/24014

Shohel, M. M., Ashrafuzzaman, M., Azim, F., Akter, T., & Tanny, S. F. (2022). Displacement and Trauma: Exploring the Lost Childhood of Rohingya Children in the Refugee Camps in Bangladesh. In R. Throne (Ed.), *Social Justice Research Methods for Doctoral Research* (pp. 244–272). IGI Global. doi:10.4018/978-1-7998-8479-8.ch012

Shohel, M. M. C. (2022). Education in emergencies: Challenges of providing education for Rohingya children living in refugee camps in Bangladesh. *Education Inquiry*, *13*(1), 104–126. doi:10.1080/20004508.2020.1823121

Shohel, M. M. C. (2023). Lives of the Rohingya children in limbo: Childhood, education, and children's rights in refugee camps in Bangladesh. *Prospects*, *53*(1–2), 131–149. doi:10.1007/s11125-022-09631-8 PMID:37197308

Shohel, M. M. C., Ashrafuzzaman, M., Babu, R., Akter, T., Tasnim, N., & Bayezid, A. (2023a). Access to Higher Education for Rohingya Refugees: Challenges, Opportunities and Future Directions. In S. Saeed & M. Zhang (Eds.), *Global Perspectives on the Difficulties and Opportunities Forced by Migrant and Refugee Students in Higher Education*. IGI Global. doi:10.4018/978-1-6684-7781-6.ch005

Shohel, M. M. C., Ashrafuzzaman, M., & Shabnam, S. (2024a). The Rohingya Crisis and the Global impact Refugee Influx. In P. Pietrzak (Ed.), *Analyzing Global Responses to Contemporary Regional Conflicts*. IGI Global.

Shohel, M. M. C., Babu, R., Ashrafuzzaman, M., Azim, F., & Bayezid, A. (2023b). Learning Competency Framework and Approach for the Displaced Rohingya Children Living in Bangladesh: A Critical Review. *Continuity in Education*, *4*(1), 50–66. doi:10.5334/cie.57

Shohel, M. M. C., Rashedujjaman, S. M., & Akter, T. (2024b). Relocation of the Rohingya Refugees to Bhasan Char: Human Rights, Government Policy and Refugee Crisis in Bangladesh. In P. Pietrzak (Ed.), *Dealing with Regional Conflicts of Global Importance*. IGI Global. doi:10.4018/978-1-6684-9467-7.ch014

Shohel, M. M. C., Roy, G., Ahsan, M. S., & Acharya, D. (2021). Access to Higher Education for Refugees in South Asia. In M. A. Prodip, G. Roy, & D. Nandy (Eds.), *Refugee Education in South Asia: Policies and Perspectives* (pp. 187–222). Nova Science Publishing House.

Sinclair, M. (2002). *Planning education in and after emergencies. UNESCO International Institute for Educational Planning*. UNESCO.

Southwick, S. M., Bonanno, G. A., Masten, A. S., Panter-Brick, C., & Yehuda, R. (2014). Resilience definitions, theory, and challenges: Interdisciplinary perspectives. *European Journal of Psychotraumatology*, *5*(1). https://doi.org/10.3402/ejpt.v5.25338

Stevens, K. E., Siraj, I., & Kong, K. (2023). A critical review of the research evidence on early childhood education and care in refugee contexts in low- and middle-income countries. *International Journal of Child Care and Education Policy*, *17*(1), 7. Advance online publication. doi:10.1186/s40723-023-00109-4

Sultana, N., Sultana, S., Saha, R., & Alam, M. M. (2023). The challenges and coping of Rohingya refugees: a comparative study of registered and non-registered Rohingya refugees in Bangladesh. Southeast Asia: A Multidisciplinary Journal, 23(2), 95-120. doi:10.1108/SEAMJ-04-2023-0033

Suma, J. T. (2022). Rohingya Refugees and Classroom Children: Cultivating a Lost Generation. In I. A. Hussain (Ed.), *Rohingya Camp Narratives*. Global Political Transitions. doi:10.1007/978-981-19-1197-2_11

Teach For All. (2019). Supporting Teachers in Emergencies: A Resource Kit. Retrieved from https://teachforall.org/news/education-emergencies-connecting-across-borders

Thomas, N., Smith, M. B., Agerhem, S., Cadesky, J., Fadel, B., & Paz, B. V. (2018). Theme Paper 1: Everyday humanitarianism. Academic Press.

Ullah, A. K. M. A. (2016). Rohingya Crisis in Myanmar: Seeking Justice for the "Stateless". *Journal of Contemporary Criminal Justice*, *32*(3), 285–301. doi:10.1177/1043986216660811

UNESCO. (2018). Education in emergencies: Policy guidelines for inclusion and equity. Retrieved from https://unesdoc.unesco.org/ark:/48223/pf0000128214

UNESCO. (2018). Improving teacher support and participation in Local Education Groups. Retrieved from.https://www.globalpartnership.org/sites/default/files/2018-07-gpe-improving-teacher-support-in-legs.pdf

UNHCR. (2021). *Refugee Education Report- Staying The Course-The Challenges Facing Refugee Education*. https://reliefweb.int/sites/reliefweb.int/files/resources/612f85d64.pdf

UNHCR. (2023). *Desperate Journeys: Rohingya Refugees in Search of Protection.* UNHCR Regional Bureau for Asia and Pacific (RBAP).

UNHCR. (2024). *Joint Government of Bangladesh - UNHCR Population Factsheet - Block Level as of March 2024*. https://data.unhcr.org/en/documents/details/107776.

UNICEF. (2018). Evaluation Office. Bangladesh - Rohingya Refugee Response: Education in Emergencies Sector Evaluation. Retrieved from https://reliefweb.int/report/bangladesh/evaluation-unicefs-response-rohingya-refugee-crisis-bangladesh

Venet, A. S. (2023). *Equity-centered trauma-informed education* (1st ed.). Routledge. doi:10.4324/9781032677989

Wallerstein, N., & Duran, B. (2010). Community-based participatory research contributions to intervention research: The intersection of science and practice to improve health equity. *American Journal of Public Health, 100*(S1), S40–S46. doi:10.2105/AJPH.2009.184036 PMID:20147663

Yosso, T. J. (2002). Toward a Critical Race Curriculum. *Equity & Excellence in Education, 35*(2), 93–107. https://doi.org/10.1080/713845283

Zaman, S., Sammonds, P., Ahmed, B., & Rahman, T. (2020). Disaster risk reduction in conflict contexts: Lessons learned from the lived experiences of Rohingya refugees in Cox's Bazar, Bangladesh. *International Journal of Disaster Risk Reduction, 50*, 101694. doi:10.1016/j.ijdrr.2020.101694 PMID:32518743

Zee, M., & Koomen, H. M. (2016). Teacher self-efficacy and its effects on classroom processes, student academic adjustment, and teacher well-being: A synthesis of 40 years of research. *Review of Educational Research, 86*(4), 981–1015. doi:10.3102/0034654315626801

# Chapter 11
# The Extraordinary Journey of AUAF Students From Despair to Hope

**Setha Fayaz**
*American University of Afghanistan, Afghanistan*

**Ahmad Suhaib Majeed**
*American University of Afghanistan, Afghanistan*

## ABSTRACT

*This chapter narrates the remarkable journey of students from the American University of Afghanistan (AUAF) following the Taliban's takeover in August 2021. The evacuation, initiated amidst the chaos at Hamid Karzai International Airport, and finally getting completed two years later, encountered numerous challenges. The narrative details the phased evacuation process, where AUAF students were dispersed to different countries in Asia, such as Kyrgyzstan, Iraq, and Qatar. The chapter delves into the mental health concerns that emerged, exploring the anxiety and uncertainty faced by students as they grappled with the abrupt regime change, the efficacy of online learning, and the challenges of continuing education amid geopolitical upheaval. The narrative also highlights the resilience of female students who faced numerous obstacles, including delays, flight cancellations, and the imposition of restrictive travel norms. The chapter captures the inspiring resilience of AUAF students, emphasizing their unwavering commitment to education amidst adversity.*

DOI: 10.4018/979-8-3693-1483-8.ch011

## INTRODUCTION

Education comes at a cost. This is particularly true in countries going through war or emergency situations. The takeover of Afghanistan by the Taliban brought a massive shift in the lives of millions of people. It affected not only those who were affiliated with the U.S. and worked for them but also half of the population who were deprived of education in the country. As of now, all of the schools are closed for girls after the elementary level (Barakat, 2024).

This was predictable in August 2021 when the Taliban came into power. Everyone who had worked for the U.S. felt the need to evacuate for the prospect of a better future and to get away from any persecution. Millions of other Afghans wanted to leave because the return of the Taliban meant that they would not be able to continue their education. Students from the American University of Afghanistan (AUAF) were particularly exposed to this risk. As an institution with a co-ed educational system and following the American curriculum, the American University of Afghanistan faced many challenges (AUAF, 2024). Having previously faced an attack that killed about 13 students and staff and left many others injured in 2016, the return of the Taliban meant that they would face the same or much worse situation under their regime (McKirdy & Popalzai, 2017).

Through the university's efforts, hundreds of AUAF students have been evacuated. They have been evacuated to three different countries in Asia. Approximately 200 students were evacuated to Kyrgyzstan, and they continued their studies at AUCA (Talant, 2021). Another group of approximately 110 students was evacuated to Iraq, where they studied at the American University of Iraq Sulaimani (Arraf, 2021). The last group of about 200 students was relocated to Qatar, where they continued their studies at the new AUAF campus, which was opened in Education City (Sharma, 2024). Though the number of AUAF students who got out of the country is much larger, these three groups only include those whom the university's efforts have directly relocated. Others have been relocated with the help of international organizations they had previously worked with. As of now, AUAF students are scattered throughout the world. They live in different countries, including Pakistan, Iran, the United States, Canada, and many European countries, where they all continue their education online.

### Evacuation Efforts in August 2021

The evacuation process for AUAF students, taking place for years and involving passage through many countries, has been extremely complicated. The efforts started in August 2021, when the Taliban took over the country. As international organizations were evacuating thousands of Afghans through the Hamid Karzai International

Airport (HKIA), a chaotic scene was created in the airport. Thousands of other Afghans would gather in the hopes of getting evacuated. This caused tumult and disorder in the area. A suicide bombing followed by an eruption of gunfire also took place at the airport on August 26, 2021, which killed at least 183 people, including 170 Afghan civilians and 13 members of the United States military (Seyler, 2022).

Despite the dire security situation at the airport, AUAF students understood the danger that the Taliban posed to them and felt the urgency to get out of the country. Some students took matters into their own hands and went to the airport to get evacuated in any way possible. Other students, having received an email from the President of the university, Ian Bickford, on the night of the takeover of Kabul by the Taliban, patiently waited for the call by the university to get themselves to the airport. The email had informed them about the possibility of evacuation through HKIA and had asked them to be ready. This possible evacuation was scheduled to happen any day between August 15 and August 30. Students were ready with bags packed. As the U.S. troop withdrawal deadline got closer, confusion grew, and the students started to get impatient. During this time, a small number of AUAF students, faculty, and staff were evacuated from HKIA after days of waiting in the airport. They were evacuated from Kabul to Qatar. From there, the students were taken to Iraq to continue their studies at the American University of Iraq in Sulaimani.

Finally, on August 29, 2023, two days before the deadline for U.S. troop withdrawal, the AUAF administration sent a mobilization email to about 600 of its students and staff (Fassihi, 2021). The email requested their immediate gathering at a provided location in order to board prepared buses and enter the airport as soon as permission is received. Some were worried about how they would be able to get into the airport and board planes. Others worried about their families, jobs, and lives they would leave behind once and for all. Many feared the failure of this attempt as the deadline for the withdrawal of international forces from HKIA was only two days away. After waiting several hours, those hundreds of students were informed that the evacuation plan was canceled due to the security situation and that they must return home. After all, it was only a few days after the deadly suicide bombing occurred at the airport, and it was not unreasonable to expect similar attacks as the airport was more packed with people during the last few days of the troops' withdrawal. Attempts made by AUAF to evacuate students failed because the U.S. military completely stopped the evacuation process from HKIA.

## Post-August Efforts

From that point onwards, it was a long waiting game for AUAF students. A group of about 200 students were contacted in September for an evacuation attempt to Kyrgyzstan, where they would continue their studies at the American University of

Central Asia (AUCA) (Crossette, 2021). This group was the first to be evacuated after the August US troop withdrawal deadline. They were evacuated in small groups. The first of these small groups had to cross the Torkham border into Pakistan, from where they would be flown directly to Bishkek, Kyrgyzstan. Crossing the border in buses and cars had its challenges. However, the students, the majority of whom were female, succeeded in getting out of the country, which was a massive relief on its own. The remaining students got into Pakistan through a flight from Kabul to Islamabad. The other process was the same for all these students. They waited about a week before boarding their flights to Bishkek, Kyrgyzstan. The completion of this process recreated hope of evacuation among the AUAF community.

One other destination that appeared on the map for AUAF students from the start was the American University of Iraq Sulaimani (AUIS). This university, located in a comparably safer part of Iraq, was in the Sulaimani region of Iraqi Kurdistan. It was one of the first universities that responded to AUAF's call for help after the fall of the republican regime. AUIS was the first destination for a small number of AUAF students and some staff members between August 15 and 30, 2021 (Shakir, 2021). This university promised to host AUAF students until a decision is made about their education and future. The second time AUAF students were relocated to AUIS was in October. Students were anticipating an update about a possible relocation process. However, a sudden email, followed by a phone call, was sent to a selected number of students about an opportunity for them to be relocated to Iraq, Sulaimani. Most students saw this as a golden opportunity and accepted it. However, there were people who had doubts about the offer to be relocated to Iraq. These doubts caused a small number of students to reject the offer. On October 17, a flight was arranged for nearly 90 individuals containing both male and female students. The new Taliban government was yet to impose restrictions on females traveling without a male legal companion (Mahram). This made the overall relocation process easier as students did not need to be relocated to a third country before being sent to their final destination. On the day of the flight, students were called upon to gather at HKIA and wait for the representative of the government of Qatar to come and coordinate them all together onto the plane bound for Qatar. After waiting for several hours, the students boarded the plane, and after a three-hour flight, they landed in Qatar. There, they were placed in a temporary location for eight days until they got their security clearance and documentation. On October 25, 2021, the green light for traveling to Iraq was given, and these students boarded a plane bound for Sulaimani, Iraq.

One of the biggest concerns of these students while traveling to Iraq was the security situation there. Almost everyone who knew about the process asked whether Iraq was safe for AUAF students. The AUAF community had a similar concern. However, they were caught between the devil and the deep sea. They faced a more significant risk back in Afghanistan, where they might have been easily identified and targeted. Hence, going to Iraq was a massive opportunity for them, and like any other opportunity, it had its own risks. Nonetheless, students accepted to be relocated to Iraq and went there.

With the evacuation of those two groups and others who got out of the country alone, another group of about 350 or so students remained in the country (Packer, 2022). About 100 students did not have a passport. Though some people were evacuated without a passport during the U.S. troop's withdrawal from August 15 to August 31, that did not seem to work after August. The complexity of the situation was worse since the Taliban were not issuing passports. Finally, the passport directorate was reopened in October 2021 (Tayeb, 2021).

Once the passport directorate was opened, the process of taking passports was the second most chaotic scene in Kabul that year. After the U.S. withdrawal from the airport, huge crowds would gather at the Kabul passport directorate to obtain passports. Thousands of people who had worked with the U.S. or other countries were still remaining in the country. However, with the passage of the withdrawal deadline, the process had become much slower. Most of these people did not have adequate documents. A passport was the most essential of all. The passport directorate had opened, and the Taliban were initially prioritizing only those who were sick and had to travel to another country to get treated and those who had scholarships.

AUAF students fell under the second category. To be able to prove that they had scholarships, they were provided with a letter by the university. However, when students went to the passport directorate, the Taliban would complicate matters and ask them to get approval from the Ministry of Education. Having cleared that process, when the students went to the passport directorate in the following days, they faced a vast mass of people. Lines were stretched from the passport directorate located in the west of Kabul, a part of the city that has been continuously attacked over the years. Cars were unable to pass because people were standing on the roads. Faced with the crowd, the Taliban did not know what to do, and they could not manage the situation well. They would threaten people with their guns. Most of them would beat people, both men and women. At some point, they had a small electric device that they were using to give electric shocks to those causing problems. Though they used it mostly on males, they would use it for females at some points. The Taliban would treat people however they pleased. With the departure of foreign powers and facing such treatment by the Taliban, the people would be even more fearful and anxious for their futures. They knew about the complexity of the process, and they all wanted to finish it quickly. In such an attempt, everyone panicked and wanted to get ahead of the line. This was a disaster, and the experience was traumatic for the students.

Most of the students had to go to the passport directorate for days and had to wait in line. When their turn did not come, they would have to return the next day earlier. Some of those students came as early as 12 or 1 in the morning. They would sit there in line and wait until 7 am in the morning when the passport directorate would be opened. It was winter, and the nights were cold. Nevertheless, they still had to endure the cold and wait hours before the line would move. When the students were able to get a stamp (which was the first step in the process) after several days of going there and waiting, they would have to come back for the other steps, which were relatively more accessible. Students would go there in groups, and most finally got their passports issued. After a few weeks, the passport directorate was temporarily closed. For those students who failed to obtain a passport, this meant that they would have to come back to the passport directorate when it opened, which would delay their evacuation process. Eventually, all the students got their passports. Nevertheless, everyone had to go through many difficulties to varying extents.

As these events unfolded, AUAF concluded a contract with Qatar Foundation, which provided that the AUAF campus would be relocated to Education City in Qatar and that about 200 students would be able to get an education there (Qatar Foundation, 2021). A hundred of those students were supposed to get relocated first by the end of 2021, and then they would be followed by the second batch of a hundred students in early 2022. Like all other processes and evacuation steps of AUAF students, many issues started dismantling the flow of efforts and progress. Of those 350 students remaining, the ones who had obtained passports were then waiting to acquire visas for Qatar, which the university administration was handling. One main problem that remained throughout the relocation process to Qatar was the delays in obtaining visas. The AUAF administration put forward its best efforts to pace the process, but the Qatari authorities capable of obtaining these visas only took action according to their own timeline. Hence, what was initially announced as an agreement to relocate AUAF students to Qatar in 2022 took nearly two years to complete. The slow pace of obtaining visas was not the only problem that AUAF faced during the process. Several other challenges elongated the process. One such problem was the COVID-19 outbreak in Qatar in early 2022. For a brief period, the number of registered cases of COVID-19 in Qatar started to rise. This created global concerns as the country would host the FIFA World Cup in less than ten months. Hence, Qatari authorities placed strict rules on people attempting to travel to Qatar during this period. This decision directly affected the pace of the relocation process of AUAF students to Qatar. As the spring semester of 2022 passed, not even one AUAF student was relocated to Qatar. Meanwhile, the students had to wait patiently in Afghanistan while continuing their education online. Though the semester was postponed for one week in the fall semester of 2021, the university was still functioning online, and the classes were being held. During this time, the students faced many issues with continuing their studies. They were mentally unprepared for their classes. With the events that unfolded, the students' family issues, and the uncertainty of their future, they could not focus properly on their classes. Security concerns were another challenge. The tragic news of the death of Zainab Abdullahi, an AUAF alumnus who was killed in Kabul at night by the Taliban, left many students in awe (Al Jazeera, 2022). However, the reason for her death was not her link with AUAF but rather the bleak security situation in the country. To add to the challenges, internet connectivity was another issue the students faced. As such, most students felt hopeless, and some of them did not do as well in their classes. This affected their GPAs, and some of them fell behind their degree plan as they had to drop or withdraw from some of the classes they could not continue. Students started to worry, and worry led to frustration.

Finally, at the end of July 2022, an optimistic update about the progress was sent to the AUAF undergraduate community. It was announced that the Qatari authorities have allowed the relocation of the first batch of students to be completed so that the visa process for the second batch is initiated. Nearly 100 students were expected to be in the first batch. The logistics and transportation of this process required it to be done at separate times and in smaller groups. Hence, the university put forward a timeframe of three weeks for the completion of the process.

The visas for the first batch of 100 students destined to be relocated to Qatar were finally issued at the end of July 2022. This meant that only the final step of the process remained, which was scheduling flights. The administration scheduled flights for the date of August 25. About a hundred students destined to go to Qatar and twenty other students destined to go to Kyrgyzstan were supposed to board that plane and reach their destinations. Students from provinces came to Kabul. Some of these students did not have a place to stay in Kabul, so they had reached the capital only hours or a day prior to the flight.

One primary concern that was highlighted by different parties as these events unfolded was the newly established rule of the Taliban government. They were not allowing girls and women to travel inside or outside the country without a Mahram (Reuters, 2022). AUAF, respecting the close relationship between the Qatari government and the Taliban authorities, requested for the government of Qatar to facilitate permission for female students to travel from Afghanistan without a Mahram. The Qatari authorities gave the assurance, and everything was expected to go as favored after that. Yet, in a delicate process like this, things almost never go as planned.

Hours before the flight, students got there on a rainy morning. Friends met and hugged each other, feeling happy they would be finally evacuated to Qatar, where they could continue their studies. However, after the students' luggage was checked and their passports were stamped, they were informed that the flight was delayed.

They had to wait six hours in the airport without knowing what was happening. Finally, some students boarded the plane, and others were in the buses, taking them to the plane. It was during this time that the airport personnel came and collected the passports of all the students. They asked the students who had boarded the plane to get out. The students felt anxious and knew that something was wrong. The male students were then given back their passports and allowed to board the plane. The female students' passports were returned, and they were told to go back home. Their justification was that females were not allowed to travel without a Mahram. Only the male students and one female student, who had her brother with her, were allowed to board the plane. The plane left Kabul almost empty that day. The female students lost the only hope that they had. They waited a few more hours for the return of their suitcases, which had already reached Qatar. Some went home with their suitcases while the others had lost them too. They returned home feeling hopeless. For some of the students who were meant to go to Kyrgyzstan, their visas were about to expire, too.

As a day full of heartbreak ended, speculation among the AUAF community started to rise about what would happen next. The first concern was the students who did not have a place to stay in Kabul. Being helpless, the administration arranged rooms for them in the Serena Hotel so they could stay for the night. These students, including about 8 to 10 individuals, stayed there for a few nights while impatiently waiting for their relocation. Since it was a group of female students without any male companions, the Taliban brought up the issue again, and these students were asked to leave the hotel. On a late evening, they were left confused under the blank sky without shelter. With nowhere else to go, they spent the night in a house belonging to the relatives of one of these students. The day after, they were guided to go to another entirely female educational establishment by the name of Moraa Educational Complex. This complex became the hosting place of AUAF female students for the coming month.

The female students who could not leave the country were assured that they would be evacuated either by issuing visas for one of their male guardians who would accompany them to Qatar and then return, or by a third country, such as Pakistan or Iran, would evacuate them. The second option was more viable. Hence, students had to decide to get themselves to either Pakistan or Iran. The university helped these students get visas. For those going through Pakistan, once they had their visas, some booked a flight to Pakistan while most went through the Torkham border crossing point into the country. One part of this plan required a place for these students to stay in for a while, in Islamabad, until the day of their flights. Hence, AUAF organized a three-bedroom flat in Islamabad where these students would stay for a few days. Flights were booked for them one or two days after they had arrived. Once one group reached their destination, the process would start for another group. This went on for several months until all of the female students who were denied at HKIA finally reached their destinations. As students still recall, upon reaching Qatar, they could not believe they had finally arrived. The journey, from start to finish, was dramatic, stressful, and highly challenging.

The next part of the plan was to relocate the second batch of students. For months, there was less communication from the university about the timing and dynamic of the relocation process for the second batch. On April 18, 2023, the administration finally gave a conclusive update to the AUAF community through an internal email. The relocation of the second batch of 100 students would happen no sooner than August. The message was clear, and it gave the students an idea of how long to wait. The email also indicated that as per the host country's request, Qatar, the majority of the 100 students included in the second batch would be female students yet again. As months passed, many students made personal attempts without the support of the university to get themselves out of the country and relocate to somewhere where they would be safe and where they would have better growth opportunities. Starting from July 10, 2023, students selected for the second batch started receiving emails and guidance about their relocation process. The first step was for the students to sign an agreement that included several clauses and conditions about the time they would be in Qatar and the dynamic of the stay. After that, students waited almost a month until they received an update about the date and time of their flight. The update was not sent to everyone; it was instead sent to smaller groups of students separately as they were about to be relocated in a divided manner. All female students had to be taken to Pakistan with a Mahram from where they would board flights to Doha. However, the male students' journey was shorter as they could fly directly from Kabul. The process had some minor difficulties. In one instance, two male students got separated from their group, missed their flight, and had to wait for several hours until the next flight. However, in general, it was a much better-planned and well-executed process than the relocation of the first batch.

## Mental Health Concerns

Along with the issues about the physical safety of the AUAF students after the fall of the republican regime, there was also concern about the mental state of these students. This concern resulted from the rise of several issues that resulted from the regime change in August 2021. As the first groups of Taliban marched into Kabul city in their minivans and pickup trucks, the first concern created among the community, along with the physical risk they were exposed to, was the stress about the future. As mentioned earlier, AUAF was a one-of-a-kind higher educational institution that followed a coeducational American curriculum inherited from the United States- the same country that the Taliban had fought against for two decades. Hence, there was a one hundred percent assurance that the university would not be physically functional in Kabul as before. This created a significant concern in the community about what the future holds for them. As a result, students were mentally unwell due to the uncertainty of their tomorrow. One female student who was relocated by AUAf to Kyrgyzstan said the following about her mental health:

*Most of my friends and family members back in Afghanistan told me how I should be happy for being amongst the first people to be relocated post-August by AUAF. To an extent, I was. However, the uncertainty that was ahead of me shadowed my happiness. I would think about my future and how all my plans, prior to the fall of the previous regime, were shattered. The main question that nobody seemed to be able to answer was what is the next step ahead? What if we fail to get immigration status to the United States? There was no authority able to answer these questions. Hence, they gave me headaches for days (2023).*

Before the fall of the republican regime, students had a sense of what they would do after graduation. Some were thinking about employment with U.N. organizations, governmental offices, or other private entities. Others planned on creating their own businesses. There were also those who planned to continue their graduate studies outside Afghanistan. However, with the takeover of the Taliban, all three of these seemed unachievable. The employment rate in the country dropped by a large percentage as the economic situation worsened. For the same reason, almost no optimism was visible in a startup company. Likewise, since no other country had an embassy in Kabul other than Pakistan, getting a visa for a master's program seemed impossible. All of these together turned the future into a dark tunnel for AUAF students, which in turn created significant stress and mental issues for them.

The bleak future was not the only factor that created mental issues for AUAF students. The ineffectiveness of the learning process at the time also created anxiety among students. Even though the online learning process was established in AUAF after the COVID-19 lockdown in March 2020, and it seemed to be the best available educational opportunity at the time, students still heavily doubted the effectiveness of the process compared to physical classes. One female student said, "I had always worked hard to maintain a good GPA. Will any of it matter though? I don't know if I will be able to find a job after finishing my studies and getting my degree" (2024).

The non-existent human connection, along with the lack of motivation among students and instructors, led to a decrease in the quality of learning. This, in turn, affected the accreditation of the AUAF degree. Since after the regime change, AUAF was a cloud university for a while and a university in exile after that, several organizations and institutions had doubts about whether to accept the AUAF degree as a credible one or not. All of these together created anxiety among AUAF students, individuals who already had traumas from the sudden and swift regime change.

## Online Classes

Throughout the evacuation process, the classes remained online. Though it was initially planned that classes would be in-person in the Fall 2021 semester, and students were very enthusiastic about starting the in-person classes after more than a year of studying online, the events that unfolded did not allow the start of in-person classes again. However, AUAF still showed enormous resilience by continuing to teach online in the face of the challenges and concurrent with the events that were unfolding in the country. Even after the reopening of its campus in Doha, Qatar, AUAF continues teaching online classes to hundreds of students worldwide (Sharma, 2024). Though the majority of these students are in Afghanistan, a large number of them are located in other countries as well. AUAF remains the only educational opportunity for Afghan students.

Though online classes have their own challenges worldwide, ranging from distractions at home to time zone differences for international students, the challenges are particularly myriad for students from Afghanistan. These challenges include weak internet connection and electricity outages. The internet connection is minimal for students in rural areas of the country. Being a full-time student and doing all of the assignments online while having a bad internet connection is very challenging for these students. Even for those in Kabul province, internet connection remains weak in parts of the city. As such, studying at AUAF remains intertwined with the challenges of studying online. Another persisting challenge is the untimely electricity outages in the country. This problem has remained for years, affecting students' ability to perform well in their studies.

As these challenges have existed and continue for students, the AUAF administration has attempted to support the students and help make online classes easier. Since the classes have been online, the university has provided data bundles for the students and provided flexibility in deadlines for the students who continue to struggle. In addition, they have tried to provide emotional support for the students through online sessions. Such support has been crucial for the students who have faced unimaginable events throughout the past years and are still resiliently continuing their education.

## CONCLUSION

The plight of AUAF students amidst the turmoil following the Taliban takeover reflects the resilience, determination, and unwavering pursuit of education in the face of adversity. Despite facing numerous obstacles and challenges, including security threats, logistical complexities, and mental health concerns, the AUAF community has displayed remarkable courage and perseverance. When faced with the endless challenges in the evacuation process, students and the administration clung to the last arrays of hope, continuing their efforts until they got to safe places to continue their education.

## REFERENCES

Al Jazeera. (2022). Taliban arrests fighter who shot dead Hazara woman at checkpoint. https://www.aljazeera.com/news/2022/1/19/taliban-arrest-fighter-who-shot-dead-hazara-woman-at-checkpoint

Arraf, J. (2021). Afghan College Students Find a New Life, and Safety, in Iraq. The New York Times. https://www.nytimes.com/2021/11/14/world/middleeast/afghan-university-students-iraq.html

Barakat, S. (2024). Taliban ban on girls' education defies both worldly and religious logic. Al Jazeera. https://www.aljazeera.com/opinions/2024/4/1/taliban-ban-on-girls-education-defies-both-worldly-and-religious-logic

Crossette, B. (2021). Afghan students reach Kyrgyzstan after a risky journey. Pass Blue. https://www.passblue.com/2021/10/13/afghan-students-reach-kyrgyzstan-after-a-risky-journey/

Fassihi, F. (2021). American University of Afghanistan students and relatives trying to flee were sent home. The New York Times. https://www.nytimes.com/2021/08/29/world/taliban-american-university-of-kabul-afghanistan.html

McKirdy, E., & Popalzai, E. (2017). American University of Afghanistan reopens after 2016 attack. https://edition.cnn.com/2017/03/28/asia/kabul-american-university-reopens/index.html

Packer, H. (2022). There's no hope: Afghanistan students still waiting for evacuation. The Pie News. https://thepienews.com/news/afghanistan-students-waiting-evacuation/

Qatar Foundation. (2021). American University of Afghanistan students to continue education in Qatar through strategic agreement with QF and QFFD. https://www.qf.org.qa/stories/american-university-of-afghanistan-students-to-continue-education-in-qatar

Reuters. (2022). Taliban ban women in Afghanistan from flying without male chaperone. https://www.reuters.com/world/asia-pacific/afghanistans-taliban-ban-women-flying-without-male-chaperone-sources-2022-03-27/

Seyler, M. (2022). Single suicide bomber killed US troops and Afghans in ISIS-K attack at Kabul airport, Pentagon finds. ABC News. https://abcnews.go.com/Politics/single-suicide-bomber-killed-us-troops-afghans-isis/story?id=82676604

Shakir, L. (2021). 26 students, professor arrive in Sulaimani from Kabul. Rudaw. https://www.rudaw.net/english/kurdistan/300820211

Sharma, Y. (2024). A Relocated University Restores Hope for Female Students. University World News. https://www.universityworldnews.com/post.php?story=20240116075152546

Talant, B. (2021). Afghan Students Make Daring Escape to Kyrgyzstan. Eurasianet. https://eurasianet.org/afghan-students-make-daring-escape-to-kyrgyzstan

Tayeb, R. (2021). Passport Department Resumes Services to Afghans. Tolo News. https://tolonews.com/afghanistan-174914

The American University of Afghanistan. (2024). About AUAF. https://auaf.edu.af/

# Chapter 12
# The Night I Fought for Life

**Marzia Azizi**
*VECINA, USA*

## ABSTRACT

*In a war-torn country, your foremost struggle is not the pursuit of wealth or riches; it is the fight for survival and then obtaining an education. In 2016, when the author enthusiastically embarked on the author's higher education journey at the American University of Afghanistan, the author had to confront death to pursue that education. It was just the beginning of the second semester when the Taliban, now in control of the country, launched a terrorist attack on my university on August 24, 2016, resulting in the tragic loss of friends, classmates, and teachers. To survive that night, the author leaped from the second floor, landing on the ground, feigning death beside a classmate who wasn't breathing and was covered in blood. The author could witness the terror and the horror right before the author's eyes, resembling a scene from a Hollywood movie. That night, not only did the author lose the author's friends, classmates, and teachers but the author also lost a piece of the author's own mental well-being. Yet, amidst the ruins, the author did not lose hope. In the author's heart, the author still carried hope and optimism for a better beginning.*

## THE NIGHT OF HORROR

It was Thursday, 4 PM on August 24th, 2016, when I left the office for the American University of Afghanistan, unintentionally leaving my phone behind. The university vibes had me so pumped that I didn't want to go back and grab the phone, fearing I would be late for class.

I had informed my family and friends multiple times about the restart of university classes and the boundless joy I felt about the new semester. The classes

DOI: 10.4018/979-8-3693-1483-8.ch012

were delightful, and the professors engaging, but little did I know that joy has its limits. Perhaps the universe couldn't handle my enthusiasm, or maybe I wasn't built to endure such excitement. That day, without my phone, I geared up for my classes. A few minutes before the class started, I went to the cafeteria to pack my dinner. After leaving the cafeteria, I entered the classroom just in time. It was a math class, and the professor started the lecture after an introduction.

Exactly at 7 PM, chaos erupted. An explosion shook the classroom, windows shattered, and panic set in. It wasn't long before the sounds of bullets and gunfire forced us to rush towards the exit. Just as we neared the stairs, a voice screamed: "Go back... go back to your classroom!" So, we all rushed back towards the classroom and locked the door behind us.

A few minutes later, one of my classmates went near the door and said, "I'll go find a way out." He slightly opened the door and looked through the corridor, and then gunfire erupted. He quickly came back and said with a shaky and dry voice, "They are here! In a few minutes, they will enter the classroom for sure." He continued, "The attacker had a military uniform, and although he aimed at me, the bullet hit the whiteboard." With a louder voice, he said, "Look at the bullseye!" We all saw that a bullet had hit the corner of the whiteboard. Hopelessness loomed as we feared an imminent assault, thinking that they would come and shoot every one of us.

The gunfire intensified; some students rushed towards the windows facing the university courtyard, and a few daring souls leaped down.

A momentary silence followed, and the students who had approached the windows were in limbo, unsure of which way to escape – inside the classroom or outside in the university courtyard!

My throat was dry. I was scared, but unlike some, I held back tears. My ears were waiting to hear the next sound. Jamshid, a student who later became one of the martyrs, was also with us. I pulled out my university ID from my bag to inform the security department that we were there. Jamshid also tried to contact security. The same boy who had approached the door a few moments ago said, "Who knows, maybe the university's security team itself is an accomplice." Thus, we gave up on this idea.

Heavy gunfire echoed in the corridor, pushing everyone away from the windows attached to the corridor wall. Some students leaped down towards the courtyard, but most of us stayed in the classroom. Silence fell again; we held our breaths and wondered what to do. At that moment, the ringing of the telephone would terrify everyone because the sound of the phone rings would amplify the gunfire. Someone whispered and asked the time. I checked my watch; it was 7:45 PM.

We were still inside the classroom, and I was praying for survival. Suddenly, gunfire roared louder this time, and all the windows towards the corridor shattered in pieces. Students ran towards the nearest outer window, and I ran towards the windows a few steps farther away. The professor was also sitting next to the window and had not dared to jump towards the courtyard yet. He said with a kind voice: "Please, my dear! Don't leap down, it is very high". I didn't say anything. I looked down; it was high, but in my heart, I believed that I wouldn't die. First, I dropped my bag, then grabbed the window tightly, hung my legs, and descended. In fact, during this time, there was no gunfire in the courtyard, giving me enough time to jump carefully. However, the gunfire from the corridor was intense.

Feet on the ground, I felt no pain. I realized that one of my shoes was missing. I didn't look for my bag, but I grabbed my shoe. My eyes fell on the small and short trees, and I ran towards them. A boy was lying motionless with his face on the ground. I sat next to him, hoping that both of us could find shelter there. However, when I looked closely, he was covered in blood and was not breathing. Fear propelled me to another tree. Another student came closer who had a mobile phone in his hand and wanted to make a call. The courtyard was dark, and all the lights were turned off, but his phone screen was very bright. I urged him to turn off his phone; otherwise, we would both be killed. He quickly left that place and ran farther away. I stayed there and sat down because it was very risky to run towards an emergency gate which was very far from where I landed. I didn't know if my feet were visible from behind the tree but I was sure my face was not.

A girl's cries broke my heart. She begged for help, but there was nothing I could do. Then, three gunshots silenced her. The gunman was standing only four steps away from me. He was wearing a military uniform, and I could see him standing tight, but his face was not visible. I could witness the sparks and flashes resulting from the gunshots. Silence descended on the university courtyard. A moment later, I saw people inside and around the Faculty Building. All of them were dressed in white, carrying guns. Most of them were speaking Pashto, but occasionally, they spoke in Dari too.

Strange events had occurred that day, stacking up one after another, allowing me to stay alive. My mobile phone was left in my office, and even though I wanted to wear my white outfit that day, I didn't know what happened, and my preference changed. I chose a black and dark outfit just like the color of the night.

Hours passed, and gunfire and explosions ceased. Chatters and murmurs of the white-clad attackers surrounded me who were running between the Faculty and Bayat buildings.

I thought they might have noticed me sitting behind the trees, but they had not showered me with bullets yet but soon they would, and tonight, would be the last night of my life.

I heard them preparing to confront the security forces. They were organized into groups, each responsible for a specific task. The scraping sounds of iron sheets being pulled to the ground were also audible. They were dragging them toward the upper floors of the Bayat Building and I was not sure what they would do with them.

When they addressed each other, I managed to memorize the names of five of them. They had a commander and a doctor. When they had the first person wounded, they called for their doctor who was also among the attackers. One of them seemed to be the commander who gave the team moral support. "Don't be afraid, friends! There's no retreat. Ensha Allah, we will defeat the enemy!" the commander screamed and encouraged the team.

From their voices, I could guess that all the white-clad attackers were probably young men aged between 25 and 35. All of them seemed to be ordinary people or university students, but that night, they were there to kill us.

Sitting in one position for a longer period, I found it hard to give myself a little shake. I wanted to look at the sky and see if God was watching us. However, I could never move myself. I resisted and remembered the lessons of Survival Strategies that a dear school teacher was teaching me. I told myself that tonight, I had to stay alive. I had to survive.

Around 1 AM, the gunfire and explosion in the courtyard became so intense that I was no longer able to hear anything. I lay down on the ground and closed my eyes as if I were already dead. This time, my feet were visible to the attackers. Trembling intensely, I was terrified that if the terrorists noticed me trembling, they might shoot me in the head. I tried my best to stop shaking, but it was impossible.

A moment of silence was restored, but this time, the cries of another girl who had fallen in the university courtyard broke the silence. She was crying and asking for help. A security guard from a distance was instructing her on how to drag herself and reach him, but the girl was crying and saying that her lower back and legs were broken and that she was not able to make any move. It hurt my heart, but there was nothing I could do. The girl sometimes needed help and sometimes fell silent.

Now that seven years have passed since that tragic night, my heart still aches for all my sisters and brothers who experienced and suffered that night, and for my Afghans who have gone through similar tragedies and confrontations with death in every corner of the country.

Perhaps it was around 3 in the morning when I heard sirens outside the university's main entrance gate, which was under the control of the terrorists. Just a few minutes later, a group of seven or eight attackers ran quickly towards the gate and I was sure they escaped. The total number of terrorists exceeded 10 people, some of them were killed with the hope of ascending to paradise (as per their beliefs), and those who survived just ran and disappeared.

At around 4 AM, some soldiers were guiding a group of students. Sometimes with a loud and harsh voice, but sometimes with a kind and soft voice. At first, I thought maybe they were taking the students hostage, but looking closely at their uniforms, I understood that they were members of special forces, saving the students and teachers stuck in the classrooms. I saw hope and salvation. One of the soldiers came closer and helped me out of the courtyard through the emergency gate.

I was trembling, and it was very cold. I saw my teacher and several of my classmates there, just outside the emergency gate. I saw Arifa, my roommate and classmate. We hugged each other and cried. Hussain, another classmate of mine, was there too, and he was glad that I was alive. He said all the students who threw themselves out of the windows were martyred, but he was happy that I was alive.

While I was writing the Dari version of this story a few months after the attack, I was in tears and shock, and that is why I never dared to share the story with the public. However, in my heart, I was sure that one day, I would tell the story to everyone, and this feeling of fear and shock would change to strength, a strength that would empower me and help me strive for a world without war and violence.

The university made me feel stronger and more hopeful when it resumed classes in 2017. No one could ever imagine that the university would reopen after such a disastrous terrorist attack, but the leadership behind this inspiring institution did not give up on us and on the goal of making education available for the citizens of Afghanistan, a conflicted-affected country.

On August 15th, 2021, when the Taliban took control of the country, another round of chaos, haplessness, and instability haunted the hopes and dreams of Afghans.

## The Day of Collapse and Haplessness

On August 15th, 2021, the news broke that the President of Afghanistan had fled the country and the Taliban were at the gates of Kabul city. As a result of the Doha agreement between the Taliban and the US government, all the US troops withdrew in May 2021. Therefore, the overall deterioration of the country's security was anticipated after that agreement. However, such a swift takeover of Kabul by a terrorist group was unexpected for Afghans.

The atmosphere in our house was filled with fear, contemplating the reality of a group responsible for violence and chaos now taking control of the country, the terrorist group who had killed thousands of Afghan men, women, and children at schools, universities, hospitals, streets, and masques.

I immediately called my father to ensure they were safe and were at home, given that the Taliban had entered the city and had seized key locations such as the presidential palace, known as "Arg." The Taliban had quickly captured renowned prisons, Pul-e-Charkhi and Bagram, releasing prisoners affiliated with ISIS, Al-Qaeda, and the Taliban. The transformation of Kabul into a city now seemingly filled with thieves, terrorists, and thugs was surreal. In the blink of an eye, the vibrant city turned into a dead one, disrupted by the absence of daily crowds, vendor noises, and vehicle commotion. People were desperate for shelter and safety, with many rushing towards Kabul International Airport in the hopes of fleeing the country and seeking safety.

In the following days, the Taliban initiated searches for former military members and individuals affiliated with the Afghan and US governments or international organizations. My husband and I collected our documents and IDs, showcasing our affiliations with the US government and international organizations. We scanned and saved soft copies, then carefully hid the hard copies making sure that the Taliban wouldn't be able to find them. While hearing about others burning their documents, I couldn't bring myself to do the same.

Despite initially planning to leave for Pakistan, the chaos at Kabul International Airport led us to stay home until it was deemed safe to head there. On August 19th, the organization promising safety and evacuation to Pakistan informed us of a charter flight the next day, allowing us only one backpack. On August 20th, in the early morning, we took a taxi first to visit my mother and then headed towards the airport. I gave my mother a tearful goodbye hug, unaware it would be our last mother-daughter hug.

Kabul, with all its flaws, held a special place in my heart. Crossing its streets, looking at trees and shops, would evoke homesickness. My heart was deeply attached to everything in the city, making it difficult for me to hold back my tears as I remembered my school, university, friends, family, and countless memories.

When I safely arrived at the Kabul International Airport, also known as Hamid Karzai International Airport, there was another horrifying sight. The airport had turned into a frightening and ruined place, with damaged machines, broken glasses, ripped billboard charts, and parked airplanes without staff. The only people walking around were a couple of cleaners attempting to clean the mess in the airport, NATO forces inside the terminals and landing field controlling charter flights, and the Taliban forces outside the terminals and entry gates holding guns and, at times, firing into the sky to control the crowd.

Unlike any normal departure, after nine hours of waiting in the terminal, where obtaining a bottle of water was impossible, we finally boarded the plane and landed in Islamabad in less than an hour. During the flight to Pakistan, my primary concern was reaching safety, with thoughts of a potential return to my country when conditions improved. A few weeks later, a friend of my husband informed us about a charter flight to Qatar and onward to the United States. The decision was difficult, weighing leaving behind our family and parents who needed us the most or risking our lives and our child's future by staying in Pakistan and potentially returning to Afghanistan. Although we were in Pakistan, we held onto the hope of finding a way for our parents to join us there. Going to the USA would further separate us. After hours of consultations with family members, we finally decided to take the flight.

On October 3rd, we landed in a military camp in Doha, joining hundreds of Afghan evacuees seeking refuge in the USA. In the camp, I could witness the haplessness and desperateness in the eyes of fellow Afghans who left their loved ones, their home, and everything behind in the quest for safety, similar to us. After forty days in Doha, we arrived in the United States on November 14th. It has been over two years since I left my home country, yet the memory of saying goodbye to my family and homeland still feels like yesterday.

Life is significantly different now, for me and for all the Afghans who fled the country. In Afghanistan, as a conflict-affected country, the primary focus of the young generation has been survival and obtaining an education. Leaving home for work or university carried the uncertainty of returning home safely, with the persistent fear of potential explosions. Like many Afghans, I had to risk my life to pursue education or work to provide for my family and finally had to leave my country to live, study, and work.

In contrast, in a developed country like the USA, people live without the constant fear and concern of attacks in public spaces. There is confidence in living a life without the fear of dying due to a suicide attack or bomb explosion on the streets. Education is readily available and, to some extent, guaranteed, eliminating the need to risk lives for education. However, in Afghanistan, many girls are currently denied access to education. Millions of school-aged girls are out of school, and working women are confined to their homes. Despite these challenges, Afghans' quest for life and education is admirable and remarkable.

## Hopes and Resilience

Despite enduring a difficult life in the country, the younger generation of Afghanistan has held onto hope, firmly believing that nothing can thwart their dreams and aspirations. The path to success, particularly for women and girls in conflict-affected countries, is undeniably challenging. Despite ongoing bans on education for Afghan girls, a growing number have expressed interest in online educational programs. Despite hurdles like internet connectivity issues and the associated costs, many persist in finding ways to overcome these challenges, driven by their belief in the envisioned future.

History has shown that regimes change, none of which endure indefinitely. The key to a generation's resilience and achievement lies in their hard work, resilience, and ability to see the bigger picture. The bare existence of the Taliban and their two rounds of ruling in Afghanistan underscores an unyielding stance on women's rights, especially access to education. Their aim to eliminate girls and women from the public sphere remains unchanged. Military confrontation and negotiations have proven futile over the past decades, highlighting the entrenched nature of the Taliban's mindset.

To counter the Taliban's oppressive stance and empower Afghan girls, the most viable option is education within their homes through online platforms. Collaborative efforts from institutions and international firms and companies are crucial. Today Afghan girls and women need more institutions like the American University of Afghanistan and the University of People. By extending a helping hand to these girls and providing online education opportunities, these institutions contribute to fostering hope and strength against extremist ideologies.

Furthermore, the involvement of major internet providers such as Starlink is essential. Their support, especially in providing free internet connections in areas with limited access, can significantly contribute to the victory of Afghan girls and women.

In conclusion, alongside the resilience and determination of Afghans, the commitment of educational institutions and the assistance of internet-providing companies play a pivotal role in shaping a brighter future for Afghan women, offering a beacon of hope and resilience against oppressive forces.

# Chapter 13
# The Russian–Ukrainian Conflict:
## Renewed Soviet Ambitions

**Habib Al Badawi**
https://orcid.org/0000-0002-6452-8379
*Lebanese University, Lebanon*

**Mohamad Zreik**
https://orcid.org/0000-0002-6812-6529
*Sun Yat-sen University, China*

## ABSTRACT

*After a long military buildup, Russian recognition of the unilaterally declared Donetsk People's Republic and the Lugansk People's Republic, and entry of the Russian Armed Forces into the Donbass region of Eastern Ukraine on February 21, 2022, the Russian forces launched a military campaign against Ukraine on February 24, 2022. Kyiv and other cities around Ukraine were bombed on February 24 after Russian President Vladimir Putin announced a military action to "disarm and de-Nazify Ukraine" in a speech. The purpose of this research is to shed light on why Russia and Ukraine are at odds and how NATO has played a part in fueling this old war. Since Ukraine was a part of the Soviet Union, this study gives some context for understanding both Russia and Ukraine. In order to better comprehend the nature of the conflict, this study employs a descriptive analytical approach based on narrative and connections between data points.*

DOI: 10.4018/979-8-3693-1483-8.ch013

## INTRODUCTION

In a dramatic escalation of the Russo-Ukrainian War that broke out in 2014, Russia invaded Ukraine on February 24th, 2022. It was made quite obvious that Kiev would meet its fate as a result of the European approach, which included a British proposition made from a podium in Washington. Amid a lack of clear positions and decisive decisions, Ukrainian President Volodymyr Zelenskyy famously declared on the afternoon of the invasion, "We are alone defending our nation." Most of the Western community sympathized with Ukraine at the beginning of the war, but, contrary to what is believed, North Atlantic Treaty Organization's forces did not move to enter the war to support the Ukrainian forces, but huge budgets were allocated to support Ukraine financially and militarily, as well as political and media support through the huge campaigns supporting it (Szporluk, 2020).

Although globalization has been occurring for millennia, the rise of International Organizations can be considered a more modern phenomenon. There was a strong and widespread desire for peace among nations after the devastation of World War II, and many international organizations were established as a result. These included the United Nations, the North Atlantic Treaty Organization, the International Monetary Fund, the European Union, and the World Bank, among others. Almost any conceivable topic is covered by one of the many international organizations currently in operation.

The two primary schools of thought in international relations, realism and liberalism, have had a contentious history of disagreement about IOs. In an anarchic world, where there is no global government, realists claim that nations, the primary actors in international affairs, are primarily motivated by their own self-interest and concern for power and security. That's why they stress the importance of states exercising caution and moderation. On the other hand, liberals tend to hold a brighter vision of humanity, which they defend as inherently good. They have faith that global prosperity, peace, and stability can be achieved via concerted international effort. Realists have been much more wary about the potential benefits of IO proliferation than liberals (Fedor et al., 2017).

Those Western politicians in power frequently adhere to liberal orthodoxy and encourage the spread and extension of IOs, making this topic important in the area of governance. However, the stabilizing effect of these institutions should not be assumed. If you want to know if IOs provide peace, democratization, and stability to global politics and inter-state interactions, look no further than the case of NATO enlargement. Policymakers should exercise caution before adopting liberal policies, and the Ukraine crisis of 2014 and that of 2022 is a telling case study of the risks expansion sometimes involves (Abbassi et al., 2022).

The original purpose of the North Atlantic Treaty Organization (NATO) was to restrain Soviet aggression and shield Western Europe against communist expansion. After the fall of the Soviet Union, the Atlantic Alliance "lost its reason for being". NATO not only made it through the end of the Cold War, but expanded in size, influence, and membership as a result of having "managed to adapt itself to a profoundly altered environment". Some academics, typically those with a liberal worldview, have seen this as a good thing, a win for the West, and a thriving example of global cooperation. However, realists, who tend to be more skeptical of international organizations, see NATO's expansionism as dangerous and frightening, claiming the alliance is a major contributor to tensions between the East and West, and notably between Russia and the West (Feklyunina, 2016).

Two issues, the expansion of NATO and the recognition of Russia's special interest in the countries of the former Soviet Union, "caused the greatest tension and have the most potential to continue," as they have, "to upset U.S.-Russian relations" beginning in 1994, during what Victor Israelyan calls the cold peace. Scholars continue to debate what the future of NATO should be, whether or not it should seek growth, and whether or not it should even exist at all (Kovacevic, 2009). It is possible to tell which of these two perspectives is more correct, or at least more pertinent in the present political situation, by looking at Russia's interests in the former Soviet bloc and its reactions to NATO's expansionist policies and activities in Eastern Europe.

The modern Ukrainian state is relatively young and is plagued by internal disputes and violence. It is easy to point the finger at the country's cultural, linguistic, regional, and ethnic schisms as a contributing factor to the state of the country's democracy. After the fall of the Soviet Union, those in the east and in Crimea, which was part of Russia from the 18th century until 1954, displayed strong Russian nationalism, while those in the west exploited nationalistic pride and pro-west impulses to rebel against their former "master." As a result, ethnic and linguistic lines are quite sharply fractured in Ukrainian politics (White & Feklyunina, 2014).

Apart from the country's internal strife, it is also squeezed between two global powers striving for influence: the United States and Russia. Both have attempted to achieve their objectives through a variety of economic, historical, political, and ideological means (Zreik, 2022). Ukraine's defensive alliance aspirations and military alliances it currently has partnerships with are of the utmost importance. The annexation of Crimea by Russia and the separatist movement in eastern Ukraine can be traced back in part to the tense relationship between the North Atlantic Treaty Organization, Russia, and Ukraine.

Although the Crimean Peninsula has historical ties to Russia due to its ethnic, cultural, and linguistic similarities to the Russian people, its strategic value is the primary reason it has always been at the center of conflicts and has been a frequent target of invasion, conquest, and annexation. Its proximity to the Black Sea is arguably the single most important strategic factor in its favor. Yanukovych and then-Russian President Dmitry Medvedev signed the Kharkov Accords in 2010 to renew and prolong the agreement allowing Russia's Black Sea Fleet to be stationed in Sevastopol from 2017 to 2042. The new government's anti-Russian bias and its support for the European Union and NATO make it likely that the Accords would have been annulled, creating a major roadblock for Russia's access to its Black Sea base in Sevastopol (Kuzio, 2018a).

Maybe it was the prospect of losing such a valuable strategic asset that prompted Russian President Vladimir Putin to seize Crimea. NATO's position in the crisis is more telling and crucial than the European Union's, despite the fact that the crisis originated over whether the Ukrainian government should sign an economic deal with the EU or Russia. This is primarily because of the character of NATO itself. NATO is a military alliance that has been Russia's foe ever since it was formed in 1949, while the European Union is a real and severe economic competitor for Russia, especially if Ukraine were to join this institution. Lacking a standing army and a history of successful military action, the European Union is ill-equipped to defend itself. A NATO-aligned Ukraine would increase this hostile military danger and bring it to one of Russia's most vital frontiers, thus Russia has good reason to worry about NATO expanding into its sphere of influence (Abbassi, 2022).

## The Beginnings of The Russian Military Campaign in Ukraine

Russian forces launched a military campaign against Ukraine on February 24, 2022. The campaign began after a long military buildup, and the Russian recognition of the unilaterally proclaimed Donetsk People's Republic and Lugansk People's Republic, followed by the entry of the Russian Armed Forces into the Donbas region in eastern Ukraine on February 21, 2022. On February 24, after a speech in which Russian President Vladimir Putin announced a military operation with the aim of "disarming and de-Nazifying Ukraine", bombings began on sites across the country, including in the capital, Kyiv (Ben Hassen & El Bilali, 2022).

Prior to the Russian surprise with a Soviet flavor, the common belief was that there would be no military confrontation between Russia and Ukraine due to the strong bonds between the two peoples, which are based on their shared Slavic heritage, belief in the same Christian faith, and shared historical experiences, as well as the fact that the Russian language is practically spoken by everyone in Ukraine.

The second, and arguably more compelling reason, is that people have learned from the bloody conflicts of the last several centuries that killing each other is pointless, and they are terrified of the prospect of a third world war in which nuclear weapons are deployed. Hiroshima and Nagasaki's atomic bombings, as horrific as they were, were executed with very rudimentary technology compared to modern arsenals.

Despite Ukraine's proximity to the territory of the Republic of Belarus, under the influence of Russian President Vladimir Putin, which can be described as the "Putin Republic," the conventional military capabilities of the Ukrainian forces show a great ability to withstand and repel supposed hordes, especially from the capital Kiev (Chortane & Pandey, 2022).

Roughly 142 million people live in the Russian Federation; of them, 69 million are working-age adults; 850,000 serve in the Russian armed forces; and another 250,000 are in reserve units. Russia's annual defense spending is estimated at 154 billion US dollars, out of a total defense budget of a little over a trillion (Jagtap et al., 2022).

There are around 1,543 military helicopters in the Russian Air Force, including 538 attack helicopters among the total of 772 fighter helicopters and 739 attack aircraft. Russian ground forces have over 30,000 armored vehicles and 12,420,000 tanks. There are more than 6,574 self-propelled weapons, roughly 7,571 field guns, and 3,991 rocket launchers. In total, there are 605 ships in the Russian navy, which includes one aircraft carrier, 15 destroyers, 11 frigates, and 48 minesweepers (Ahmed et al., 2022).

There are 43.7 million people in Ukraine, but only 22.3 million working-age men. The Ukrainian army has up to 200,000 soldiers, with another 250,000 in reserve. The Ukrainian Ministry of Defense spends almost $1.1 billion annually on defense.

The Ukrainian Air Force possesses a total of 318 aircraft, comprising 69 fighters, 29 attack aircraft, 32 military transport aircraft, 71 training aircraft, and 112 military helicopters, of which 34 are attack helicopters. About 2,596 tanks, 12,303 armored vehicles, 1,067 self-propelled guns, almost 2,000 field guns, and 490 rocket launchers make up the ground troops. Ukraine's naval fleet is made up of 38 decommissioned Soviet warships, giving it a respectable amount of seagoing might (Steffen & Patt, 2022).

Kiev, the capital and a symbol of independence, sovereignty, and patriotism, is the clear goal of the invasion corps, and the Russian all-out offensive has rendered these inflexible figures ineffective.

## The Era of Ukraine Abandoning Its Nuclear Shield in the 1994 Budapest Treaty

The globe awoke to the sounds of war on February 24, when Russia launched a full-scale airstrike against Ukraine's defenses and hordes of ground troops charged across the border. The question was about the alleged deterrent from a country that inherited the Soviet nuclear weapons and its enemy, the Russian Federation, when the conventional military power balance was tilted in Moscow's favor.

When the Soviet Union disintegrated in 1991, four new states—Russia, Ukraine, Belarus, and Kazakhstan—each acquired nuclear weapons. Kyiv now has the third-largest nuclear arsenal in the world, behind only Moscow and Washington. More than a thousand long-range winged nuclear missiles and 1,240 warheads were carried by the 176 intercontinental ballistic missiles and 44 heavy bombers (Pickyance, 1994).

In the post-Soviet era, Ukraine became a nuclear state with a Soviet legacy, and a strategic arsenal of 1,900 nuclear warheads and a tactical arsenal of 2,500 nuclear weapons. Ukraine's control over the nuclear weapons on its territory was undoubtedly more theoretical than operational, as the power to control their launch remained in Moscow's hands. Nonetheless, this was of little short-term concern, as the country is both uranium-rich and home to remarkable technological and, especially, production capabilities. In 1962, the Soviet Union supplied Cuba with ballistic missiles that originated from bases and manufacturers in Ukraine (Fitzgerald, 2019).

More than 50 countries met at a meeting in Budapest in 1994, and there the "Budapest Memorandum of Security Assurances" was signed and it now applies to Kyiv. The memorandum aimed to restart nuclear disarmament and complete the removal of nuclear weapons from Ukraine. When Kiev gives up its weaponry, it will obtain a formal promise of its territorial sovereignty. This was a major distinction, but in the naive and optimistic atmosphere of the immediate post-Cold War era, it seems to have been largely ignored.

A Ukrainian envoy told his American counterparts that Ukraine was "never under the misconception that the Russians would honor the accords they had signed" prior to the signing of the Budapest memorandum. Ukraine knew that its former status as an empire would not allow it to survive so easily, and so its government hoped only to "obtain agreements that would allow it to seek help in international forums when the Russians violate" these agreements, a childish approach that does not amount to dealing with the destinies of nations and peoples (Bussas et al., 2017).

Ukraine agreed to remove all nuclear weapons from its territory by the deadline set forth in the Budapest Memorandum of 1994 (Jewell, 2015). According to the OSCE Final Document, the Russian Federation, the United States, and the United Kingdom committed to the following:

- Respect Ukraine's sovereignty, existing borders, and independence in accordance with the OSCE Final Document.
- No weapons shall be used against the political independence and territorial integrity of Ukraine, unless for self-defense purposes and in other cases under the Charter of the United Nations.
- Refrain from economic coercion, which is aimed at subjecting Ukraine's exercise of the inherent rights of its sovereignty to its own interests and thus securing any advantages for itself.

- Demand that the United Nations Security Council take immediate action if, as a State party to the Treaty on the Non-Proliferation of nuclear weapons (NPT), Ukraine becomes the subject of a threat or victim of aggression using nuclear weapons.
- Nuclear weapons shall not be used against Ukraine except in response to attacks by that country on the parties to the Memorandum, their territories, and their allies.
- Providing advice if there is a disagreement about the obligations.

On Thursday morning, February 24, 2022, the head of Ukraine's National Security Council stated that "international guarantees were empty words," commenting on the international nominal support, so far, for his country amid a seemingly all-out Russian invasion. "All aircraft capable of transporting nuclear bombs have been taken away from us as well. Ukraine suffered the loss of strategic bombers and cruise missiles, and the country's missile mines and fuel were damaged. The treaty signed in Budapest is just that, a document, and does not contain any meaningful mechanisms for the execution of safeguards. What this suggests is that the disarmament of Ukraine is a done deal, and the guarantees are just that, words" he added (Abbassi et al., 2022).

The secretary of Ukraine's National Security Council warned that "the Budapest memorandum proved the absurdity of disarmament, showed the destruction of the system of collective security, and transformed the new century into a medieval style (the law of the jungle and the logic of force) that poses a potential threat to everyone."

## The Religious Roots of the Russian-Ukrainian Confrontation

Russian and Ukrainian forces are currently engaged in battle across the Donetsk and Lugansk regions, as well as in Kharkov, Kherson, Berdyansk, the Crimea, and the outskirts of Kiev. After hearing reports of "thousands of Chechen warriors currently deployed in southern Ukraine," "hundreds of Chechen fighters arrive in the forest before heading to the Ukrainian front," and "dozens of Chechen special forces carry cards with the names and photos of their targets inside Ukraine," Chechen leader Ramadan Kadyrov made the announcement amidst Islamic songs, religious prayers, and mass prayers. In response, the Russian Ministry of the Interior claimed that they had prevented a terrorist attack by ISIS in the Kaluga region, which is located about 200 kilometers west of Moscow (Skinner, 2009).

Therefore, the following questions follow from logic:

Does the powerful Russian army really need a few dozen soldiers to win a war that's already in the bag?

Was this bizarre emergency really necessary given the lack of religious and ethnic tensions between Russians and Ukrainians?

Why are the Chechens, who have a long history of fighting against the Russians in this civil war between two Slavic peoples who share a common racial heritage and a common faith in Orthodox Christianity, fighting against one other?

For what reason is "Delivery - On Demand" the standard operating procedure for ISIS's terrorist acts?

Despite decades of Soviet repression, Orthodox Christianity is the dominant religious doctrine in both Russia and Ukraine, with over 71% of Russians and 78% of Ukrainians identifying as poor Christians (Richters, 2012).

Historically, the Orthodox churches have been organized on national foundations, with patriarchs enjoying autonomy within their respective churches despite their shared belief in the Orthodox faith; the Greek Ecumenical Patriarch of Constantinople, currently based in Turkey, is widely regarded as the most influential of these patriarchs. He has a high standing among peers but, unlike the Catholic Pope, is not at the top of the church's hierarchy.

Unlike the Catholic Church, which is ruled by a single spiritual leader—the Pope—the Orthodox Church is made up of 14 separate churches with their own leaders and canons. The Russian Church is the largest of these, followed by the Roman Church, the Greek Church, the Serbian Church, and finally the Bulgarian Church (Marzuki et al., 2023).

Historically, the Slavic people, especially those located in what is now Ukraine, have been loyal to the Russian Orthodox Church, which is now known as the Moscow Patriarchate. The government estimates that there were more than 12,000 dioceses under this church's jurisdiction in 2018 (Ponomariov, 2019).

In April of 2018, then-President of Ukraine Petro Poroshenko attempted to form an independent Ukrainian Orthodox Church, with the blessing of the Ecumenical Patriarch of Constantinople, Bartholomew I, whose position as the Cape Bishop of the Old Capital of the Byzantine Empire is essential to his continuing legitimacy. Constantinople, Alexandria, Greece, and Cyprus were the first to recognize the newly formed Orthodox Church of Ukraine, but the Moscow Church and the Orthodox Church in the United States of America were swift to reject it.

A clear and severe challenge to the hegemony of Moscow's political authorities, the Orthodox Church of Ukraine enjoys full autonomy as the result of decades of effort by Ukrainian nationalists Yin-Yang in cooperation with the West. Their stated goal was to establish their own national church, independent of the authority of the Russian Church, to express the spiritual independence of the Ukrainian Church of Ukraine.

Ukrainian Orthodox constitute 30 percent of all Christians associated with the Moscow Church, hence the Ukrainian Church's announcement of its departure from the Russian Church was "a significant blow to the objectives of the Kremlin and the Russian Patriarchate on numerous levels" (Demydova, 2019).

The Moscow Patriarchate, the "mother and foundation," affirms that Russians and Ukrainians are one people who must be unified by one church, and this affirmation is at the heart of the current conflict between the Moscow Patriarchate and the Kiev Church. Putin has used this slogan publicly, saying the radical Nazi in Ukraine is damaging the "spiritual union" of the two peoples (although he is pushing Chechen "Sufi" fighters into Ukraine).

In a public speech rejecting divorce and homosexuality and promoting traditional family values, President Putin is portraying himself as a champion of conservative traditions and the authority of the church. Putin is rumored to have been baptized in secret in the early 1950s, when the Iron Curtain and the prospect of the Gulag camps threatened any little religious smell. His mother is supposed to have given him a cross necklace on the occasion (Kuzio, 2018b).

The Patriarch of Moscow, Kirill, has spoken out in support of the Kremlin and the Russian military, expressing pride in the country and its military, and highlighting the Church's importance in protecting the system of "Orthodox values that the sacred Russian civilization holds for the world" in his homily.

However, the current president of Ukraine, a Jew named Volodymyr Zelenskyy, has not placed the same emphasis on religious nationalism as his predecessors have. He made an announcement on the stormy Saturday of the Russian bombardment, saying that he had spoken to the spiritual leaders of all major religions in Ukraine, including the Orthodox Church, the Catholic Church, the Muslim faith, and the Jewish faith.

The Ukrainian Orthodox Church backed the president, while they were a staunch opponent of the Russian imperial tradition of epiphany, writing that "as a Ukrainian people they will not have a unique culture," and that "Ukraine has its own church" that backs the country's sovereignty and independence from the Russian Federation.

In the midst of this religious-ethnic tangle in its cultural-civilizational connotation, it is important to note that 75 percent of the two peoples share commonalities in doctrine, ethnicity, and history. This leads us to revisit our original questions, to which we have not yet received satisfactory answers (Suslov, 2014).

## Indications for The Use of Decisive Russian Military Force Against Ukraine

Even though previous Russian military assessments and Western expectations led to the capital not holding even for two days, Ukraine held its ground and Kiev did not fall despite fierce fighting. The European media's gloomy outlook was justified by the close proximity of Kiev to the Belarusian border, through which the initial waves of the direct attack had swept.

From a previous estimate of 100,000, it is now believed that the invading forces number closer to 200,000, and that they are supported by combat formations of tanks, artillery, coverage, and naval support (Zreik, 2023a). Moscow was compelled to contribute additional troops to support the Russian army's war effort after the initial attack consisted of a missile strike followed by ground advances with ineffective air cover. This included sectors of the National Guard and internal security units. Some 60 percent of Russia's total war capacity, across all branches, has reportedly been amassed on the fronts of the Russian-Ukrainian war, according to leaks from British sources. Approximately two-thirds of the attacking forces are effectively spread out along multiple axes, with a depth of up to 500 kilometers inside Ukrainian territory, according to the leaks, while the remaining one-third is deployed in tactical depth inside Russian territory to ensure access to logistical support (Associated Press, 2022).

The main support units needed for the military campaign have moved to their positions alongside the fighting forces, according to reports from the United States. Russian and allied forces suffered heavy casualties, necessitating the deployment of tank repair facilities, field hospitals stocked with blood supplies, and mud removal machinery.

Here, we must define the "spring phenomenon of Rasputitsa," in which the melting of ice turns the earth into a sea of clay. This occurred around the middle of March in the countries of Russia, Ukraine, and Belarus, and led some to conclude that "it seems that the scenario of a blitzkrieg has been completely ruled out with the continued resistance of the Ukrainians to the Russian attack as best they can." The Rasputitsa phenomenon, where the ground is covered with a layer of clay that makes you feel your weight is getting heavier as you try to move through it, may soon confront the Russian armed forces, which has previously defeated those who have tried to invade Russia in the past (Al Shaher & Zreik, 2022). During the ongoing operations, Russian forces will be able to overcome this phenomenon with means and equipment that have been specifically invented to counter it. Alternatively, Russian forces may try to avoid it by concentrating on the major cities rather than the rural areas where farms and mud-land abound (ECB, 2022).

When the reserve forces were called into action, some supporting units traveled nearly 4,000 miles from the Russian Far East, and heavy armor was transported by rail from the Armenian-Azerbaijani border (Zreik, 2019). Many special battalions came in via land, while others entered through Kursk, which is only about 80 miles from the Ukrainian border. Karachev must be traversed in order to get to the Bryansk area (Reuters, 2022).

Massive Russian forces, likely including infantry and airborne forces, have arrived on the Crimean Peninsula, with some units already reportedly engaged in operations deep within Ukraine. In the Black Sea and the Sea of Azov, both to the east and west of Crimea, Russia has stationed amphibious assault ships.

Moscow has deployed the strategic threats of the Iskander short-range ballistic missile launcher, Spetsnaz special operations forces, and substantial air defenses to prevent any irresponsibility on the part of NATO. From the Atlantic to the Pacific, the Russian Navy routinely holds military exercises. As many as nine Russian frigates in the Black Sea area are armed with cruise missiles, and another four Russian vessels with cruise missiles are stationed in the Caspian Sea.

Since the letter "Z" does not appear in the Russian alphabet, its presence on Russian military vehicles during the early days of the conflict has been viewed as a mystery. However, like other letters such as (O), (X), (A), and (V), the "Z" was likely an abbreviation for (запад) "Zapad," which means "West" in Russian, or (за победу) "Za pobedu," which means "to the victory" (Tosun & Eshraghi, 2022).

## Russia's War of Attrition on the Ukrainian Front

According to the principles of theoretical military reasoning, the conflict between Russia and Ukraine is not a fair fight between equal opponents. The Russian Federation has a significant advantage over the Ukrainian resistance in terms of infantry, armored forces, armed vehicles, and attack aircraft. The successor states to the Soviet Union were equipped with cutting-edge weapons, many of which had been tested in Syria, as well as superior military might and extensive experience projecting force across national boundaries (Greene & Robertson, 2022).

Russian forces have been repelled at the gates of the Ukrainian capital Kiev, the focal point of their initial advance to assume a pro-Moscow puppet government, despite the fact that the Russian hordes have been stuck there for weeks despite their overwhelming numerical superiority. Russian forces have been unable to maintain control over strategic positions inside Ukraine due in large part to their inability to perform even the most fundamental of logistical tasks, such as ensuring that military vehicles have enough fuel, soldiers have food, and wounded military personnel receive treatment.

The Russian security system has reportedly been riddled with resignations and firings among high-ranking officials as a result of the country's inability to properly study and analyze the Ukrainian interior, adding insult to injury after the disastrous military defeat. Estimates from Russia before the war stated that the Ukrainian people—with their Slavic ancestry, Orthodox religious ties, and Soviet historical memories—were unhappy with their current leadership and skeptical of their country's westward trend and, by extension, joining NATO. Therefore, it appears that Russia's primary invasion plan was based on this superficial assessment, with the assumption that Ukrainian resistance would be marginalized by some elements of the Nazi far right. "Russia has made big assumptions about its ability to reach Kiev within 48 hours, and most of its decisions have been shaped on the subject.

It was a strategic choice, shaped by bias and assumption, which tried a crazy dash that failed." While traditionally in such a conflict, militarism calls for the heavy use of so-called "common weapons," the common attack elements of the military strike force, such as armored tanks, mechanized infantry, attack aircraft, and cruise missiles, to launch a full-scale, coordinated attack (Bounguo & Yatié, 2022).

The Russian leadership, angered by the Ukrainian people's resolve, has ordered the use of "some" of Russia's most devastating weapons and tactics, such as the widespread bombing of cities. Efforts were made to avoid employing such weapons in order to maintain the "surgical military operation" veneer applied to this war by the powers that be. A large amount of firepower is unnecessary for this purpose.

With the help of NATO member states and US intelligence, the Ukrainian forces put up an impressive defense and fierce resistance, making the Russians pay dearly for their careless advance and their weak supply lines. Despite suffering heavy damage in the initial Russian bombardment, Ukrainian air defenses have been holding strong for the past few weeks, denying the Russians air superiority and thus preventing them from achieving the decisive factor necessary to claim victory. Russian supply lines were long and emaciated, but the Ukrainians used a Turkish drone called Bayrakdar to effectively slash them (Abbasssi et al., 2022).

It is too soon to declare Ukraine the victor in the conflict and the protector of its independence and sovereignty; armies typically adapt during conflict; Russia may change tactics and adopt a more appropriate strategy to destroy the fierce Ukrainian resistance, thus the massive tactical bombardment and siege of Ukrainian cities.

Not only has Russia failed to deploy its elite teams, its most capable fighting force, but it has also become mired in conflicts fought by conscripts, who lack the training and dedication of professional soldiers. With each passing day of fighting (and ambush), however, Russia suffers greater human cost, economic hardship, and diplomatic pressure. As a result, it's crucial that the battle be won by the best forces possible (Reuters, 2022).

The Kremlin is wagering that Russia can quickly defeat Ukraine's military if it can quickly adjust its approach and withstand its genuine power. However, NATO made an early statement on its stance toward direct military intervention, "Due to the fact that NATO is not a military party to the conflict, US President Joe Biden has stated categorically that 'When Russian weapons clash with the American,' NATO is not a party to the conflict and therefore Article V of the Charter of Establishment of the Alliance does not apply to Ukraine. If this happened, it would be the third global war ", which he and his European allies do not want. To wear down the Russian military, economy, and state, the West is instead engaged in a "proxy war of attrition" (Zreik, 2023b).

Ukraine's military had potent defensive weapons, including US-made Javelin portable anti-armor missiles and Stinger portable air defense missiles, which prevented Russian forces from enjoying air sovereignty and the surge of armored corps (Associated Press, 2022).

Is it time for direct negotiations between Russian President Vladimir Putin and Ukrainian President Volodymyr Zelenskyy to end the bloodshed, considering that five months have passed since the start of the war, there is no chance of a quick Russian victory, the morale of the Ukrainian resistance is high, and US and NATO support for the Kiev government is on the rise. Will Slavic brotherhood prevail over global tensions? Or is it inevitable that the United States and the Russian Federation will go to war?

## CONCLUSION

As the conflict between Russia and Ukraine has escalated over the past few months, we have witnessed huge military, economic, and military changes. To contradict the premise upon which Moscow is so adamant, radical nationalism in Ukraine is not the only driving force, or even the dominant cause, behind the obstinate Ukrainian resistance. Even while there are significant differences between the two faiths in Russia and Ukraine, the significance of the current religious renaissance in both nations should not be overstated, and hence this conflict cannot be viewed as a religious one. It's not just a border disagreement though, though regional concerns have been and continue to be a major roadblock to peace talks between Moscow and Kiev.

The root of the current dispute is found in the fact that the two countries' social and political systems are fundamentally incompatible. A fundamental enmity exists between the two points of view on the nature of the international system and the globe today. Two competing theories explain how a state makes decisions and how power is distributed throughout its institutions. The criteria for the efficacy of state authority are distinct from more fundamental considerations about the nature and sources of legitimacy.

Ukraine is now an exemplar of a successful liberal democracy in the West. Contrarily, Russia has departed more and more from the liberal democratic paradigm, especially during the past two decades. Ukrainian society is organized from the bottom up, in contrast to the top-down structure of Russian society and politics. It's hard to put a finger on what caused such a dramatic divide to open up between the two major groups of Russian people and the future prospects for either. But for the time being, the fundamental mismatch between the two social organization models appears to have not only turned into a military clash in the heart of Europe, but also to substantially determine the logic of the behavior of the participants in this battle.

There have been two seemingly opposite stages of the Russian-Ukrainian conflict so far, and the world is on the cusp of a third, the nature of which is hotly debated.

Since February 24th, 2022, the Russians went on the offensive. The Donbas region, the northern and western outskirts of Kharkiv, the southern city of Kherson, and the western capital of Kyiv have all been targeted in an attempt to replicate the blitzkrieg strategy. Troops were attacked and pushed from three directions: to the north and east of Russia, to the west of Belarus, and to the south of Crimea.

Weakness in command and control, poor supply plans, underestimation of Russian capabilities, and the spread of misinformation are all contributing to the current state of affairs. As a result of these severe setbacks, the Russian military has lost a significant amount of ground.

The war entered its second phase, it has consisted of traditional military conflicts between two adversaries. The Russians reverted to their more conventional fighting tactics, pulling back from Kiev and the surrounding area to concentrate on the east (Donbass province) and holding on to the south while they continued their offensive to seize the rest of the coastline along the Sea of Azov.

A slow and natural advance of the forces occurred in relation to the capabilities of the forces and the size of the losses. However, they are working to take advantage of their mistakes in the first stage by concentrating efforts and not distracting them, and by paying attention to the transfer and securing of logistics and the mobilization and concentration of means of fire.

The Russians have a great opportunity to weaken the Ukrainian resistance as we enter a new phase, which is now in full swing. Attention is diverted to the south in an effort to capture Mykolaiv and Odessa and gain control of all of Ukraine's Black Sea coast (Slovanisk, Kramatorsk, and Constantinev, Yeka). Second, expanding southwards from the seized territory and shifting focus to the north in an attempt to capture the remaining cities in the Donbass region, with a particular emphasis on the cities to the west of the region all the way to the capital. Finally, conceivable outcomes for Russia's "special military operation" inside Ukraine's territory may be presented, each with far-reaching geopolitical ramifications.

It is the Western position that a return to the "unipolar moment" in international affairs, when the United States was the dominant power, is possible if Moscow is beaten in this conflict.

For Ukraine, the fact that the current government has been able to hold on despite losing certain border regions represents a "taming" of Putin's Soviet-dreamy Russia. The final resolution of the battle between the Russian and Ukrainian development models would be again postponed if, as the second scenario suggests, the war ends with an incomplete but mutually acceptable political compromise between Moscow and Kiev as well as between Russia and the West. It's possible that a short-term peace agreement between the West and Russia will be followed by a continuation of the core conflict between the two social organization paradigms. The agreement between Russia and the West, calls for a significant investment of time, effort, and political flexibility on all sides. The agreement's results, however, will alter the global landscape by, among other things, reshaping the International Monetary Fund, the World Trade Organization, and other multilateral institutions, modernizing the rules of public international law, and bringing the United Nations system to the forefront of global governance.

The third possibility is that no agreements will be reached regarding the Ukraine conflict anytime soon, and that the escalation will continue in accordance with the following mechanisms: attrition of the parties, a fragile truce, the accumulation of troops, and a new escalation. If this occurs, the conflict in Ukraine could become a catalyst for the final collapse of the modern international order. There will be a dramatic increase in regional conflicts, a proliferation of nuclear weapons will set off a domino effect of other dangerous weapons, and international terrorism will thrive as ineffectual international and regional institutions lose their footing in world politics. Such shifts will bring about widespread unrest and possibly even chaos in the years to come; it is only after this unrest and chaos has subsided that the first shoots of the new global order will appear.

## REFERENCES

Abbassi, W., Kumari, V., & Pandey, D. K. (2022). What makes firms vulnerable to the Russia–Ukraine crisis? *The Journal of Risk Finance*.

Ahmed, S., Hasan, M. M., & Kamal, M. R. (2022). Russia–Ukraine crisis: The effects on the European stock market. *European Financial Management*.

Al Shaher, S., & Zreik, M. (2022). Sino-Iranian Relations: More Cooperation Despite Sanctions. *Resmilitaris,* 12(2), 637-649.

Associated Press. (2022, February 26). Ukraine crisis: UNHCR predicts 4 mn will flee to neighbouring countries. Business Standard India. https://www.businessstandard.com/article/international/ukraine-crisis-unhcr-predicts-4-mn-will-flee-toneighbouring-countries-122022600772_1.html

Ben Hassen, T., & El Bilali, H. (2022). Impacts of the Russia-Ukraine war on global food security: Towards more sustainable and resilient food systems? *Foods*, *11*(15), 2301. doi:10.3390/foods11152301 PMID:35954068

Boungou, W., & Yatié, A. (2022). The impact of the Ukraine–Russia war on world stock market returns. *Economics Letters*, *215*, 110516. doi:10.1016/j.econlet.2022.110516

Bussas, V., Sharma, A., & Shouche, Y. (2017). IP and The Budapest Treaty—Depositing Biological Material for Patent Purposes. In *Microbial Resources* (pp. 275–292). Academic Press. doi:10.1016/B978-0-12-804765-1.00014-X

Chortane, S. G., & Pandey, D. K. (2022). Does the Russia-Ukraine war lead to currency asymmetries? A US dollar tale. *Journal of Economic Asymmetries, 26*, e00265. doi:10.1016/j.jeca.2022.e00265

Demydova, V. (2019). Independence of the Ukrainian Church: How Russia Is Losing Its Soft Power Tool in Ukraine. *Akademik Hassasiyetler, 6*(11), 251–274.

ECB. E. C. (2022). Economic Bulletin Issue 1, 2022. https://www.ecb.europa.eu//pub/economicbulletin/html/eb202201.en.html

Fedor, J., Lewis, S., & Zhurzhenko, T. (2017). Introduction: War and Memory in Russia, Ukraine, and Belarus. In War and Memory in Russia, Ukraine and Belarus (pp. 1-40). Palgrave Macmillan.

Feklyunina, V. (2016). Soft power and identity: Russia, Ukraine and the 'Russian world (s)'. *European Journal of International Relations, 22*(4), 773–796. doi:10.1177/1354066115601200

Fitzgerald, T. (2019). Biological deposits for patenting purposes under the Budapest Treaty. *Microbiology Australia, 40*(3), 114–116. doi:10.1071/MA19031

Greene, S. A., & Robertson, G. (2022). Affect and autocracy: Emotions and attitudes in Russia after crimea. *Perspectives on Politics, 20*(1), 38–52. doi:10.1017/S1537592720002339

Huang, H. (2018). The pathology of hard propaganda. *The Journal of Politics, 80*(3), 1034–1038. doi:10.1086/696863

Jagtap, S., Trollman, H., Trollman, F., Garcia-Garcia, G., Parra-López, C., Duong, L., Martindale, W., Munekata, P. E. S., Lorenzo, J. M., Hdaifeh, A., Hassoun, A., Salonitis, K., & Afy-Shararah, M. (2022). The Russia-Ukraine conflict: Its implications for the global food supply chains. *Foods, 11*(14), 2098. doi:10.3390/foods11142098 PMID:35885340

Jewell, C. (2015). WIPO's Budapest Treaty facilitates biotech patenting. *WIPO MAGAZINE*, (4), 2–7.

Kovacevic, A. (2009). *The impact of the Russia–Ukraine gas crisis in South Eastern Europe*. Oxford Institute for Energy Studies.

Kuzio, T. (2018a). Russia–Ukraine crisis: The blame game, geopolitics and national identity. *Europe-Asia Studies*, *70*(3), 462–473. doi:10.1080/09668136.2018.1443643

Kuzio, T. (2018b). Euromaidan revolution, Crimea and Russia–Ukraine war: Why it is time for a review of Ukrainian–Russian studies. *Eurasian Geography and Economics*, *59*(3-4), 529–553. doi:10.1080/15387216.2019.1571428

Marzuki, S. Z., Zreik, M., Buyong, S. Z., Goyipnazarov, S., Mirzaliev, S., & Bakhramovna, I. A. (2023). An Importance Satisfaction Analysis of Hosting Events in Hangzhou, China: A Pilot Study. *Advances in Social Sciences Research Journal,* 10(6.2), 107–117. doi:10.14738/assrj.106.2.14996

Pickvance, C. (1994). Housing privatization and housing protest in the transition from state socialism-a comparative-study of Budapest and Moscow. *International Journal of Urban and Regional Research*, *18*(3), 433–450. doi:10.1111/j.1468-2427.1994.tb00277.x

Ponomariov, A. (2019). Ukrainian church autocephaly: The redrawing of the religious borders and political identities in the conflict between Ukraine and Russia. *Russian Analytical Digest*, *231*(25), 2–9.

Reuters. (2022, February 15). Gas supply shock would cut value of Europe's economy, ECB says. Reuters. https://www.reuters.com/article/gas-europe-growth-idUSL1N2UQ0EV

Richters, K. (2012). *The Post-Soviet Russian Orthodox Church: Politics, Culture and Greater Russia*. Routledge. doi:10.4324/9780203115732

Skinner, B. (2009). The Western front of the Eastern church: Uniate and Orthodox conflict in 18th-century Poland, Ukraine, Belarus, and Russia. Northern Illinois Univ. Press.

Steffen, B., & Patt, A. (2022). A historical turning point? Early evidence on how the Russia-Ukraine war changes public support for clean energy policies. *Energy Research & Social Science*, *91*, 102758. doi:10.1016/j.erss.2022.102758

Suslov, M. (2014). The Utopia of 'Holy Russia' in Today's Geopolitical Imagination of the Russian Orthodox Church: A Case Study of Patriarch Kirill. *Plural. History, Culture. Society*, (1-2), 81–97.

Szporluk, R. (2020). *Russia, Ukraine, and the breakup of the Soviet Union*. Hoover Press.

Tosun, O. K., & Eshraghi, A. (2022). Corporate decisions in times of war: Evidence from the Russia-Ukraine conflict. *Finance Research Letters, 48*, 102920. doi:10.1016/j.frl.2022.102920

White, S., & Feklyunina, V. (2014). Identities and foreign policies in Russia, Ukraine and Belarus. *New York. Palgrave Macmillan, 10*, 9781137453112.

Zreik, M. (2019). Geopolitical Situation in Northeast Asia: Development of Sino-Russian Relations. *RUDN Journal of Political Science, 21*(3), 450–459. doi:10.22363/2313-1438-2019-21-3-450-459

Zreik, M. (2022). US-China Relations in The Era of Multi-Polarism: Trade War Between the Two Economic Giants. *Eurasian Research Journal, 4*(2), 43–65. doi:10.53277/2519-2442-2022.2-03

Zreik, M. (2023a). Navigating the Dragon: China's Ascent as a Global Power Through Public Diplomacy. In S. Kavoğlu & E. Köksoy (Eds.), *Global Perspectives on the Emerging Trends in Public Diplomacy* (pp. 50–74). IGI Global. doi:10.4018/978-1-6684-9161-4.ch003

Zreik, M. (2023b). Analytical study on Foreign Direct Investment Divestment Inflows and Outflows in Developing Economies: Evidence of China. *Chinese Economy, 56*(6), 1–16. doi:10.1080/10971475.2023.2193118

# Chapter 14
# When Extraordinary Circumstances Call for Mutual Aid:
## The Arrival of Afghan Academics in the U.S.

**Stephen A. Rosenbaum**
*University of California, Berkeley, USA*

**Davida Finger**
*Loyola University New Orleans, USA*

**Negina Khalili**
*Loyola University New Orleans, USA*

**Ghazi Hashimi**
*Northwestern University, USA*

## ABSTRACT

*In this chapter, the authors use the narrative essay as a device for providing insight into the experiences of legal scholars, with a spotlight on the personal journeys of two Afghan scholars whose lives and livelihood were put in jeopardy when their country's law professors and lawyers were abruptly thrust into a struggle for academic freedom, vigorous advocacy, and personal safety. They also share the reflections of two US-based legal educators who helped to secure the engagement of Afghans with American law schools at the time their own government was in the midst of a chaotic retreat and the Taliban had regained total control. This series of vignettes examines challenges faced by educators, lawyers, and students in Afghanistan and the United States and the profound impact of the Taliban's totalitarian takeover on academic freedom, human rights, and the pursuit of education.*
DOI: 10.4018/979-8-3693-1483-8.ch014

## PROLOGUE

In this chapter, the authors discuss efforts to facilitate the arrival of Afghan scholars at-risk after the August 2021 Taliban takeover of Afghanistan. To add to the existing literature on so-called "scholar rescue" is a multi-perspective view on the role and response of American academic institutions, bringing together the contemporary voices of two Afghan scholars and two United States faculty members in a unique exploration of the collaboration that resulted from the crisis. Academy in Exile's founder Kader Konuk has written:

*"Historicizing the figure of the 'refugee scholar,' a figure that has been referred to as exile, émigré, refugee, expatriate, displaced scholar, pariah, or, as is the trend now, simply person 'at risk,' is helpful in identifying the intellectual traditions that inform current international aid efforts in higher education." (Konuk, 2020, p. 277).*

The aid effort chronicled here includes at-risk Professors Ghazi Hashimi and Negina Khalili, both Afghan lawyers, who eventually were welcomed at U.S. law schools as visiting scholars after evacuating from Afghanistan. The other two authors are Professor Stephen Rosenbaum, University of California, Berkeley, who helped to lead efforts to bring Afghan scholars to the U.S., and Professor Davida Finger, Loyola New Orleans, who responded to Rosenbaum's outreach and worked to secure a position at her own institution for Professor Khalili. Although scholar rescue programs in the United States and Europe "remain the most important and enduring response to safeguarding and restoring the scholarly freedoms of academics whose intellectual rights have been threatened" (Adebayo, 2022, p. 1818), the effort reported here falls outside the domain of these formalized initiatives.

The four authors' perspectives are woven together to describe the initial period of the urgent evacuation of Afghan scholars starting in August 2021. Taken together, these perspectives provide a novel view of the crisis and the role of U.S. academic institutions and faculty by bringing together Afghan and American. academic perspectives. Throughout the chapter, each author offers their own viewpoint as the evacuation process unfolded during a time period when many in the United States were seeking a way to lend support to Afghan lawyers, judges, and scholars, and some academic institutions were able to find ways to open their doors to Afghans forced to flee.

After an overview of the Afghan crisis situation that prompted the effort to create visiting positions for Afghan scholars at-risk, Professor Rosenbaum anchors the chapter by providing context for his engagement with Afghan scholars. Professors Hashimi and Khalili share brief remarks about their own escapes from Afghanistan. Professor Finger offers her perspective as a faculty member following Professor

Rosenbaum's lead to place scholars at-risk as visitors, which was a significant step toward securing evacuation and arrival in the United States. The account of the Afghan scholars' departures and arrivals highlights issues, obstacles and a path forward. By centering on the perspectives of the scholars at-risk, together with those of U.S. faculty during this challenging time, and drawing from the literature on the historic and contemporary émigré scholar experience, the chapter reflects on collective lessons learned, at both the individual and institutional levels, especially in light of the likely need to replicate similar situations for other scholars in the future.

## Taliban 2.0 Takeover

In August 2021, many Americans were riveted to their screens–on TV, social media and other platforms-–in disbelief. Multiple messages were arriving via email, Facebook, WhatsApp and later via Telegram Messenger and Signal. Our former Afghan students and contemporary law colleagues were in danger. The Taliban was steadily advancing on Kabul and it was only a matter of days before the U.S. military would evacuate. (Mobasher & Qadam Shah, 2021). With the undemocratic rulers' return to power and their ultra-orthodox interpretation of Islam, the ongoing reconstruction and reform process was halted. A challenging, unclear path started to take form and tens of thousands of Afghans attempted to leave their country out of fear of the Taliban's totalitarian rule. (Qadam Shah & Rosenbaum, 2023, pp. 221-22).

The Taliban returned to reign with neither a clean slate nor clean hands. There were many incentives for reprisals against secular, Western-educated, and American-affiliated intellectuals, given their public profiles and past or present associations with the U.S. or international NGOs. Several had published in English or Dari during the previous decade on subjects deemed anti-Islamic or pro-Western. Many Afghan legal academics had addressed issues such as political participation of women and national development in a post-Taliban era. Some were on the law faculty at the American University of Afghanistan, an institution modeled on American higher educational standards–and the target of a lethal bomb attack by suspected Taliban operatives at the time the former self-styled emirate government was ploddingly plotting its return to power.

## G. HASHIMI

The United States played a key role in the political and economic development of Afghanistan from 2001 to 2021. Since the fall of the Taliban regime, "impressive amounts of moneyha[d] been spent in the effort to 'restore' or 'develop' Afghanistan…" (Eddy, 2009, p.1). The February 2021 Agreement between the U.S.

and the Taliban increased concerns among my compatriots, given horrific memories of the Islamic emirate's earlier governance from 1996 to 2001. We expected the United States would not leave us without support and would avoid unilaterally handing power over to the mullahs. The agreement led to the Taliban's swift return to power, erasing the progress made during the previous 20 years. On August 14, while conducting an interview of a prospective employee with the Civil Service Commission, my colleagues and I received news that the Taliban insurgents had entered Kabul, leading to cancellation of the interview as we rushed to find safe places to hide. My colleagues and I quickly embarked on our journeys homeward. Due to road congestion, the typical 40-minute drive from the Commission to my residence turned into an agonizing four hours.

## S. ROSENBAUM

For more than ten years, a critical mass of Afghan legal scholars had devoted their professional lives in teaching, training, research, and advocacy with new methods and on new causes. They had graduated from University of Washington or Ohio Northern University with hands-on knowledge, technical understanding of teaching methodology, and an esprit de corps developed during their time together in the United States. They had since designed interactive course plans for subjects ranging from lawyering skills and civil procedure to international criminal law, women's rights, and ethics. The curriculum was being normalized. They were joined by legal academics, lawyers, and jurists also educated in the U.S., United Kingdom, Australia, India or in Europe.

A decade earlier, I had been part of a team of UW clinical faculty who conducted a comparative law tutorial for LLM students from Afghanistan, under the auspices of a U.S. State Department-funded legal education support program, LESPA. Ghazi Hashimi, a Kabul University law professor and criminal defense attorney, was one of my students. Our goal was to lay the groundwork through skills-based, interactive education, to reform the law and Shari'a faculties pedagogy that had changed little since the days before the Soviet invasion. (Choudhury, 2014, pp. 259-61; Swenson & Sugerman, 2011). We wanted to ensure that clinical education—broadly defined and decisively organic—would not be divorced from the rest of the curriculum, but rather be endorsed and embraced by their faculty colleagues, university administration, and Ministry of Higher Education. Our students in the tutorial were the people responsible for piloting this new curriculum and they needed help to do so. (Rosenbaum, 2016).

Upon its return to power, the clear motivation of the Taliban directorate was to root out any instruction, examination or vulgarization of material and messages that fell outside the confines of a fundamentalist Islamic sect.

## G. HASHIMI

During my time in the United States, I had observed the active engagement of women in both the public and private sectors. During my youth in a Nangarhar province village, education for girls was not encouraged. When I returned to Afghanistan in 2013, I was motivated to prioritize the empowerment of women, particularly in the area of education. I undertook a number of projects, with a focus on female students, many of which were made possible by support from NGOs such as the Asia Foundation and International Legal Foundation and from the U.S. and other Western governments.

These activities included a capacity building program for both male and female students at Kabul University's law and Shari'a faculties. I was also an Associate Professor in the University's legal clinic, where I taught various litigation skills and helped female students with professional client interviews in family law cases. I served as a moot court judge and prepared teaching materials for academic Legal English courses conducted by LESPA, the legal education support program which had sponsored my graduate studies at UW. Many of the participants in moot court and LESPA were women.

As a senior specialist in the Independent Administrative Reform and Civil Services Commission, from 2018 until the collapse of the Government in August 2021, I helped in hiring national and provincial administrative directors who were instrumental in mitigating corrupt practices. During my two years with the Commission, I noticed that many young people were keen on entering government service through open competition and that the number of individuals who held master's and doctorate degrees in Afghanistan was increasing.

## N. KHALILI

After obtaining my bachelor's degree in Shari'a law from Herat University, I served as the first female prosecutor in my native Ghor province, prosecuting high profile gender-based cases involving death by stoning, honor killing, child marriages, domestic violence, extreme cases of violence against women, and homicide. The threats to me and my family from the Taliban, and targeted killings against prosecutors, impeded our progress, as the Taliban insurgency and political control had never actually subsided in many parts of the country after the demise of the Islamic Emirate in 2001. Nevertheless, I continued to provide a window for women to pursue perpetrators of crimes against them.

I not only wanted to seek justice, but to encourage girls and women to feel safe and to study and work outside the home. After the collapse of the Emirate, people in Ghor province and elsewhere were witnessing the blossoming of democracy and the incorporation of human rights values. They saw the first school for all boys and girls and the reopening of universities. A new generation in Afghanistan got a chance to get a peek into the modern world.

In 2017, I was selected for a scholarship funded by the U.S. Department of State at Ohio Northern University Pettit College of Law to pursue an LLM in democratic governance and rule of law. This was the chance of a lifetime, although it was very much against the custom for a young woman to go abroad for education and be accepted back into society afterwards. This was particularly true in Ghor province, despite having the support of my family. With a master's degree in hand, I returned to Afghanistan in 2018 with a more informed understanding of human rights—and had a chance to make a difference.

Appointed as the Head of Prosecution for Juvenile Protection, I partnered with UNICEF to draft a training curriculum for juvenile prosecutors, as most prosecutors were not fully trained under the new law to differentiate between juvenile and adult offenders. Later, while on staff at the Attorney General's office, I was selected to develop a capacity building plan. I also lectured part-time on human rights courses at Rana and Fanoos Universities in the faculty of law, which informed my development of a curriculum for training prosecutors. In 2020, I was asked to serve as director of a State Department-funded uniform case management system, so that any information needed by prosecutors or clients could easily be provided with the click of a mouse.

## DEPARTURES AND ARRIVALS

### S. Rosenbaum

The listservs for the Association of American Law Schools (AALS) Clinical Legal Education Association faculty and for international human rights clinicians were networks that I used widely to share information, as I anchored the effort to facilitate the evacuation of Afghan scholars who had trained in law and democracy. Many law clinic faculty responded and ultimately, scores of U.S. universities became host schools for Afghan scholars at-risk. Typically, these were law schools, sometimes with a joint appointment in another campus department or school, or sometimes in partnership with other law schools in the vicinity. This was a convenient way to share costs and spread the benefits. My own school's Dean, Erwin Chemerinsky, an influential legal academic, lawyer and newly elected President of the AALS, put out an email to all law school deans urging them to take on an Afghan "visiting scholar" or a "fellow" or some other specially designated, short-term faculty or staff member.

As a contingent faculty member, I was not immersed in the ways of institutional leadership. I figured that law school administrators are constantly approached about a number of worthy causes and that this particular "ask" would compete for limited resources. (Lässig, 2017, pp. 788-89). Or perhaps it would be met with polite indifference. It soon became apparent that some prospective schools were not appreciating the need to act quickly and were applying standard protocols for a permanent, tenure-track hire. The tension between humanitarian and utilitarian responses to attacks on academic freedom has been noted by earlier scholars, i.e. the need to respond to the persecution of scholars balanced against establishing a given scholar's potential value to the host institution. (Konuk, 2020, p. 272). My pitch was to convince would-be hosts that expediency, flexibility and safety were our paramount concerns, tilting the effort in a pragmatic and humanitarian direction. As we moved further from the withdrawal of American troops, with borders closing and visas delayed, I realized that departure from Afghanistan or a third country was not imminent for many scholars, and their arrival was not subject to a predictable timetable. With the passage of time, I had the impression that Russia's invasion of Ukraine in early 2022, as the conflict *du jour,* was beginning to eclipse the crisis and chaos in Afghanistan and the continued need for American material, financial and moral support for Afghans in-country and out-of-country.

In terms of our resources, it helped to have assembled a national team of *pro bono* immigration law firms and to suggest that costs could be offset by local in-kind assistance, or joint arrangements, as discussed above. The role played by State Department alumni and Karen Hall, Ohio Northern University's former director of the Democratic Governance and Rule of Law LLM Program, was also critical. FPJRA was able to compensate Hadley Rose and Habib Rahmani for their efforts in helping get placements, get people out of Afghanistan and provide some financial assistance for travel and incidental costs.

## D. Finger

As Law Clinic Director at Loyola New Orleans and an active member of the national law clinic faculty community, I had been considering the situation in Afghanistan from a distance and trying to determine whether there were ways that clinic faculty in the U.S. could provide some support. I had never been involved in related justice issues, but felt compelled to consider ways to support Afghan colleagues, although I knew that I did not have my own established networks. The Jewish concept that anchors my work, *tikkun olam*—repairing the world—continued to come back to me as a guidepost. The news was getting worse every day for Afghan women. On August 14, 2021, I received the following e-message from Professor Stephen Rosenbaum, which he sent to law clinic faculty throughout the U.S.:

Dear Colleagues,

*Our fellow legal clinicians and quasi-clinical faculty in various Afghanistan universities are trapped. Their past/current affiliation with US or other Western universities, NGOs and IGOs taints them in the eyes of Taliban operatives and other insurgents.*

*These law faculty and NGO activists do not qualify for the much-publicized Special Immigrant Visa (SIV). Despite bureaucratic and logistical hurdles, many are seeking a way out of country. I'm writing to see if your law school or university has a liaison for possible placement of Afghan evacuees (or those from Myanmar, Syria, Yemen, Ethiopia…) in positions like Visiting Scholar or some other low-cost post that allows them and their families to escape a fate that ranges from stigmatization to punishments to death. The Council for At-Risk Academics (www.cara.ngo) does its own vetting and has limited placements.*

*I'm interested in joining forces with others who have leads and/or fresh ideas. Please contact me off-line at above email address.*

*In solidarity,*

*Steve*

That email was the catalyst I had been waiting for, and I wanted very much to pursue the proposed action in collaboration with Steve whom I knew had long standing international experience. I forwarded Professor Rosenbaum's message to the Loyola New Orleans Dean, Madeleine Landrieu, who responded quickly with a message of support; we soon engaged other University leadership to confirm institutional support. From there, Loyola began what became an eight-month process to secure the arrival of an Afghan at-risk scholar. Professor Rosenbaum grounded this effort for our institution and many others; we were in touch daily for months as we worked through visa issues, evacuation logistics, and more. We also worked alongside the Fragomen law firm, which volunteered its considerable expertise on immigration matters on a *pro bono* basis. Despite access through many networks, the process and policies for travel to the U.S. were anything but clear. Most notably, with family members remaining in Afghanistan, the scholars had difficult decisions about the best immigration path forward for themselves.

Professor Rosenbaum facilitated the framework for discussions with the institution by creating and sharing a punch list of "Commitments of the Host Institutions" that enabled me to have increased clarity in discussing expectations with College of Law and University Leadership.

Our vetting of visitors gave priority to candidates who were: (1) Western-trained (usually U.S.) legal academicians, jurists, or NGO advocates; (2) holders of LLM, JSD, or PhD degrees issued by an accredited Western higher education institution; (3) at risk for stigmatization, punishment, economic sanctions, and/or death; (4) able to travel solo (with other arrangements for family, if any); and (5) willing to seek supplemental means of financial or in-kind support. We also prioritized candidates who had a justice, rule of law, or access to justice orientation and had made a *bona fide* application (i.e., not subject to sham or cronyism).

With this structure in place, my discussions with the Loyola leadership team followed a reasonably concrete track of what was required of us with the vetting of visitors in mind. This structure further enabled streamlined discussions that ultimately resulted in a commitment of funding for one year in a partnership by the Loyola New Orleans College of Law and the College of Arts and Sciences.

It took only one week after Professor Rosenbaum first messaged us to move the process sufficiently forward for engaging a visiting position. We received the CVs of five candidates. The challenge of this process was not lost on me; I was keenly aware that the scholars at-risk were not necessarily able to choose the location, institution or community that they would be moving to. I convened a group of faculty members from departments across the University to engage with the selection process that resulted in our institution's match with Professor Negina Khalili. This collaborative process was key to making the decision, which I was not equipped to do alone and perhaps more importantly, to developing highly interested stakeholders especially given that our institution did not have a pre-existing scholar at-risk committee. The faculty members from these different departments were early, supportive connections for Professor Khalili once she arrived and the process of building a community became a priority.

Many questions remained on how we would provide sufficient resources, physical and emotional, to our newly arriving colleague. However, given the exigency of the situation, we moved forward to secure Professor Khalili's arrival to Loyola New Orleans without having answers to everything immediately.

## N. Khahili

When the government of the Islamic Republic of Afghanistan collapsed in 2021, I was working at the Attorney General's Office, in charge of the elimination of violence and harassment against women. It was as if someone had just turned off the lights and there was sudden darkness, as insurgents were roaming the streets of Kabul with their guns pointing and shooting at innocent people. My journey from the Kabul Airport to the U.S. started on August 21, 2021 when, in the span of two weeks, the Taliban took control of the entire country and began to zealously dismantle all democratic institutions.

With the help of Loyola University and Professor Davida Finger, and the mobilization of law schools abroad, I was able to evacuate to safety after three long and desperate nights on the Afghan airport tarmac. Those days were truly a disaster–without food, sanitation,or and safety. Like many of my fellow citizens who were trying to escape the wrath of the mullahs, I found safe passage at first in a third country. I recall seeing the pile of hijabs at the airport left behind as we were able to board flights out of country. I spent seven months in Qatar and then was in limbo in Albania for months before a United States visa was issued to me. Without the support of people like my Ohio Northern mentors, State Department leadership, and Members of Congress, many would have lost their lives in the ensuing violence wrought by Taliban militants.

## G. Hashimi

The evacuation process was poorly managed and lacked a clear plan, leaving many Afghans waiting for months in hotels for flights. Along with other members of the Afghan US law alumni association (AALA), I maintained on-line contact with our U.S. counterparts at FPJRA (Friends of the Public-Private Partnership for Justice Reform in Afghanistan). After I fled my office on August 21, I received an instruction a couple days later to converge near the Kabul airport and attempt entry through the American military base. I opted for a taxi, urging the driver to avoid Taliban checkpoints. Despite venturing through side streets, the trip to the airport was harrowing. Upon arrival, I encountered fellow professors from Kabul University, as well as other AALA members, all summoned to the airport. The surroundings were chaotic, with thousands of Afghans vying for entry to the airport despite the daunting presence of Taliban militants. The struggle to reach the military base proved nearly insurmountable for myself and a fellow professor. We were advised by FPJRA to return home and await information on the next flight. Many weeks later, on October 23, a potential flight failed to materialize, due to airport congestion and technical glitches and we received a FPJRA directive to attempt a risky journey to Mazar-i- Sharif in the north.

The Taliban continued to obstruct scheduled evacuation flights. As the Russian-Ukrainian war unfolded, a U.S. contact informed us of a temporary halt in flights and suggested relocating to a third country. Obtaining visas was challenging and expensive. Neighboring countries like Uzbekistan, Tajikistan, and India had stopped issuing visas for Afghan citizens; the only options were Iran and Pakistan. I eventually arrived in Pakistan in February 2022. However, administrative processes at the

U.S. Consulate prolonged my visa application for another 15 months, with little explanation for the delay. Struggling to sustain myself in Pakistan with an expired visa, I reluctantly returned to Afghanistan to secure a new visa before re-entering. The experience with Pakistani police was marked by unwarranted demands for bribes from Afghan citizens. I did not arrive in Chicago until July 2023 to begin my fellowship at Northwestern University, which had early on selected me as one of their visiting scholars. My personal experience resonates with what has been documented in the literature to the extent that "[v]isas raise hope for oppressed and suppressed scholars, but visa politics can just as easily shatter hope and life." (Adebayo, 2022, p. 1829).

## D. Finger

The need for clear expectations throughout this process cannot be overstated. In this crisis context that included linguistic, cultural, social, and a myriad of other barriers, it was of utmost importance that expectations be managed through timely communications that remained as clear as possible to secure the eventual arrival of Professor Khalili. Steve Rosenbaum and I remained in close touch for many months to track events as they unfolded with Professor Khalili's arrival and to troubleshoot, as we encountered barriers and discovered new needs. I also spoke with other U.S. law faculty who had responded to Steve and were embroiled in similar institutional and immigration issues that we considered together, sharing notes and ideas for immigration issues, onboarding upon arrival, and building communities of support. It appears that even the formal scholar at-risk programs are subject to this "academic immobility conundrum" insofar as physical movement is hampered between countries predominantly in the Global South and Global North, due to travel restrictions and changes in visa governance and politics. (Adebayo, 2022, p. 1829). There are also difficult legal and personal decisions faced by émigré scholars in adjusting their immigration status. (Yarar & Karakaşoğlu,, 2022, p. 1633; Baser & Öztürk, 2022, p. 9).

To further the process for Negina Khalili, Loyola provided her with a written contract with a start date that eventually was pushed back to track her eventual arrival date in the U.S. from Albania. I was mindful that even the ability of Professor Khalili to access documents via email had its own challenges. Moreover, she was accepting an appointment sight unseen at an unfamiliar institution and in an unknown location. I was aware of those painful challenges throughout the process to secure Professor Khalili's arrival.

Loyola had arranged for a position that focused on speaking at the College and in the community. The semester after her arrival, Professor Khalili also taught a human rights course to undergraduate honors students, which she spent considerable time designing and planning during the time she was also acclimating to New Orleans.

While my institution had committed to a paid, year-long visitor position, it was clear that broader support needs would be required well beyond that timeframe. There was much to consider in terms of basic needs, housing, medical, and integration in the University and community. I was able to confirm with the local Catholic Charities office that Professor Khalili would be accepted as a refugee client and as such, afforded the benefits of case-management through that non-profit agency, which included a critical piece: access to an immigration attorney without cost.

## S. Rosenbaum

I hesitate to use the word "rescue." (Adebayo, 2022, p. 1832). In the end, this kind of effort is driven largely by personal and professional connections. (This is what historian Simone Lässig referred to as "a sense of professional solidarity on the part of their [at-risk scholars'] foreign colleagues and the wide respect for the academic system that had trained them." (Lässig, 2017, p. 787). For me, my time as a visiting faculty member at UW allowed me to bond with Afghan students who became peers once they returned home. Unfortunately, this was not enough to motivate almost all of my UW colleagues, much less the administration, for whom the now-defunct legal education support program was more of a distant memory than an urgent cause.

In the midst of *ad hoc* recruitment of hosts and writing letters of support, I became aware of the niche of organizations supporting at-risk scholars with financial grants and suitable out-of-country academic placements. I was profoundly moved in reading about the genesis of these organizations, many of which had saved the careers and lives of Jewish professors who had been dismissed, ostracized, and eventually expelled and/or murdered by the Nazi regime. That resonated with me deeply, given my personal background and identity as a Jew.

Sociologist Kudus Adebayo has written about the importance of "solidarising politically with exiled academics," criticizing "the limited engagement of rescue organisations with the political undercurrents that led to academic threat and displacement in the first place." (Adebayo, 2022, p. 1827). In the same vein, political philosopher Seçkin Sertdemir Özdemir has urged that "[h]umanitarian policies based on the compassionate approach must be superseded by acts of solidarity grounded in the sense of belonging to a common world. Only then can the political problems that give rise to such exile be confronted." (Sertdemir Özdemir, 2021, p. 950).

When there is a crisis or tragedy in the community of international clinical law teachers, my first reaction has been to convey solidarity with faculty and lawyers in places where I have some connection. As advocates and activists, clinicians may find ourselves "more *of* the Academy—than *in* the Academy occupied by our fellow doctrinal faculty members—having to distance ourselves from the institution, with the need to immediately react to real world events on issues that touch our professional lives." (Rosenbaum, 2022, p. 55). I have since been called upon to reflect on the possible negative impact of the "victim savio[u]r" narrative. (Sertdemir Özdemir, 2019, pp. 943-46; Adebayo, 2022, pp. 1829-30 & 1832). Following a politically motivated dismissal by the Erdoğan regime from her university post, one exiled Turkish academic wrote of the media's characterization of her flight to Western Europe:

*"I was and am not a refugee. . . . The western gaze needs new victims, but am I that victim? Well, I am a 'scholar at risk', that much is true enough. ... Europe needs me as a victim to assure itself that it is indeed 'the saviour' that it has imagined itself to be." (Sertdemir Özdemir, 2019, p. 944).*

Inasmuch as we were not eager to position ourselves as saviors of victims of an authoritarian regime, we felt compelled, in our appeal to US university administrators, to rely on stories of desperation for those remaining in Taliban territory. (Sertdemir Özdemir, 2021, pp. 942-43; Adebayo, 2022, p. 1825).

## Adjusting to New Realities

Political and international security issues in Afghanistan continue to be important to the Afghan diaspora and the population living under a renewed restrictive government, and also to strategic analysts, journalists, and historians throughout the globe. Emigrant multilingual, multicultural academics, already familiar with the literature and politics, will have a head start in informing academic peers, media outlets, and international and regional policy makers. Hopefully, some of the analyses and messaging will also find their way into clandestine on-line or broadcast information accessible to compatriots remaining in their homeland.

Transitions and relocation are not always easy. After her flight from Nazi Germany, historian and philosopher Hannah Arendt described the plight of exiled academics as having "lost" their "occupations" and "confidence" that they were "of some use in this world." The loss includes language and "the naturalness of reactions, the simplicity of gestures, the unaffected expression of feelings..." These are left behind on the other side of the "rupturing" of one's private and public-professional life due to displacement. (Newman, 2020, p. 286 (citation omitted)). In more concrete terms, "scholars do not all arrive in their respective places of relocation with the same level of preparedness and readiness to allow for proper integration into research life. Some may have lost their means of identity and life's work… without any hope of recovering them." (Adebayo, 1830).

## D. Finger

Once Professor Khalili arrived in New Orleans, it became clear early on that we would need to extend support beyond one year, especially while her asylum petition was still pending. With many Afghans arriving in the U.S. simultaneously in a relatively short period of time, a permanent job search for her would not be easy. Moreover, ensuring a meaningful scope of work at our institution took on increased importance as Professor Khalili's long-term career options and life plan more generally in the United States became more of a focus. For her second academic year, we successfully applied for a grant that focused on supporting newly arrived Afghans; this covered part of Negina's salary. In committing to working with scholars at-risk, institutions must consider whether multi-year support is possible and provide clarity about that as early on as possible.

Ideally, institutions would establish a scholar at-risk committee to delineate policies to keep expectations clear and known for all involved, with established parameters for available funding and the timing and focus of visiting appointments. This seems to be in keeping with the recommendation of Exile Studies scholars to develop "pre-crisis networks" through the Academy, personal contacts, professional associations and foundations. (Lässig, 2017, pp. 784-86). A pressing question that comes back to me again and again is how are Afghan refugees securing permanent employment in the U.S. and how will they find work in their chosen field.

## N. Kahili

From the start, my vision has been to promote human rights with a particular focus on the most vulnerable, i.e., women and children. Loyola has provided me with a chance to work towards my goal and connect with people working towards the same. In addition to lecturing on human rights subjects at the University, I also enrolled in the LLM program. I am also on staff at Loyola's Women's Resource Center, partnering with wonderful people to advocate for women's rights and gender-based violence prevention in New Orleans.

Living as a refugee and away from family comes with its difficulties and I hope one day I will have them join me in the United States. Adjusting to a new life in a new place, and finding new friends and social support, are the challenges any refugee faces. But, with the help of Loyola and friends I am grateful to minimize the difficulties and to settle. I have not witnessed much Islamophobia here and have found America to be welcoming of all faiths. Catholic Charities has assisted me in many ways without asking my religion. After my arrival I stayed at first with Professor Finger and her family. While they celebrated Passover, I was fasting during Ramadan and praying at a local mosque.

It is still disheartening for me as a woman who fought for twenty years for democracy and rights in Afghanistan to see my people, particularly girls and women, in this deplorable condition. I don't have any plans to return to my country as long as the Taliban is in power. Maybe I will go back when there is a government that supports women's rights and other civil and political rights.

## G. Hashimi

Having had prior experience in the United States and the University of Washington academic culture and maintaining a network with my Afghan peers made the transition a little smoother. (Yarar & Karakaşoğlu, 2022, p. 1632; Lässig, 2017, p. 796). But, it is easy to underestimate the emotional toll that comes with emigration, leaving behind one's home and extended family, a culture and a way of life. Even for myself, who was accustomed to living in the United States. In effect, I am "starting over" as a lawyer and an academic. Currently, I am a Clinical Fellow with the Bluhm Legal Clinic at Northwestern University and a Visiting Scholar with the University's Buffett Institute for Global Affairs. Grateful for the opportunity and for my mentors, including Professor Juliet Sorensen, I am conducting research on embezzlement and other corrupt banking practices in Afghanistan and the degradation of law in a Taliban 2.0 regime.

While promising, the resettlement of my family in America requires a period of adjustment to American culture and the environment—particularly for my wife and children who had never been to the United States. At the beginning, we encountered difficulties in finding an affordable apartment due to high rents and my lack of a rental history and credit score. Despite these challenges, the fellowship program and the integration of my children into local schools have been valuable. School authorities have accommodated our religious values and for the most part we feel safe as Muslims to freely practice our faith.

English language proficiency poses a significant challenge. My wife misses her relatives in Afghanistan, adding an emotional layer to adaptation. She has since found solace and companionship in a new Afghan friend. The road ahead for language acquisition and cultural assimilation may be long, but perhaps traveled more easily with a supportive friend. While we remain hopeful about a prosperous and free Afghanistan, we acknowledge that the timeline for improvement may be long. My family and I do not contemplate returning any time soon. I can understand how for some academics who receive a scholarship, the migration decision process can instill some confidence. Rather than leading to a "disrupted" or "disordered" life, a scholarship might bring "some hope for a life in 'harmony' or 'order'." (*Id.*). I maintain optimism about overcoming any obstacles in our pursuit of a stable and fulfilling life in the United States.

Many Afghan legal scholars who had returned to the Islamic Republic after their education abroad were at ease with languages and a global lifestyle. It is these qualities which made them a particularly attractive target for the Taliban and other rogue insurgents, but also allow them to adapt to life outside Afghanistan, explore newer research media and venues, and to contribute to the diversity of a new academic institution. Ghazi reminds us, however, that it is easy to "gloss[] over the profound difficulties encountered by scholars upon arrival in their respective host countries." (Konuk, 2020, p. 276). In addition to the difficult emotional and practical adjustment, there are "disciplinary differences, the loss of access to the archives and libraries they left behind, ongoing uncertainties regarding citizenship, the barriers presented by an unfamiliar language" and possibly cultural or religious prejudice or isolation. (*Id.*)

Neither Ghazi nor Negina has experienced much animus against them as Muslims or as immigrants or people of color. Yet, they may have Afghan or other Global South emigrant peers in the U.S. or elsewhere in the West "who fall into the category of the politically exiled, rather than the internationally defined legal categories of migrants, refugees, displaced or stateless persons or asylum seekers." (Sertdemir Özdemir, 2019, p. 946). Perhaps for them, their focus in exile is "not on the hope of building a new life in the hostland, but on return to the homeland." These peers may also be subject to anti-immigrant populism or other hostile environments and sense that their status is temporary and insecure. (*Id.*).

## The Scholar Rescue Phenomenon

The literature about prior periods of scholar rescue reveals trends that are both similar to, and distinct from, our own experience, especially in terms of the scale and aftermath. Earlier accounts of academic migration, under the rubric of Exile Studies, concentrated heavily on intellectual elites. Researchers "were generally less interested in the many 'nameless' émigrés" such as the Afghan scholars who are the subject of this chapter. (Lässig, 2017, pp. 770-71). Professor Konuk has reminded us that "[t]he collective exile, imprisonment, or killing of dissident intellectuals has been used as a tool by authoritarian and fascist regimes since the early twentieth century." (Konuk, 2020, p. 270). This was the case with the arrest and mass deportation of Armenian intellectuals in Istanbul in 1915 who were eventually deported to Ottoman Syria, ushering in the months-long genocidal killing. (*Id.*, pp. 270-71).

Organized efforts to come to the aid of academics abruptly dismissed or forcibly retired in 1933 from their university positions after the Nazis consolidated power in Germany, are considered to be the inspiration or roots for today's so-called scholar rescue initiatives. (Konuk, 2020, p. 270). The Academic Assistance Council, founded in England, was one such organization. Later renamed the Council for At-Risk Academics (CARA), its objective was to assist scholars who were victims of political and so-called "racial" persecution to relocate outside the reach of the Third Reich. In the United States, the Institute of International Education initiated the Emergency Committee in Aid of Displaced Foreign Scholars, the precursor of today's IIE-Scholar Rescue Fund. Philipp Schwartz, a Jewish professor of pathology, founded the *Notgemeinschaft deutscher Wissenschaftler im Ausland* (Aid Organization for German Academics Abroad) and coordinated rescue efforts from Switzerland, where he took refuge after dismissal from a German university. The organization also convinced the newly secularizing and Westernizing Turkish state to accept a number of purged professors. (Konuk, 2020, p. 273-74; Newman, 2020, p. 289).

In contrast to the *ad hoc* and individualized initiatives by academics and administrators undertaken to place Afghan scholars in the United States, the 1930s migration relied largely on self-aid organizations, private foundations and colleges and universities (at the institutional level), and to a lesser extent on the Roosevelt Administration and establishment media outlets. (Lässig, 2017, pp. 776-78). In both instances, the sought-after university placements were meant to be temporary, but not necessarily for the same reasons. Anti-semitism and nativism in the U.S. was still prevalent in academia in the earlier part of the 20th century and there was the belief that the expelled scholars would soon return to their posts after a quick defeat of Hitler. For the Afghan émigrés, however, it was almost exclusively a question of institutions' financial capacity and willingness to deviate from standard curricular planning, recruitment and hiring practices.

Relocation from Nazi Germany, Austria and other annexed territories was motivated only in part by humanitarian instincts. It was also an opportunity for fledgling faculties, including historically black colleges, to fill their ranks in the sciences and humanities with "scientific capital" and for more established universities to hire European luminaries (Konuk, 2020, p. 271; Newman, 2020, p. 294; Lässig, 2017, pp. 772 & 779). This utilitarian or strategic "brain gain" motivation, which often obscured any empathetic appeal, was absent in the case of the Afghan scholars or of only incidental interest to university administrators. (Lässig, 2017, pp. 770-72). The latter émigrés who were considered for positions in the U.S. were neither well-known, nor necessarily filling curricular gaps in the American professoriate and were offered only temporary positions. At best, some of the Afghans had an alumni connection with the University or were able to teach a class in Islamic law, comparative law or Central Asian studies. Xenophobia and anti-immigrant sentiment notwithstanding, humanitarian arguments in support of scholar rescue continue to resonate in the West. "A memory culture shaped by a sense of responsibility imposed by the Holocaust" allows for humanitarianism to prevail over concern about excellent scholarly qualifications. (*Id.*, p. 794).

In a celebrated initiative, the president of the maverick New School of Social Research founded the University in Exile as a home for the displaced German Jewish and anti-fascist scholars and their families, invoking the image of Byzantine scholars fleeing from the Ottomans. With the help of philanthropists, the school facilitated assimilation in the U.S. of the "faculty as a whole," and not simply an individual scholar. Eventually, the school became a sanctuary for academics from other countries. (Konuk, pp. 273-74; Newman, p. 290). The work of these intellectuals had a profound impact on the shape of the United States Academy in the 20th century. (Newman, p. 291). Some of the professors remained on the faculty for years and others moved on to other American universities. In recent years, the University in Exile morphed into a globally expanding consortium of colleges and universities in more than 15 nations, mainly in the Global North. Each institution commits to hosting at least one threatened scholar every year. More than 150 exiled scholars from 24 countries have been assisted. (The New University in

Exile, https://newuniversityinexileconsortium.org/history/mission/our-mission/, 2024).

There is no Academy in Exile *per se* in the United States, populated by expatriates, to serve as a faculty-in-waiting for the day the Taliban is once more ousted. There is, however, a network of Afghan émigré professors, lawyers and judges who are part of an Online Education Academy offering Afghan women access to education. Originally formed in 2019 as the Afghanistan Law, Shari'a, and Political Science Association, ALPA is also committed to holding annual conferences for Afghan law and political science professors. The courses, which span a variety of disciplines, including law, human rights and journalism, are "solidifying ALPA's position as a beacon of learning for Afghan women." (ALPA (in Exile), www.alp-a.com, 2024). Another iteration of a home for displaced faculty is the American University of Afghanistan. For many years the premier (private) university in Afghanistan, since its founding in 2006, AUAF quickly relocated to Qatar after the Taliban 2.0 takeover. Its law faculty has included a number of Western-educated graduates. The University continues to educate hundreds of students in Afghanistan and in the diaspora, on-line and in person–with a strong commitment to critical thinking, civic engagement and equal opportunity for women. In its vision statement, the University declares: "Until the day we return to our campuses in Kabul, AUAF lives in the minds of its students and the accomplishments of its alumni – who prove what we have always believed, that education will prevail." (AUAF, (https://auaf.edu.af/, 2024).

Complementing the Academy in Exile model is the Academy in Exile (AiE), consortium in Germany, co-founded in 2017 by Professor Konuk. Like the New University in Exile, it is a joint initiative of several universities, institutions and foundations hosting scholars at-risk because of their academic work and/or civic engagement, "enabling persecuted scholars to collaborate with each other" through long-term fellowships, emergency stipends, residencies and guest professorships. (AiE, https://www.academy-in-exile.eu/, 2024). Other German higher education institutions and foundations have regularized placements and compensation and have timely reacted to crises in an organized fashion–whereas most of their U.S. counterparts have responded to irregular and informal appeals to place and fund at-risk scholars. Would-be hosts in Germany sometimes take the initiative to apply for stipends on behalf of their endangered or colleagues abroad. In other instances, exiled scholars have utilized their own personal contacts in Germany or were aided by bilateral academic channels or self-organized solidarity networks. (Yarar & Karakaşoğlu, 2022, pp. 1617 & 1631). Even for those scholars who have no personal network or clear support abroad, "the academic capital appears as the life saver to hold on." (*Id.*, p. 1632).

It is ironic, or rather, fitting, that Germany, which once expelled its vilified and censured professoriate en masse has become an important migration destination for émigré academics. (Lässig, 2017, p. 794). As for Turkey, there is only irony: The nation that took in many dismissed German Jewish scholars in the 1930s has been purging its own enemies from the Academy decades later.

## Going Forward: The Afghan Education Vaccuum and the Legal Academy in Exile

Only a year after the public burning in Berlin of university library books written by Jews and "un-German" authors, and the wholesale dismissal from academia of Jews and regime opponents, the Third Reich Education Minister is reported to have asked his dinner companion how was his university mathematics department now that it was free of "Jewish influence." The professor allegedly replied, "There is no mathematics [there] any more." (Newman, 2020, p. 288 (citation omitted)). Decades later, more than two years after the return of Taliban 2.0, the higher education ministry's Department of Invitation and Guidance directed private universities to remove all books from the "Republican era" and purge their libraries of any books that were contrary to the Hanafi (Sunni Islamic) school of jurisprudence. (*Kabul Now*, 2023). Around the same time, the Acting Minister of Higher Education announced, incongruously, that the emirate was committed to improving the quality of higher education and "plans to enhance the capacity of university professors." He assured Afghans pursuing doctoral or master's degrees (at home or abroad) that there were now funds available to pay them to teach. Meanwhile, as an indication of the wide scale faculty exodus, a student told a reporter, "We want experienced professors because the professors who have been appointed for these two-year periods have very little teaching experience." (*TOLOnews*, 2023)

The ranks of Western-trained academics in Afghanistan's faculties of law and Shari'a have been depleted, whether through dismissal, resignation and/or exile. Kader Konuk of the Academy in Exile has "inquire[d] into the ways in which knowledge may be preserved and transferred in exile" and what can we learn from past experiences to counter "the great forgetting wrought by totalitarian and fascist governments?" (Konuk, 2020, p. 270). It is important that the body of legal knowledge and teaching methodologies brought to America with the Afghan legal scholars not be lost. And, to the extent feasible, the learning appetite and skills-building capacity of faculty and students trapped in their totalitarian-governed homeland be "nourished" from abroad in any way possible.

## D. Finger

Girls continue to face extensive barriers in terms of access to education in Afghanistan and there has been a deterioration in the quality of instruction and infrastructure. (Special Inspector General for Afghan Reconstruction, 2023, pp. 5-15 & 19). Whether or not the Taliban genuinely laments the loss of the Western-educated law and Shari'a professoriate, it is clear that the university faculty ranks are shrinking in many fields and educational quality has declined since August 2021. Many professors fled after the Taliban takeover and those who remained have been subject to threats and intimidation. Universities have been forced to decrease the number of secular courses. Administrators and teaching staff are leaving their jobs and enrollments are declining, in part due to restrictions on secondary and higher education of girls and women. All of this contributes to a reduction in teaching capacity. (*Id*., pp. 9-12).

## G. Hashimi

As Afghans navigate the uncertain path ahead, it is crucial to recognize the potential within the country, and extend a supportive hand. The Taliban have prevented girls from secondary and higher education. The future looks bleak, especially for women. The uncertain future of Afghanistan depends on the international community, which bears moral responsibility for applying pressure on the Taliban to change their approach to governance and human rights. Without this change, there is no future for the younger generation.

The presence of educated Afghans, both within and outside the country, reflects a collective aspiration for knowledge and skills development and presents an opportunity for positive change. Utilizing the material resources and intellectual capital amassed over the past two decades, I am dedicated to contributing to a community of collaboration and assistance among my peers as we try to rebuild the nation and foster long-term development. Together with Afghan colleagues in ALPA, I am teaching on-line legal writing courses in the Dari language. Through ALPA's Online Education Academy, we aim to empower women students with essential legal skills. I am also writing recommendation letters for Kabul Law School graduates who are aspiring to pursue opportunities abroad.

## N. Khalili

Education is key in developing societies, but the Taliban have banned women and girls from attending high school and universities, despite 20 years of investment in the educational sector, including scholarship, and opportunities for Afghan women to continue education in the U.S. and around the world. Although a lot of funding was allocated for women's right to development and welfare, the male-dominated society and over two decades of war make it impossible for a fruitful outcome in the short term. Poverty, insecurity, lack of jobs, and restriction on women and girls made the situation worse. Much more needs to be done.

Through individual mentoring and teaching, I am trying to do my part in preserving and transferring knowledge and skills. Under the aegis of the Afghanistan Prosecution Association, I am assisting prosecutors still in-country and helping those who wish to leave with their visa applications. And, in interviews with national and international media, I can continue to speak freely about the dire situation in Afghanistan and help ensure that events there are not not overshadowed by other world crises. I can speak more loudly now that I am in the diaspora. (Baser & Öztürk, 2022, p. 4 (citation omitted)).

## S. Rosenbaum

A couple years after the Afghan scholar exodus, I joined a network, in partnership with Monash University, to offer free on-line lessons to displaced, dislocated or ostracized populations (Classrooms Without Walls, 2024). For Afghan girls and women, the mission is to create a space to exercise skills in critical thinking, where they can feel free to speak about the constraints they are experiencing under an authoritarian, fundamentalist regime. Notwithstanding Professor Konuk's message about the importance of preserving knowledge in exile and resisting totalitarian edicts on educational form and substance, I asked myself how does on-line teaching change the status quo? What is the objective of a class like this in the context of the Taliban regime when so many opportunities have been curtailed and agency and identity repressed? After agreeing to teach this course *pro bono*, I put this question to some of my former Afghan law students who had returned to Afghanistan to teach in the law schools under the Islamic Republic and then had to flee when the Taliban took over. Here are some of their comments:

These women are (or may be) mothers, and they are sisters and daughters. If they are able to respect and teach the rest of the family (or their children) to respect each other's human rights, that in itself is enough to bring a change in society. Individuals constitute a family and a family is a crucial part of a society.

Understanding human rights and the rule of law can also offer emotional resilience. If they're aware of their rights, these women and girls may feel less isolated and more connected to a global community of individuals who believe in, and fight for, those same rights. Having an awareness of global standards and rights can give them a glimmer of hope that change might someday come or that there are parts of the world where their rights are recognized.

Societies change, and circumstances can shift. If these women ever find themselves in situations where their worlds expand, having a strong foundation in human rights and rule of law can prepare them to be leaders and advocates in their communities.... When opportunities arise, either in Afghanistan or if they find themselves outside its borders, these women will be better equipped to advocate for change.

## EPILOGUE

As noted above, there was not really any tension between humanitarian and utilitarian motivations in support of the at-risk academics from Afghanistan. In placing endangered scholars, few school administrators were incentivized by the potential benefit to the host institution. Still, it must be recognized that these "scholars bring knowledge from the home countries to enrich the intellectual environment of their host institutions." (Adebayo, 2022, p. 1822).

In her case study of academics escaping the autocratic regime in Turkey, on temporary fellowships or grants in Europe, researcher Seçkin Sertdemir Özdemir has written that "politics based in compassionate humanitarianism" may not be "appropriate... when we examine close up current campaigns designed to help persons displaced by political persecution." (Sertdemir Özdemir, 2019, p. 936). While she does not "devalue the social importance of emotions such as pity and compassion and their power to assist our capacity to empathize with others," she has asserted that such "welcoming policies" may lead to "exiled academics experienc[ing] a form of anonymization and marginalization which demands that they acquiesce to their normatively assigned roles as victims to be saved." (*Id.*, p. 937).

Migration and diaspora studies scholar Kudus Adebayo has cast the marginalization or victimization as a product of Eurocentrism or bias against academics from the Global South. (Adebayo, 2022, pp. 1825-26 & 1830). This may be indicated by an "institutional insistence that one's nationality should define the border of one's research and public engagement in the host country…thereby robbing them of their agency to freely choose what to research." (*Id.*, p. 1830 (citations omitted)). It is further reflected in the inaccessibility of scholars publishing in English-language peer-reviewed journals "given the pre-conditions of the higher education system where they originally built their careers" (*id.*, p. 1825) or in "how the publications of at-risk scholars in their mother tongue are dismissed, and in the devaluation of academic titles they carry." (*Id.,* p. 1830).

This is not to suggest, however, that the victim-savior narrative, othering, anonymization or marginalization is the inevitable lot of émigré scholars at their respective host institutions. Nor is it manifested to date in the experience of the scholars profiled in this chapter. In fact, it is reasonable to fill a void on the faculty in the areas of the scholars' present research and draw upon their expertise and lived experience. There are also some college and university positions where teaching capacity is valued as much or more than research potential. Moreover, the emigrant's desires, ambitions and expectations will vary–as will those of their hosts and sponsors.

Scholars Bahar Baser and Ahmet Erdi Öztürk have described the array of definitions and complexities surrounding the terms *exile* and *refugee* as typically tied to "experiences of longing and belonging, nostalgia and sorrow, advocacy, and resistance." (Baser & Öztürk, 2022, p. 7 (citations omitted)). They cite an essay by literary critic and post-colonial studies pioneer Edward Said, describing the unsurmountable but "essential sadness" of the exilic situation and as something "compelling to think about but terrible to experience." Baser and Öztürk ask, however: "Does exile always need to have a negative connotation?" (*Id*).

AiE founder Kader Konuk posits that the most effective scholar rescue model in our times is one with emphasis on the knowledge—rather than on the individual scholar as the agent of that knowledge. It is a model "that foregrounds the conditions and modes of transforming knowledge in exile" insofar as forming cohorts of refugee scholars "facilitate[s] the concentration, absorption, and transformation of knowledge at host institutions." (Konuk, 2020, p. 276). In effect, the creation of academies in exile, whether an independent bricks-and-mortar residency, consortium or on-line institution, it is the dissident intellectuals' knowledge or scientific capital that must be preserved in the host country. And the knowledge is not static.

The New School of Social Research still serves as a model insofar as it "continues to attract distinguished and socially active faculty who challenge long-held theories and push scholarship and social discourse in new directions." (https://www.newschool.edu/nssr/history, 2024). Moreover, the various scholar rescue programs should continue to investigate areas of overlap and particularized expertise "towards the creation of a coordinated response at scale." (Adebayo, 1832). Whether constituted as a single institution, or more likely a consortium, the academy in exile should conscientiously attract scholars who bring existing knowledge but are also amenable to innovation and transformation.

When war, disaster, or authoritarian regime change is visited on places where we have professional partnerships, we must continuously ask ourselves, in our capacity as justice educators: What are the options—and obligations—for those of us who teach and otherwise engage with colleagues abroad, to support their institutional or other political struggles, particularly when there are threats to academic freedom, livelihood or lives? For those of us endangered by the regime change, we must ask: What are our expectations of support from our colleagues abroad and what can we do to assure that in exile at host institutions we will continue to survive and thrive?

## REFERENCES

Adebayo, K. O. (2022). The state of academic (un)freedom and scholar rescue programmes: A contemporary and critical overview. *Third World Quarterly*, *43*, 1817.

Baser, B., & Öztürk, A. E. (2022). From Exit to Voice: Reflections on Exile through the Accounts of Turkey's Intelligentsia. *Middle East Critique*, *31*(4), 1. doi:10.1080/19436149.2022.2132193

Choudhury, N. (2014). Pluralism in Legal Education at the American University of Afghanistan. *Suffolk Transnational Law Review*, (37), 249.

Eddy, J. (2009). Rule of Law in Afghanistan: The Intrusion of Reality. *Journal of International Cooperation Studies* (17), 1.

Konuk, K. (2020). Academy in Exile: Knowledge at Risk. In V. Agnew, K. Konuk, K. & J.O. Newman, (Eds.), *Refugee Routes: Telling, Looking, Protesting, Redressing*. (269-284). The Academy in Exile Book Series. doi:10.14361/9783839450130

Lässig, S. (2017, Winter). Strategies and Mechanisms of Scholar Rescue: The Intellectual Migration of the 1930s Reconsidered. *Social Research*, *84*(4), 769–807. doi:10.1353/sor.2017.0051

Mobasher, B., & Qadam Shah, M. (2021, December 14). Examining the Taliban's Words, Thoughts, and Deeds, Part I: The Myth of Taliban 2.0. *The Diplomat*. https://thediplomat.com/2021/12/examining-the-talibans-words-thoughts-and-deeds-part-i-the-myth-of-taliban-2-0/

Newman, J. O. (2020). Scholar Rescue: The Past of the Future. In V. Agnew, K. Konuk, K. & J.O. Newman, (Eds.), *Refugee Routes: Telling, Looking, Protesting, Redressing*. (285-297). The Academy in Exile Book Series. doi:10.14361/9783839450130

Qadam Shah, M. & Rosenbaum, S. A. (2023). Development Disrupted: The Case of Afghanistan Taliban 2.0 and Lessons Learned About Foreign Aid Management. *Washington International Law Journal*, (32), 207.

Rosenbaum, S. A. (2016). A New Day: Prime Time to Advance Afghan Clinical Education. *Asian Journal of Legal Education*, *3*(1), 1–21. doi:10.1177/2322005815607131

Rosenbaum, S. A. (2022). From A Distance: A "Disciplined Democracy" Comes Undone in Myanmar. *UCLA Pacific Basin Law Journal*, (39), 47.

Sertdemir Özdemir, S. (2019). Pity the Exiled: Turkish Academics in Exile, the Problem of Compassion in Politics and the Promise of Dis-exile. *Journal of Refugee Studies*, 936.

Special Inspector General for Afghan Reconstruction (SIGAR). (2023). *Status of Education in Afghanistan: Taliban policies Have Resulted in Restricted Access to Education and a Decline in Quality* (SIGAR 24-01 Evaluation Report).

Swenson, G., & Sugerman, E. (2011). Building the Rule of Law in Afghanistan: The Importance of Legal Education. *Hague Journal on the Rule of Law*, (3), 130.

Taliban Orders Purge of Private University Libraries in Afghanistan. *Kabul Now* (2023, December 21). https://kabulnow.com/2023/12/taliban-orders-purge-of-private-university-libraries-in-afghanistan/

We Are Committed to Raising Quality of Higher Education. *TOLO News* (2023, December 11). https://tolonews.com/afghanistan/provincial-186442

Yarar, B., & Karakaşoğlu, Y. (2022). "Scholars at Risk" in Germany: Forced Migration and Agency in Forced Migration Decision-Making. *Journal of Refugee Studies*, *35*(4), 1616–1637. doi:10.1093/jrs/feac044

# Compilation of References

Abawe, Z., Daud, B., Daudzai, H., Jabarkhail, M., & Yousaf, F. (2023). Afghanistan and the way forward: Incorporating indigenous knowledge into policymaking. *Global Policy, 14*(1), 192–198. doi:10.1111/1758-5899.13144

Abbasi, K., & Monsutti, A. (2023). From Muhājir to Āwāra: Figures of Migration and Exile Among Afghans. In *Migration in South Asia: IMISCOE Regional Reader* (pp. 187–200). Springer International Publishing. doi:10.1007/978-3-031-34194-6_13

Abbassi, W., Kumari, V., & Pandey, D. K. (2022). What makes firms vulnerable to the Russia–Ukraine crisis? *The Journal of Risk Finance*.

Abdallah, A. K. (2023). Teacher-Led, Student-Focused, and Unleashing the Power of Teacher Empowerment for School Improvement and Success. In A. Abdallah & A. Alkaabi (Eds.), *Restructuring Leadership for School Improvement and Reform* (pp. 1–21). IGI Global. doi:10.4018/978-1-6684-7818-9.ch001

Abdallah, A. K., & Abdallah, R. K. (2023). Achieving academic excellence: The intersection of teacher development, quality education, and entrepreneurship. In S. Chakravarti (Ed.), *Innovations in Teacher Development, Personalized Learning, and Upskilling the Workforce* (pp. 136–158). IGI Global. doi:10.4018/978-1-6684-5518-0.ch007

Abdallah, A. K., & Al-Kaabi, A. M. (2023). Induction Programs' Effectiveness in Boosting New Teachers' Instruction and Student Achievement: A Critical Review. International Journal of Learning. *Teaching and Educational Research, 22*(5), 493–517.

Abdallah, A. K., & Alkaabi, A. M. (2023). Role of teachers in reinforcing students cultural and heritage awareness at Abu Dhabi schools to meet global challenge. *Cogent Social Sciences, 9*(1), 1. doi:10.1080/23311886.2023.2194734

Abdallah, A. K., AlKaabi, A. M., & Ramadan, R. S. (2023). The critical role of principals in leading effective inclusive schools. In E. Efthymiou (Ed.), *Inclusive Phygital Learning Approaches and Strategies for Students With Special Needs* (pp. 256–281). IGI Global. doi:10.4018/978-1-6684-8504-0.ch012

Abdallah, A. K., & Farhan, A. F. (2023). Breaking Barriers and Empowering Women Leaders to Drive School Improvement. In A. Abdallah & A. Alkaabi (Eds.), *Restructuring Leadership for School Improvement and Reform* (pp. 399–419). IGI Global., doi:10.4018/978-1-6684-7818-9.ch020

Abdallah, A. K., Ismail, O., Abdallah, R. K., & Alkaabi, A. M. (2023). Perceptions of Students About the Use of Webinars in Classrooms: A Case of Abu Dhabi University. *International Journal of Information and Communication Technology Education, 19*(1), 1–17. doi:10.4018/IJICTE.322793

Abdallah, A. K., & Musah, M. B. (2023). Principal and Teacher Licensing as a Tool to School Improvement. In A. Abdallah & A. Alkaabi (Eds.), *Restructuring Leadership for School Improvement and Reform* (pp. 278–300). IGI Global. doi:10.4018/978-1-6684-7818-9.ch014

Adebayo. (2022). *The state of academic (un)freedom and scholar rescue programmes: a contemporary and critical overview*. Academic Press.

Admiraal, W., Schenke, W., De Jong, L., Emmelot, Y., & Sligte, H. (2021). Schools as professional learning communities: What can schools do to support professional development of their teachers? *Professional Development in Education, 47*(4), 684–698. doi:10.1080/19415257.2019.1665573

*Afghanistan civilian deaths hit record high in 2018: UN*. (2019, February 24). Aljazeera. https://www.aljazeera.com/news/2019/2/24/afghanistan-civilian-deaths-hit-record-high-in-2018-un

*Afghanistan security situation update: country of origin information report: September 2021*. (2021). European Union Agency for Asylum. https://euaa.europa.eu/publications/coi-report-afghanistan-security-situation-update-september-2021

*Afghanistan's refugees: forty years of dispossession*. (2019, June 20). Amnesty International. https://www.amnesty.org/en/latest/news/2019/06/afghanistan-refugees-forty-years/

Aguilar, P., & Retamal, G. (2009). Protective environments and quality education in humanitarian contexts. *International Journal of Educational Development, 29*(1), 3–16. doi:10.1016/j.ijedudev.2008.02.002

Ahearn, M. C. (2005). Professional development for teachers in crisis and post-crisis contexts: A review of the literature. *Journal of Refugee Education, 12*(2), 189–210. http://file.snnu.net/res/20126/18/018526a6-3cbf-4c9d-ac0f-a0740094aa75.pdf

Ahmadi, F. (2022). Higher education opportunities and challenges in Afghanistan: Afghan women's perspective (Doctoral dissertation, BRAC University).

Ahmadi, M. H. (2022). Higher Education of Afghanistan under the Taliban Rule. Academic Press.

Ahmad, M. M., & Nusrat, R. (2022). Human Trafficking and Crimes in the Rohingya Refugee Camps in Bangladesh. In E. D. Essien (Ed.), *Handbook of Research on Present and Future Paradigms in Human Trafficking* (pp. 26–42). doi:10.4018/978-1-7998-9282-3.ch003

Ahmed, K. (2021). Educating Rohingya Children and Youth in Emergencies: A Reflection From the Field. In R. Gordon, K. Ahmed, & M. Hosoda (Eds.), *Evolving Multicultural Education for Global Classrooms* (pp. 1–20). IGI Global., doi:10.4018/978-1-7998-7649-6.ch001

Ahmed, S., Hasan, M. M., & Kamal, M. R. (2022). Russia–Ukraine crisis: The effects on the European stock market. *European Financial Management*.

Akpovo, S. M., Nganga, L., & Acharya, D. (2018). Minority-World Preservice Teachers' Understanding of Contextually Appropriate Practice While Working in Majority-World Early Childhood Contexts. *Journal of Research in Childhood Education*, *32*(2), 202–218. https://doi.org/10.1080/02568543.2017.1419321

Akramy, S. A. (2021). Implementation of outcome-based education (obe) in Afghan universities: Lecturers' voices. *International Journal of Quality in Education*, *5*(2), 27–47.

Akter, T. 2023. Implementation of Myanmar Curriculum (MC) in the Rohingya Refugee Camps of Bangladesh: A Case Study on Implementation Scope and Challenges. [unpublished master's dissertation]. Nottingham, University of Nottingham.

Al Jazeera. (2022). Taliban arrests fighter who shot dead Hazara woman at checkpoint. https://www.aljazeera.com/news/2022/1/19/taliban-arrest-fighter-who-shot-dead-hazara-woman-at-checkpoint

Al Shaher, S., & Zreik, M. (2022). Sino-Iranian Relations: More Cooperation Despite Sanctions. *Resmilitaris,* 12(2), 637-649.

Alam, A. (2022). Investigating sustainable education and positive psychology interventions in schools towards achievement of sustainable happiness and wellbeing for 21st century pedagogy and curriculum. *ECS Transactions*, *107*(1), 19481–19494. doi:10.1149/10701.19481ecst

Alam, M. S. (2022). Rohingya Refugees in Bangladesh: Exploitations and Challenges for Host Community. *Global Journal of Human Social Science*, *22*(H2), 67–72. doi:10.34257/GJHSSHVOL22IS2PG67

Alamyar, M. (2018). Education in Afghanistan: A Historical Review and Diagnosis. *College and University*, *93*(2), 55–60.

Alariqi, A. A., Najafi, M., Abdulrab, M., Murray, C., & Slimanzai, H. (2019, October). Factors affecting e-learning effectiveness in a higher learning institution in Afghanistan. *Proceedings of the 11th International Conference on Education Technology and Computers* (pp. 176-181). 10.1145/3369255.3372275

AlbarosaE.ElsnerB. (2023, January 06). *Forced Migration and Social Cohesion: Evidence from the 2015/16 Mass Inflow in Germany*. Social Science Research Network. https://doi.org/doi:10.2139/ssrn.4319003

Aldhahyani, D. F. Q., & Al-Eryani, D. A. A. (2023). An International Peer-Reviewed Open Access Journal.

Alemi, Q., Panter-Brick, C., Oriya, S., Ahmady, M., Alimi, A. Q., Faiz, H., Hakim, N., Sami Hashemi, S. A., Manaly, M. A., Naseri, R., Parwiz, K., Sadat, S. J., Sharifi, M. Z., Shinwari, Z., Ahmadi, S. J., Amin, R., Azimi, S., Hewad, A., Musavi, Z., ... Ventevogel, P. (2023). Afghan mental health and psychosocial well-being: Thematic review of four decades of research and interventions. *BJPsych Open*, *9*(4), e125. doi:10.1192/bjo.2023.502 PMID:37424447

Al-Naabi, I., Kelder, J., & Carr, A. (2021). Preparing teachers for emergency remote teaching: A professional development framework for teachers in higher education. *Journal of University Teaching & Learning Practice*, *18*(5), 52–72. Advance online publication. doi:10.53761/1.18.5.4

Alqodsi, E. (2023). Teacher Civil Liability in the Case of Breaching Educational and Control Obligations. In A. K. Alkaabi (Ed.), *Restructuring Leadership for School Improvement and Reform* (pp. 327–338). IGI Global. doi:10.4018/978-1-6684-7818-9.ch016

Alqodsi, E. M., Jadalhaq, I. M., & El Maknouzi, M. E. (2023). *Technology-enhanced legal education: A study of its impact on student learning outcomes in the UAE.* IGI Global. doi:10.4018/978-1-6684-5518-0.ch004

Alqodsi, E., & Aljahoori, S. A. (2023). Legal protection of the right to education for people with special needs: Zayed higher organization for people of determination as a model. In E. Efthymiou (Ed.), *Inclusive Phygital Learning Approaches and Strategies for Students With Special Needs* (pp. 199–213). IGI Global. doi:10.4018/978-1-6684-8504-0.ch010

Altınay, A. G., & Pető, A. (2022). Women's courageous resistance to gender apartheid in Afghanistan: A conversation with Shaharzad Akbar. *European Journal of Women's Studies*, *29*(4), 489–505. doi:10.1177/13505068221130406

Amiri, R., & Jackson, A. (2021). Taliban Attitudes & Policies Towards Education. Working Paper 601.

Amiri, R., & Jackson, A. (2021). Taliban Attitudes & Policies Towards Education. Working Paper 601. Available at: https://cdn.odi.org/media/documents/taliban_attitudes_towards_education.pdf

Amiri, M. H. (2023). Education and Minority Groups in Afghanistan: The Case of Hazara People. *Journal of Asian Social Science Research*, *5*(1), 33–48. doi:10.15575/jassr.v5i1.70

Amouzeshi, A., Abedi, F., Amiri, M., Amouzeshi, Z., Jamshidi, A., & Salehi, F. (2021). Afghan students' willingness to continue their studies abroad and the related factors: A study of Afghan students studying at selected Afghan universities of medical sciences in the academic year 2019-2020. *Educational Development of Judishapur*, *12*(1), 248–254.

Anderson, C. D. (2016). The Role of Teachers in Restoring Normalcy in Conflict-Affected Communities: A Longitudinal Study. *The Journal of Educational Research*, *109*(4), 425–437.

Anderson, R. E. (2019). The Psychological Impact of Conflict Zones on Children: Implications for Education. *Child Development Perspectives*, *13*(3), 143–148.

Arar, K., Kondakci, Y., Kaya Kasikci, S., & Erberk, E. (2020). Higher education policy for displaced people: Implications of Turkey's higher education policy for Syrian migrants. *Higher Education Policy*, *33*(2), 265–285. doi:10.1057/s41307-020-00181-2

Arar, K., Kondakci, Y., & Streitwieser, B. (2020). Higher education for forcibly displaced migrants, refugees and asylum seekers. *Higher Education Policy*, *33*(2), 195–202. doi:10.1057/s41307-020-00184-z

Arar, K., Kondakci, Y., Streitwieser, B., & Saiti, A. (2021). *Higher education in the era of migration, displacement and internationalization*. Routledge. doi:10.4324/9780429345180

Arifani, D. (2021). The Taliban & Afghanistan: Conflict & Peace in International Law Perspective. *International Journal of Law Reconstruction*, *5*(2), 306–316. doi:10.26532/ijlr.v5i2.17704

Arraf, J. (2021). Afghan College Students Find a New Life, and Safety, in Iraq. The New York Times. https://www.nytimes.com/2021/11/14/world/middleeast/afghan-university-students-iraq.html

Associated Press. (2022, February 26). Ukraine crisis: UNHCR predicts 4 mn will flee to neighbouring countries. Business Standard India. https://www.businessstandard.com/article/international/ukraine-crisis-unhcr-predicts-4-mn-will-flee-toneighbouring-countries-122022600772_1.html

Atakpal, H. (2019, June 14). *Afghanistan Ranked World's Least Peaceful Nation*. Tolo News. https://tolonews.com/afghanistan/afghanistan-ranked-world%E2%80%99s-least-peaceful-nation

Avci, Z. Y., O'Dwyer, L. M., & Lawson, J. (2019). Designing effective professional development for technology integration in schools. *Journal of Computer Assisted Learning*, *36*(2), 160–177. https://doi.org/10.1111/jcal.12394

Axyonova, V., Kohstall, F., & Richter, C. (2022). Academics in Exile. *The Academy in Exile Book Series*, *2*, 11–30. doi:10.14361/9783839460894-002

Aydin, H., Gundogdu, M., & Akgul, A. (2019). Integration of Syrian refugees in Turkey: Understanding the educators' perception. *Journal of International Migration and Integration*, *20*(4), 1029–1040. doi:10.1007/s12134-018-0638-1

Azami, A. S. (2021). Warlords, the United States, and the state of anarchy in Afghanistan. *Central European Journal of Politics*, *7*(1), 46–75. doi:10.24132/cejop_2021_1

Bannova, O., & Mayorova, V. 2(019, November). Building trans-national and multi-disciplinary academic curricula through adaptation of a project-based approach. In AIP Conference Proceedings (Vol. 2171, No. 1). AIP Publishing. doi:10.1063/1.5133292

Baquedano-López, P., Alexander, R. A., & Hernandez, S. J. (2013). Equity Issues in Parental and Community Involvement in Schools: What Teacher Educators Need to Know. *Review of Research in Education*, *37*(1), 149–182. https://doi.org/10.3102/0091732X12459718

Barakat, S. (2024). Taliban ban on girls' education defies both worldly and religious logic. Al Jazeera. https://www.aljazeera.com/opinions/2024/4/1/taliban-ban-on-girls-education-defies-both-worldly-and-religious-logic

Barbora P. (2021). The Teacher's Wellbeing in Teacher-Student Interactions. Lifelong Education: the XXI century, 33, 103-115. Doi.org/ doi:10.15393/j5.art.2021.6691

Baser, B., & Öztürk, A. E. (2022). From Exit to Voice: Reflections on Exile through the Accounts of Turkey's Intelligentsia. *Middle East Critique*, *31*(4), 1. doi:10.1080/19436149.2022.2132193

Beltman, S. (2020). Understanding and examining teacher resilience from multiple perspectives. Cultivating Teacher Resilience, 11-26. https://doi.org/ doi:10.1007/978-981-15-5963-1_2

Ben Hassen, T., & El Bilali, H. (2022). Impacts of the Russia-Ukraine war on global food security: Towards more sustainable and resilient food systems? *Foods*, *11*(15), 2301. doi:10.3390/foods11152301 PMID:35954068

Bigsby, E. (2018). *The SAGE Encyclopedia of Communication Research Methods*. SAGE Publications, Inc. doi:10.4135/9781483381411

Bird, L. (2009). Promoting resilience: developing capacity within education systems affected by conflict. *Think piece commissioned for the EFA Global Monitoring Report 2011: The Hidden Crisis–Armed Conflict and Education.*

Boroujerdi, M. (2023). The Plight of Threatened Afghan and Iranian Scholars and Students. Academic Press.

Boungou, W., & Yatié, A. (2022). The impact of the Ukraine–Russia war on world stock market returns. *Economics Letters*, *215*, 110516. doi:10.1016/j.econlet.2022.110516

Boyd, K. J., Gold, A. U., & Littrell, M. K. (2021). Teaching practices around natural hazards and community resilience in Colorado. *Journal of Geoscience Education*, *70*(4), 460–473. https://doi.org/10.1080/10899995.2021.1964319

Bozkurt, A., Jung, I., Xiao, J., Vladimirschi, V., Schuwer, R., Egorov, G., ... Paskevicius, M. (2020). A global outlook to the interruption of education due to COVID-19 pandemic: Navigating in a time of uncertainty and crisis. *Asian Journal of Distance Education*, *15*(1), 1–126. doi:10.5281/zenodo.3778083

Braithwaite, A., Salehyan, I., & Savun, B. (2019). Refugees, forced migration, and conflict: Introduction to the special issue. *Journal of Peace Research*, *56*(1), 5–11. doi:10.1177/0022343318814128

Brenner, M.E. & Kia-Keating, M. (2016). Psychosocial and academic adjustment among resettled refugee youth. *Annual review of comparative and international education 2016*, 221-249.

Brown, S. L. (2017). Pedagogical Approaches in Conflict Zones: Strategies for Teacher Development. *International Journal of Educational Development*, *60*, 101779.

Brunzell, T., Stokes, H., & Waters, L. (2019). Shifting teacher practice in trauma-affected classrooms: Practice pedagogy strategies within a trauma-informed positive education model. *School Mental Health, 11*(3), 600–614. doi:10.1007/s12310-018-09308-8

Burde, D. (2014). *Schools for Conflict or for Peace in Afghanistan.* Columbia University Press.

Burridge, N., Maree Payne, A., & Rahmani, N. (2016). 'Education is as important for me as water is to sustaining life': Perspectives on the higher education of women in Afghanistan. *Gender and Education, 28*(1), 128–147. doi:10.1080/09540253.2015.1096922

Bussas, V., Sharma, A., & Shouche, Y. (2017). IP and The Budapest Treaty—Depositing Biological Material for Patent Purposes. In *Microbial Resources* (pp. 275–292). Academic Press. doi:10.1016/B978-0-12-804765-1.00014-X

Bužinkić, E. (2023). *Schooling aftershocks: Perpetual displacements of refugee youth amid multiple pandemics and social (b) orders* (Doctoral dissertation, University of Minnesota).

Campbell, S., Greenwood, M., Prior, S., Shearer, T., Walkem, K., Young, S., Bywaters, D., & Walker, K. (2020). Purposive sampling: Complex or simple? Research case examples. *Journal of Research in Nursing, 25*(8), 652–661. doi:10.1177/1744987120927206 PMID:34394687

Cannon, S. R., Davis, C. R., & Fuller, S. C. (2020). Preparing for the Next Natural Disaster: Understanding How Hurricanes Affect Educators and Schooling. AASA Journal of Scholarship & Practice, 17(2), 6-16. https://eric.ed.gov/?id=EJ1259714

Caritas Switzerland. (2019). *Teachers in Crisis Contexts: Cox's Bazar.* Retrieved from https://inee.org/resources/promising-practices-teacher-well-being-management-and-school-leadership https://inee.org/resources/promising-practices-teacher-well-being-management-and-school-leadership.

Castellanos, F. (2018). Teaching refugee children: increasing teacher self-efficacy through trauma-informed trainings *(Doctoral dissertation).* http://hdl.handle.net/2152/65961

Cerna, L. (2019). *Refugee education: Integration models and practices in OECD countries.* Academic Press.

Cervantes-Duarte, L., & Fernández-Cano, A. (2016). Impact of armed conflicts on education and educational agents: A multivocal review. Revista Electrónica Educare, 20(3), 1. https://doi.org/10.15359/ree.20-3.12

Chortane, S. G., & Pandey, D. K. (2022). Does the Russia-Ukraine war lead to currency asymmetries? A US dollar tale. *Journal of Economic Asymmetries, 26*, e00265. doi:10.1016/j.jeca.2022.e00265

Cho, S., Crenshaw, K. W., & McCall, L. (2013). Toward a Field of Intersectionality Studies: Theory, Applications, and Praxis. *Signs (Chicago, Ill.), 38*(4), •••. https://doi.org/0097-9740/2013/3804-0001$10.00

Choudhury. (2014). *Pluralism in Legal Education at the American University of Afghanistan.* Academic Press.

Chowdhury, M. J. I. (2023). Teachers' preparedness to implement Myanmar curriculum for the Rohingya children in the Rohingya camps of Cox's Bazar (Doctoral dissertation, Brac University). http://hdl.handle.net/10361/20106

Clemens, V., von Hirschhausen, E., & Fegert, J. M. (2020). Report of the intergovernmental panel on climate change: Implications for the mental health policy of children and adolescents in Europe—a scoping review. *European Child & Adolescent Psychiatry*, 1–13. PMID:32845381

Connah, L. (2021). US intervention in Afghanistan: Justifying the Unjustifiable? *South Asia Research*, *41*(1), 70–86. doi:10.1177/0262728020964609

Coombes, A., & Ponta, O. (2022). Building community engagement and teacher support in education: Qualitative findings from process evaluations in two exceptional settings. *Journal of Development Effectiveness*, *15*(3), 302–319. doi:10.1080/19439342.2022.2067889

Corrales, A., Goldberg, F., Price, E., & Turpen, C. (2020). Faculty persistence with research-based instructional strategies: A case study of participation in a faculty online learning community. *International Journal of STEM Education*, *7*(1), 1–15. doi:10.1186/s40594-020-00221-8

Couch, D. (2019). *The Policy Reassembly of Afghanistan's Higher Education Sector* (Doctoral dissertation, ResearchSpace@ Auckland).

Cox's Bazar Education Sector. (2018). Joint Education Needs Assessment: Rohingya Refugee in Cox's Bazar. https://reliefweb.int/report/bangladesh/joint-education-needs-assessment-rohingya-refugee-cox-s-bazar-june-2018

Crespo, N. M., & Cooperativa, S. (2021). Women and Peacebuilding from the diaspora and exile in Europe. International Catalán Institute for Peace, 19, 91-96.

Crompton, H., Burke, D., Jordan, K., & Wilson, S. W. (2021). Support provided for K-12 teachers teaching remotely with technology during emergencies: A systematic review. *Journal of Research on Technology in Education*, *54*(3), 473–489. doi:10.1080/15391523.2021.1899877

Crossette, B. (2021). Afghan students reach Kyrgyzstan after a risky journey. Pass Blue. https://www.passblue.com/2021/10/13/afghan-students-reach-kyrgyzstan-after-a-risky-journey/

Darawsheh, S. R., Al-Shaar, A. S., Alshurideh, M., Alomari, N. A., Elsayed, A. M., Abdallah, A. K., & Alkhasawneh, T. (2023). The Relation Between Creative Leadership and Crisis Management Among Faculty Members at Imam Abdulrahman Bin Faisal University in Light of the Corona Pandemic from the Perspective of Department Heads, The Effect of Information Technology on Business and Marketing Intelligence Systems (Vol. 1056). Springer. doi:10.1007/978-3-031-12382-5_83

Darling-Hammond & Gardner. (2017). *Effective Teacher Professional Development*. Retrieved March 15, 2024, from https://learningpolicyinstitute.org/sites/default/files/product-files/Effective_Teacher_Professional_Development_REPORT.pdf

Demydova, V. (2019). Independence of the Ukrainian Church: How Russia Is Losing Its Soft Power Tool in Ukraine. *Akademik Hassasiyetler*, *6*(11), 251–274.

Deo, N. (2014). The politics of education in Afghanistan. *Pakistan Journal of Women's Studies: Alam-e-Niswan, 21*(1).

Dinakhel, M. A., Bakhshali, S. Z., & Ali, I. (2023). An Overview of Political Suppression of Legal Political Parties in Afghanistan: 1940-2010. *Khairulummah, 2*(02), 12–22.

Dore, R. (1984). *Education in Tokugawa Japan*. Athlone Press, Center for Japanese Studies. The University of Michigan.

Dupree, L. (1984). Afghanistan in 1983: And Still no Solution. *Asian Survey, 24*(2), 229–239. doi:10.2307/2644442

Dushkevych, M., Barabashchuk, H., & Hutsuliak, N. (2020). Peculiarities of student distance learning in emergency situation condition. Revista Romaneasca pentru Educatie Multidimensionala, 12, 71-77. https://doi.org/10.18662/rrem/12.1sup2/248

Easar, F., Azizi, H., Rahmani, K., Moradi, M., Taieb, R., & Faqiryar, W. N. (2023). *Education in Afghanistan since 2001: Evolutions and Rollbacks*. MISC.

ECB. E. C. (2022). Economic Bulletin Issue 1, 2022. https://www.ecb.europa.eu//pub/economicbulletin/html/eb202201.en.html

Eddy. (2009). *Rule of Law in Afghanistan: The Intrusion of Reality*. Academic Press.

Ehrkamp, P., Loyd, J. M., & Secor, A. J. (2022). Trauma as displacement: Observations from refugee resettlement. *Annals of the American Association of Geographers, 112*(3), 715–722. doi:10.1080/24694452.2021.1956296

El-Serafy, Y., Adam, T., & Hassler, B. (2022). The effectiveness of technology-supported teacher professional learning communities in emergency settings. In S. Gravett & N. Petersen (Eds.), *Future-Proofing Teacher Education: Voices from South Africa and Beyond*. Routledge. doi:10.4324/9781003185499-13

Etzold, B., Isikozlu, E., Christ, S., Morosanu, L., Kraler, A., Pastore, F., . . . Al Jarrah, R. (2022). *Nothing is more permanent than the temporary: Understanding protracted displacement and people's own responses TRAFIG Synthesis Report* (Doctoral dissertation, TRAFIG).

Faisal, M. M., & Ahmed, Z. (2022). Stateless Rohingya Refugees in Bangladesh: Life and Livelihood Challenges. *Journal of Social and Development Sciences, 13*(4), 11–23. doi:10.22610/jsds.v13i4(S).3326

Falk, D., Varni, E., Johna, J. F., & Frisoli, P. (2019). Landscape review: Teacher well-being in low resource, crisis, and conflict-affected settings. Education Equity Research Initiative.https://www.edu-links.org/sites/default/files/media/file/TWB%20Landscape%20Review_June%202019.pdf

Fani, A. (2022). Disciplining Persian Literature in Twentieth-century Afghanistan. *Iranian Studies, 55*(3), 675–695. doi:10.1017/irn.2021.14

Farahani, H., Langari, M. N. M., Eliasi, L. G., Tavakol, M., & Toikko, T. (2023, April 18). "How Can I Trust People When They Know Exactly What My Weakness Is?" Daily Life Experiences, and Resilience Strategies of Stateless Afghans in Iran. *Journal of Immigrant & Refugee Studies*, 1–16. doi:10.1080/15562948.2023.2199252

Farahi, K., & Guggenheim, S. E. (2020). *Pathways for post-peace development in Afghanistan*. United States Institute of Peace.

Farzana, T. (2023). Humanism and Skill Development among the Rohingya Refugees through Content-based Instruction. *International Journal For Multidisciplinary Research*, *5*(1), 1–18. doi:10.36948/ijfmr.2023.v05i01.1388

Fassihi, F. (2021). American University of Afghanistan students and relatives trying to flee were sent home. The New York Times. https://www.nytimes.com/2021/08/29/world/taliban-american-university-of-kabul-afghanistan.html

Fedor, J., Lewis, S., & Zhurzhenko, T. (2017). Introduction: War and Memory in Russia, Ukraine, and Belarus. In War and Memory in Russia, Ukraine and Belarus (pp. 1-40). Palgrave Macmillan.

Feklyunina, V. (2016). Soft power and identity: Russia, Ukraine and the 'Russian world (s)'. *European Journal of International Relations*, *22*(4), 773–796. doi:10.1177/1354066115601200

Fitzgerald, T. (2019). Biological deposits for patenting purposes under the Budapest Treaty. *Microbiology Australia*, *40*(3), 114–116. doi:10.1071/MA19031

Flamant, A., Fourot, A. C., & Healy, A. (2022). city network activism and the governance of migration. *Local Government Studies*, *48*(6), 1017–1026. doi:10.1080/03003930.2022.2045277

Folven, I. (2022). Education in Crises. Education for What?-The Understood Purpose of Education for Rohingya Refugees in Cox's Bazar, Bangladesh, and Implications on Their Education in Times of Crisis.

Frieder, M. L. (2020). The Rohingya Artolution: Teaching Locally Led Community-Based Public Art Educators in the Largest Refugee Camp in History. Doctoral dissertation, Teachers College, Columbia University. doi:10.7916/d8-n1dz-c651

Gelmez-Burakgazi, S. (2020). Curriculum adaptation and fidelity: A qualitative study on elementary teachers' classroom practices. *Issues in Educational Research*, *30*(3), 920–942.

*Germany: Immigrants made up over 18% of 2022 population*. (2023, April 20). Deutsche Welle. https://www.dw.com/en/germany-immigrants-made-up-over-18-of-2022-population/a-65383249

Ghasemi-Tafreshi, S., & Lafortune, G. (2023). The Schooling of Forced Immigrant Afghan Youths in Iran: A Study of the Factors Leading to Exclusion. *Comparative and International Education*, *52*(1). Advance online publication. doi:10.5206/cie-eci.v52i1.14714

Ghosh, A. (2019). Afghan migration and pakistan's policy response: Dynamics of continuity and change. *Public Policy Research in the Global South: A Cross-Country Perspective*, 215-230.

Gibson, K. E., Barnett, J., Haslam, N., & Kaplan, I. (2020). The mental health impacts of climate change: Findings from a Pacific Island atoll nation. *Journal of Anxiety Disorders*, *73*, 102237. doi:10.1016/j.janxdis.2020.102237 PMID:32485590

Gindeel, R., & Etim, J. S. (2014). Education in War Affected Areas (Internally Displaced Persons communities) in Eastern Sudan. *Journal of Studies in Education*, *4*(4), 108–121. doi:10.5296/jse.v4i4.6540

Global Campus of Human Rights. (2023). *Scholarship Programme for Afghan Scholars and Students at Risk.* https://gchumanrights.org/research/projects/afghan-scholarship-programme.html

Göçer, A. P. D. D. (2020). *What education builds as nation in afghanistan: a case of nation-building* (Doctoral dissertation, Middle East Technical University).

Gonchar, O., Kryvuts, S., Petukhova, O., & Tokar, M. (2019). Organization of Pedagogical Relationships between Learners and Teachers in the Framework of Inclusive Education. *Romanian Journal for Multidimensional Education/Revista Romaneasca pentru Educatie Multidimensionala,* 11(2).

Grawert, E., & Mielke, K. (2018). *Coping with protracted displacement: How Afghans secure their livelihood in Afghanistan, Iran and Pakistan. Academic Press.*

Greene, S. A., & Robertson, G. (2022). Affect and autocracy: Emotions and attitudes in Russia after crimea. *Perspectives on Politics*, *20*(1), 38–52. doi:10.1017/S1537592720002339

Greenfield. (2021). Afghanistan in the Era of Fentanyl. Academic Press.

Gu, Q., & Day, C. (2007). Teachers resilience: A necessary condition for effectiveness. *Teaching and Teacher Education*, *23*(8), 1302–1316.https://doi.org/10.1016/j.tate.2006.06.006

Gutierrez, E. D. (2020). The paradox of illicit economies: Survival, resilience, and the limits of development and drug policy orthodoxy. *Globalizations*, *17*(6), 1008–1026. doi:10.1080/14747731.2020.1718825

Habib, I. N. (2020). *The Effects of International Child Development Program (ICDP) to Improve Teachers Attitude Towards the Rohingya Refugee.* Retrieved from https://www.duo.uio.no/handle/10852/80864

Habib, M. R., Chowdhury, A. R., & Uldanov, A. (2023). Creating pathways to opportunity: Non-formal educational 'Inclusion' for Rohingya refugee children in Bangladesh. *Journal of South Asian Development*. Advance online publication. doi:10.1177/09731741231202872

Hamedani, M. Y. G., & Markus, H. R. (2019). Understanding culture clashes and catalyzing change: A culture cycle approach. *Frontiers in Psychology*, *10*, 700. doi:10.3389/fpsyg.2019.00700 PMID:31031669

Hanifi, S. M. (2021). Local Experiences of Imperial Cultures. Academic Press.

Haque, S., Dow, A., & Hasan, M. (2023, September). ERICC Research Agenda for Cox's Bazar, Bangladesh. https://inee.org/sites/default/files/resources

Hascher, T., Beltman, S., & Mansfield, C. (2021). Teacher wellbeing and resilience: Towards an integrative model. *Educational Research*, *63*(4), 416–439. doi:10.1080/00131881.2021.1980416

Hayward, F. M., & Karim, R. (2019). The Struggle for Higher Education Gender Equity Policy in Afghanistan: Obstacles, Challenges and Achievements. *Education Policy Analysis Archives*, *27*(139), n139. doi:10.14507/epaa.27.3036

Heamatzai, M. N., & Rafiqzad, A. J. (2023). Unleashing the Potential of Afghan Women: Strides, Challenges, and Pathways to Prosperity in Economic Development. Challenges, and Pathways to Prosperity in Economic Development.

Hedayati, A. (2023). Women's education from the perspective of Islam and consequences of restrictions on women education in Afghanistan. *Sprin Journal of Arts. Humanities and Social Sciences*, *2*(08), 1–9.

Heinemann, A. M. B. (2018, March 22). The making of 'good citizens': German courses for migrants and refugees. *Studies in the Education of Adults*, *49*(2), 177–195. doi:10.1080/02660830.2018.1453115

Heltne, U. M., Dybdahl, R., Elkhalifa, S., & Breidlid, A. (2019). Psychosocial Support and Emergency Education: An Explorative Study of Perceptions among Adult Stakeholders in Sudan and South Sudan. *Sustainability*, *12*(4), 1410.https://doi.org/10.3390/su12041410

Himat, A. N., Takal, M. H., & Hakimi, M. F. (2021). Afghan students' satisfaction from online learning during COVID-19 at Kandahar University, Kandahar, Afghanistan. *American International Journal of Social Science Research*, *6*(1), 16–29. doi:10.46281/aijssr.v6i1.931

Howard, P. L., Auwarter, A., & Gallimore, R. (2016). Teacher professional development in complex emergencies: Building capacity for social-emotional learning. *Peabody Journal of Education*, *91*(2), 226–243. https://journals.sagepub.com/doi/10.1177/00224871231153084

Howard, S., & Johnson, B. (2004). Resilient teachers: Resisting stress and burnout. *Social Psychology of Education*, *7*(4), 399–420. doi:10.1007/s11218-004-0975-0

Huang, H. (2018). The pathology of hard propaganda. *The Journal of Politics*, *80*(3), 1034–1038. doi:10.1086/696863

Hu, D., Yuan, B., Luo, J., & Wang, M. (2021). A review of empirical research on ICT applications in teacher professional development and teaching practice. Knowledge Management & E-Learning. *International Journal (Toronto, Ont.)*, *13*(1), 1–20. https://doi.org/10.34105/j.kmel.2021.13.001

Imran, M., Chen, Y., Wei, X. M., & Akhtar, S. (2020). Veiled courage: Inside the women's resistance against violence through their writings. *Asian Journal of Women's Studies*, *26*(1), 74–93. doi:10.1080/12259276.2020.1718392

Inayatullah, S. (2022). Women's Education in Afghanistan-Women's Education in Afghanistan: Disparities in Education Under the Taliban Affecting Women's Capabilities Development.

Instefjord, E. J., & Munthe, E. (2017). Educating digitally competent teachers: A study of integration of professional digital competence in teacher education. *Teaching and Teacher Education*, *67*, 37–45.https://doi.org/10.1016/j.tate.2017.05.016

Inter-agency Network for Education in Emergencies (INEE). (2021). *Teacher Wellbeing Resources Mapping & Gap Analysis | INEE*. (2021, August 19). https://inee.org/resources/teacher-wellbeing-resources-mapping-gap-analysis

Inter-agency Network for Education in Emergencies. (2022) A Guidance Note for Teacher Wellbeing in Emergency Settings. https://inee.org/resources/guidance-note-teacher-wellbeing-emergency-settings

International Alert. (2019). *Education and Peacebuilding: A Guide for Policymakers and Practitioners*. International Alert.

International Committee of the Red Cross (ICRC). (2018). *Education in Armed Conflict: Law, Policy, and Practice*. ICRC.

International Institute for Applied Systems Analysis (IIASA). (2018). Education and the sustainable development agenda. International Institute for Applied Systems Analysis (IIASA). https://www.jstor.org/stable/resrep24554

International Organization for Migration (IOM). (2018). *Education in Emergencies: Responding to the Needs of Displaced Populations*. IOM.

Islam, M. S., & Şahin, A. (2023). Impact of Forcibly Displaced People's Identity Crisis on Host State's Security: The Case of Rohingya Refugees in Bangladesh. Saudi Journal of Humanities and Social Sciences. doi:10.36348/sjhss.2023.v08i04.002

IslamA.MozumderT. A.RahmanT.ShatilT.SiddiqueA. (2022, March 15). Forced Displacement, Mental Health, and Child Development: Evidence from the Rohingya Refugees. https://doi.org/doi:10.31235/osf.io/b4fc7

Islam, I., & Naing, Z. (2023). Five years lost: Youth inclusion in the Rohingya response. *Journal on Migration and Human Security*, *11*(1), 99–108. doi:10.1177/23315024231160765

Islam, M. S. (2019). Communal peace in Bangladesh after Rohingya crisis: An assessment of different approaches to community development. *Journal of Prevention & Intervention in the Community*, *48*(3), 256–271. doi:10.1080/10852352.2019.1625604 PMID:31244395

Islam, M. S., & Kaiy, M. (2022). Provision of education to Rohingya refugee children in Bangladesh: Exploring the forms of discrimination and intersectionality. *Journal of Refugee Studies*, 1–22. doi:10.1093/jrs/feac022

Islam, S., & Mozumder, M. K. (2021). Challenges to providing quality mental health services for forcibly displaced populations: The case of Rohingya refugees in Bangladesh. *Asian American Journal of Psychology, 12*(3), 234–240. doi:10.1037/aap0000195

Ismail, A. O., Alriyami, R., & Alhosani, M. (2023). The art of assuring quality education: Internal approaches and best practices. In A. K. Abdallah & A. M. Ahmed (Eds.), *Restructuring Leadership for School Improvement and Reform* (pp. 182–201). IGI Global. doi:10.4018/978-1-6684-7818-9.ch010

Jackson, A. (2021). *Negotiating survival: Civilian-insurgent relations in Afghanistan.* Oxford University Press. doi:10.1093/oso/9780197606179.001.0001

Jafari, B. A., & Schuster, L. (2019). Representations of exile in Afghan oral poetry and songs. *Crossings: Journal of Migration & Culture, 10*(2), 183–203. doi:10.1386/cjmc_00002_1

Jagtap, S., Trollman, H., Trollman, F., Garcia-Garcia, G., Parra-López, C., Duong, L., Martindale, W., Munekata, P. E. S., Lorenzo, J. M., Hdaifeh, A., Hassoun, A., Salonitis, K., & Afy-Shararah, M. (2022). The Russia-Ukraine conflict: Its implications for the global food supply chains. *Foods, 11*(14), 2098. doi:10.3390/foods11142098 PMID:35885340

Jandigulov, A., Abdallah, A. K., Tikhonova, Y., & Gorozhanina, E. (2023). Management and leadership in online learning. *Education and Information Technologies, 28*(10), 13423–13437. Advance online publication. doi:10.1007/s10639-023-11699-4

Jazayery, L. (2003). The Migration-Development Nexus: Afghanistan Case Study. *International Migration (Geneva, Switzerland), 5*(40), 231–254. doi:10.1111/1468-2435.00218

Jenkins, G. (2020). Teacher agency: The effects of active and passive responses to curriculum change. *Australian Educational Researcher, 47*(1), 167–181. doi:10.1007/s13384-019-00334-2

Jewell, C. (2015). WIPO's Budapest Treaty facilitates biotech patenting. *WIPO MAGAZINE*, (4), 2–7.

Johnson, P. D. (2017). The Role of Cultural and Ethnic Factors in Education in Conflict Zones. *Comparative Education Review, 61*(2), 280–302.

Joint education needs assessment: Rohingya refugee in Cox's Bazar. (2018, June 1). ReliefWeb. https://reliefweb.int/report/bangladesh/joint-education-needs-assessment-rohingya-refugee-cox-s-bazar-june-2018

Karim, A., & Hussain, F. (2019). 'When Will You Start Teaching the REAL Curriculum?': Challenges and Innovations in Education for Rohingya Refugees. In P. Nielsen & H. C. Kimaro (Eds.), Information and Communication Technologies for Development. Strengthening Southern-Driven Cooperation as a Catalyst for ICT4D. ICT4D 2019. IFIP Advances in Information and Communication Technology (Vol. 551). Springer. https://doi.org/10.1007/978-3-030-18400-1_55

Karim, M. U. H. F. (2023). Refugee Education in Bangladesh: A Nation-State Paradox. *Journal of Population and Social Studies, 31*, 762–782. http://doi.org/10.25133/JPSSv312023.042

Karlsson, P., & Mansory, A. (2008). *Islamic and modern education in Afghanistan: Conflictual or complementary*. Institute of International Education, Stockholm University.

Katawazai, R. (2021). Implementing outcome-based education and student-centered learning in Afghan public universities: The current practices and challenges. *Heliyon*, *7*(5), e07076. doi:10.1016/j.heliyon.2021.e07076 PMID:34095574

Kayen, H. S. (2022). Improvements and Setbacks in Women's Access to Education: A Case Study of Afghanistan. *Muslim Education Review*, *1*(1), 19–36. doi:10.56529/mer.v1i1.5

Khan, A. S. (2016). The Resilience of Education in Conflict Zones: A Comparative Analysis of Strategies. *Comparative Education*, *52*(3), 334–352.

Khan, A., Ahmad, A., & Gul, A. (2020). Constitution of Afghanistan: An analysis of educational and linguistic provisions. *Liberal Arts and Social Sciences International Journal*, *4*(1), 66–75. doi:10.47264/idea.lassij/4.1.7

Khan, S. U., Khan, S., & Mukherjee, T. (2024). The "Subaltern" Will Speak: Investigating Portrayals in the Acclaimed OTT Series of India and Pakistan. In N. Kalorth (Ed.), *Exploring the Impact of OTT Media on Global Societies* (pp. 1–24). IGI Global. doi:10.4018/979-8-3693-3526-0.ch001

Khwajamir, M. (2016). History and problems of education in Afghanistan. *In SHS Web of Conferences (Vol. 26, p. 01124).* EDP Sciences.

Kirk, J., & Winthrop, R. (2013). Teaching in contexts of emergency and state fragility. In J. Kirk, M. Dembélé, & S. Baxter (Eds.), *More and better teachers for quality for all: Identity and motivation, systems and support* (pp. 221–139).

Klasen, S., & Lamana, F. (2009). 'The impact of gender inequality in education and employment on economic growth: New evidence for a panel of countries. *Feminist Economics*, *15*(3), 91–132. doi:10.1080/13545700902893106

Kölbl, M. (2021). Ethnomusicology, Fieldwork, and the Refugee Experience: Notes on Afghan Music in Austria. *Music & Minorities, 1.*

Kong, Y. (2021). The Role of Experiential Learning on Students' Motivation and Classroom Engagement. *Frontiers in Psychology*, *771272*(12), doi:10.3389/fpsyg.2021.771272

Konuk, K. (2020). Academy in Exile: Knowledge at Risk. Academic Press.

Konuk, K. (2020). *Academy in Exile* (Vol. 269). The Academy in Exile Book Series.

Kovacevic, A. (2009). *The impact of the Russia–Ukraine gas crisis in South Eastern Europe*. Oxford Institute for Energy Studies.

Kumar, R. (2023). A Study of Refugee Crisis in Mohsin Hamid's Exit West in the Backdrop of Current Afghanistan Refugee Crisis. Academic Press.

Kurniawati, N., & Fauziyah, F. (2022). Fostering students' engagement in synchronous learning using interactive web-based media. Indonesian EFL Journal, 8(1), 13-22. https://garuda.kemdikbud.go.id/documents/detail/2570935

Kurtzer, J., Vaidya, S., & Islam, I. (2022, September 21). *Four Years On: An Update on the Rohingya Crisis*. https://www.csis.org/analysis/four-years-update-rohingya-crisis

Kuzio, T. (2018a). Russia–Ukraine crisis: The blame game, geopolitics and national identity. *Europe-Asia Studies*, *70*(3), 462–473. doi:10.1080/09668136.2018.1443643

Kuzio, T. (2018b). Euromaidan revolution, Crimea and Russia–Ukraine war: Why it is time for a review of Ukrainian–Russian studies. *Eurasian Geography and Economics*, *59*(3-4), 529–553. doi:10.1080/15387216.2019.1571428

Larasati, R., & Kuswandono, P. (2023, April 2). Enhancing Teachers' Resilience through Teacher Professional Development. *Language Circle: Journal of Language and Literature*, *17*(2), 391–402. doi:10.15294/lc.v17i2.42673

Lässig, S. (2017, Winter). Strategies and Mechanisms of Scholar Rescue: The Intellectual Migration of the 1930s Reconsidered. *Social Research*, *84*(4), 769–807. doi:10.1353/sor.2017.0051

Laxton, D., Cooper, L., Shrestha, P., & Younie, S. (2020). Translational research to support early childhood education in crisis settings: A case study of collaborative working with Rohingya refugees in cox's Bazar. Education 3-13, 49(8), 901-919. doi:10.1080/03004279.2020.1813186

Lee, J. Y. (2018). The Impact of Teacher–Student Relationships on Resilience in Conflict-Affected Schools. *Journal of School Psychology*, *56*, 1–15. doi:10.1016/j.jsp.2018.06.002 PMID:30340698

Leufer, N., Prediger, S., Mahns, P., & Kortenkamp, U. (2019). Facilitators' adaptation practices of curriculum material resources for professional development courses. *International Journal of STEM Education*, *6*(1), 1–18. doi:10.1186/s40594-019-0177-0

LSE. (2023). *Sahar Halaimzai*. https://www.lse.ac.uk/ideas/people/Sahar-Halaimzai

Mahdi, M. I., & Wani, N. U. H. (2021). Factors affecting e-learning adoption in Afghanistan: Empirical evidence from technical and vocational education and training authority. *Kardan Journal of Economics and Management Sciences*, *4*(2), 1–22. doi:10.31841/KJEMS.2021.1

Mahmud, E., & Nalifa, M. (2020). Barriers to Institutional Education for Refugees: A Case Study of Rohingya Community in Bangladesh. Business, economics and sustainable development.

Mäkinen, A. (2020). *Memory and Exile: Spatiotemporal Narrations of Displacement in Khaled Hosseini's The Kite Runner and And the Mountains Echoed* (Master's thesis).

Mansfield, C. F., Beltman, S., Broadley, T., & Weatherby-Fell, N. (2016, February). Building resilience in teacher education: An evidenced informed framework. *Teaching and Teacher Education*, *54*, 77–87. doi:10.1016/j.tate.2015.11.016

Marshall, E. A., Butler, K., Roche, T., Cumming, J., & Taknint, J. T. (2016). Refugee youth: A review of mental health counselling issues and practices. *Canadian Psychology*, *57*(4), 308–319. doi:10.1037/cap0000068

Marzuki, S. Z., Zreik, M., Buyong, S. Z., Goyipnazarov, S., Mirzaliev, S., & Bakhramovna, I. A. (2023). An Importance Satisfaction Analysis of Hosting Events in Hangzhou, China: A Pilot Study. *Advances in Social Sciences Research Journal,* 10(6.2), 107–117. doi:10.14738/assrj.106.2.14996

McBrien, J. L. (2005). Educational needs and barriers for refugee students in the United States: A review of the literature. *Review of Educational Research*, *75*(3), 329–364. doi:10.3102/00346543075003329

McKirdy, E., & Popalzai, E. (2017). American University of Afghanistan reopens after 2016 attack. https://edition.cnn.com/2017/03/28/asia/kabul-american-university-reopens/index.html

McNamara, E. (2023). Calling Triple A: Providing a Path From Humanitarian Parole for Afghan Refugees. Academic Press.

Mebert, L., Barnes, R., Dalley, J., Gawarecki, L., Ghazi-Nezami, F., Shafer, G., Slater, J., & Yezbick, E. (2020). Fostering student engagement through a real-world, collaborative project across disciplines and institutions. *Higher Education Pedagogies*, *5*(1), 30–51. doi:10.1080/23752696.2020.1750306

Meera, M., & Yekta, K. (2021). The Challenges to Political Participation of Women in Afghanistan: Historical and Cultural Perspectives. *Asian Studies (Quezon City, Philippines)*, *9*(1), 65–91. doi:10.4312/as.2021.9.1.65-91

Meininghaus, E., & Mielke, K. (2019). Beyond Doha and Geneva: Peacemaking engagement of Afghans and Syrians in North Rhine-Westphalia and Germany. Academic Press.

Mendenhall, M. (2018). Education for Peacebuilding in Post-Conflict Contexts: Recommendations for Policy and Practice. *Comparative Education*, *54*(3), 349–368.

Mentes, O. & Talas, M. (2021). The international humanitarian aid delivered for afghanistan on primary and secondary education after 2001. *Journal of World of Turks/Zeitschrift für die Welt der Türken,* 13(1).

Metni, E. (2022). *Exploring Lebanese Teachers' Engagement in a Low-Cost, Technology-Enhanced, Problem-Solving, Orientated Learning Intervention with Refugee Children* (Doctoral dissertation, UCL (University College London)). Retrieved from https://discovery.ucl.ac.uk/id/eprint/10151196/

Mielke, K., & Meininghaus, E. (2021). Engaging refugees and migrants as peace advocates: how Germany can contribute to peace in Syria and Afghanistan. Academic Press.

Miller, E., Ziaian, T., de Anstiss, H., & Baak, M. (2022). Practices for inclusion, structures of marginalisation: Experiences of refugee background students in Australian secondary schools. *Australian Educational Researcher*, *49*(5), 1063–1084. doi:10.1007/s13384-021-00475-3

Miszak, N. & Motwani, N. (2021). *Afghanistan Case Study*. Academic Press.

Mitu, K., Jones, N., Vintges, J., & Devonald, M. (2022). Climate risks and truncated opportunities: How do environmental challenges intersect with economic and social disadvantages for Rohingya adolescents in Bangladesh? *Sustainability (Basel), 14*(8), 4466. doi:10.3390/su14084466

Mobasher & Shah. (2021). *Examining the Taliban's Words, Thoughts, and Deeds, Part I: The Myth of Taliban 2.0.* Academic Press.

Mohiuddin, M., & Molderez, I. (2023). *Rohingya influx and host community: A reflection on culture for leading socioeconomic and environmental changes in Bangladesh.* European Journal of Cultural Management and Policy. doi:10.3389/ejcmp.2023.11559

Molnar, B. E., Bradshaw, C. P., & Buka, S. L. (2015). Building resilience in Teachers of Traumatised Students in the Classroom. *School Psychology Quarterly, 30*(4), 381–394. https://www.researchgate.net/publication/304010563_Classroom_Ready_Building_Resilience_in_Teacher_Education

Morales, L., Rajmil, D., & Andreosso-O'Callaghan, B. (2023). The Global Economic and Political Landscape through the Lenses of Multiple and Multifaceted Crises. Academic Press.

Moravej, M. (2022). Higher Education, Gender, and Empowerment in Afghanistan: A Nation in Conflict (Doctoral dissertation, Lehigh University).

Morrice, L. (2022). Will the war in Ukraine be a pivotal moment for refugee education in Europe? *International Journal of Lifelong Education, 41*(3), 251–256. doi:10.1080/02601370.2022.2079260

Muram, S., & Wani, N. U. H. (2020). Linkage between International political relations and foreign direct investment: A case study of Afghanistan, *Kardan. Journal of Social Sciences and Humanities, 3*(1), 1–32.

*Nadim: We Are Committed to Raising Quality of Higher Education.* (2023). https://tolonews.com/afghanistan/provincial-186442

Naseh, M., Potocky, M., Stuart, P. H., & Pezeshk, S. (2018, October 09). Repatriation of Afghan refugees from Iran: A shelter profile study. *Journal of International Humanitarian Action, 3*(13), 1–12. doi:10.1186/s41018-018-0041-8

Natarajan, A., Moslimani, M., & Lopez, M. H. (2022, December 16). *Key facts about recent trends in global migration.* Pew Research Centre. https://www.pewresearch.org/short-reads/2022/12/16/key-facts-about-recent-trends-in-global-migration/

Nazari, M., Miri, M. A., & Golzar, J. (2021). Challenges of second language teachers' professional identity construction: Voices from Afghanistan. *TESOL Journal, 12*(3), e587. doi:10.1002/tesj.587

Nedelsky, N. R. (2023). A New German 'We'? Everyday Perspectives on Germanness and its Boundaries. *Nationalities Papers, 51*(6), 1215–1234. doi:10.1017/nps.2022.106

Nehan, N. (2022). The Rise and Fall of Women's Rights in Afghanistan. *Afghanistan,* 111.

Newman, J. O. (2020). Scholar Rescue: The Past of the Future. Academic Press.

Nicolai, S. (2003). *Education in Emergencies: A tool kit for starting and managing education in emergencies, Save the Children.* Retrieved from https://inee.org/sites/default/files/Staffing_Building_Team_Supporting_Staff.PDF.

Nilsen, M. J., Olney, J., Maung, K., Karim, L., Ahmad, S., Haque, N., & Mubarak, H. R. (2023). Community-Led Education among Rohingya Refugees and the Politics of Refugee Education in Bangladesh. *Journal of Refugee Studies*, *36*(4), 712–735. doi:10.1093/jrs/fead037

Noora, M., & Wani, N. U. H. (2021). Linkage between imposed career study, career planning, and employability perceptions among higher education students: An empirical investigation of Afghanistan, *Kardan. Journal of Social Sciences and Humanities*, *4*(2), 1–18.

Nussbaum, C. (2000). Women's Capabilities & Social Justice. *Journal of Human Development*, *1*(2), 219–247. doi:10.1080/713678045

O'Connor, K., & Seager, J. (2021). Displacement, Violence, and Mental Health: Evidence from Rohingya Adolescents in Cox's Bazar, Bangladesh. *International Journal of Environmental Research and Public Health*, *18*(10), 5318. doi:10.3390/ijerph18105318 PMID:34067724

Oates, L., Goger, L.K., Hashimi, J. & Farahmand, M. (2019). *An early stage impact study of localised OER in Afghanistan.* Academic Press.

Orfan, S. N., Akramy, S. A., Noori, A. Q., & Hashemi, A. (2021). Afghan lecturers' perception of problem-based learning: A case study of takhar university. *Journal of Problem-Based Learning*, *8*(2), 62–68. doi:10.24313/jpbl.2021.00059

Otukile-Mongwaketse, M., Mangope, B., & Kuyini, A. B. (2016). Teachers' understandings of curriculum adaptations for learners with learning difficulties in primary schools in B otswana: Issues and challenges of inclusive education. *Journal of Research in Special Educational Needs*, *16*(3), 169–177. doi:10.1111/1471-3802.12069

Özdemir, S. (2019). Pity the Exiled: Turkish Academics in Exile, the Problem of Compassion in Politics and the Promise of Dis-exile. *Journal of Refugee Studies*, 936.

Packer, H. (2022). There's no hope: Afghanistan students still waiting for evacuation. The Pie News. https://thepienews.com/news/afghanistan-students-waiting-evacuation/

Palik, J., & Østby, G. (2023). Interventions to improve refugee children's access to education and quality learning: A scoping review of existing impact evaluations. *International Review of Education*, *69*(1-2), 227–247. doi:10.1007/s11159-023-10004-2 PMID:37313292

Pherali, T., Abu Moghli, M., & Chase, E. (2020). Educators for Change: Supporting the Transformative Role of Teachers in Contexts of Mass Displacement. *Journal on Education in Emergencies*, *5*(2), 147. doi:10.33682/gtx6-u5q8

Pickvance, C. (1994). Housing privatization and housing protest in the transition from state socialism-a comparative-study of Budapest and Moscow. *International Journal of Urban and Regional Research*, *18*(3), 433–450. doi:10.1111/j.1468-2427.1994.tb00277.x

Pillai, V. S. (2020). No-tech solutions to continue learning: An example from Afghanistan. *Education.*

Pincock, K., Betts, A., & Easton-Calabria, E. (2021). The rhetoric and reality of localisation: Refugee-led organisations in humanitarian governance. *The Journal of Development Studies, 57*(5), 719–734. doi:10.1080/00220388.2020.1802010

Ponomariov, A. (2019). Ukrainian church autocephaly: The redrawing of the religious borders and political identities in the conflict between Ukraine and Russia. *Russian Analytical Digest, 231*(25), 2–9.

Powell, C. (2014). Women and girls in the Afghanistan transition. Council on Foreign Relations® Incorporated.

Powell, C. G., & Bodur, Y. (2019). Teachers' perceptions of an online professional development experience: Implications for a design and implementation framework. *Teaching and Teacher Education, 77*, 19–30.https://doi.org/10.1016/j.tate.2018.09.004

Prodip, M. A. (2023). Education for what? The politics of education for Rohingya refugees in Bangladesh. *Politics & Policy, 51*(6), 1028–1045. Advance online publication. doi:10.1111/polp.12541

Qatar Foundation. (2021). American University of Afghanistan students to continue education in Qatar through strategic agreement with QF and QFFD. https://www.qf.org.qa/stories/american-university-of-afghanistan-students-to-continue-education-in-qatar

Ra'ees, W. (2023). Geostrategic and Internal Dynamics of the Afghan Conflict. *Editorial Board*, 37.

Rahman, M. M., Shindaini, A. J., & Abdullah, A. B. (2022). undefined. *Asia Pacific Education Review, 24*(3), 433–445. doi:10.1007/s12564-022-09770-9

Rahman, M. M., Shindaini, A. J., & Husain, T. (2022). Structural barriers to providing basic education to Rohingya children in the Kutupalong refugee camp, Cox's Bazar, Bangladesh. *International Journal of Educational Research Open, 3*, 100159. doi:10.1016/j.ijedro.2022.100159

Rahman, M. S., & Siddik, M. A. (2022). Need for Teachers' Professional Development in a Low-Resource Context During and After COVID-19: A Bangladesh Perspective. *International Journal of Teacher Education and Professional Development, 5*(1), 1–14. doi:10.4018/IJTEPD.295547

Ramadan, R. S., & Ismail, O. A. (2023). Building a better future: The role of school inspection in driving educational excellence. In *Restructuring Leadership for School Improvement and Reform* (pp. 202–221). IGI Global. doi:10.4018/978-1-6684-7818-9.ch011

Rapley, T. J. (2001). The art(fulness) of open-ended interviewing: Some considerations on analysing interviews. *Qualitative Research, 1*(3), 303–323. doi:10.1177/146879410100100303

Razzaque, A. (2023). *Identity Ruptures and Community Fragmentation: Afghans Resettling in the US After the Taliban Takeover* (Doctoral dissertation, Georgetown University).

Rehman, W. (2023). Afghanistan's parallel crises under the Taliban. Academic Press.

Reidy, K. (2020, February 10). Expanding education for Rohingya refugee children in Bangladesh. UNICEF. https://www.unicef.org/bangladesh/en/stories/expanding-education-rohingya-refugee-children-bangladesh

Reuters. (2022). Taliban ban women in Afghanistan from flying without male chaperone. https://www.reuters.com/world/asia-pacific/afghanistans-taliban-ban-women-flying-without-male-chaperone-sources-2022-03-27/

Reuters. (2022, February 15). Gas supply shock would cut value of Europe's economy, ECB says. Reuters. https://www.reuters.com/article/gas-europe-growth-idUSL1N2UQ0EV

Richardson, J., Karam, C. M., & Afiouni, F. (2019). The global refugee crisis and the career ecosystem. *Career Development International*, *25*(1), 1–13. doi:10.1108/CDI-04-2019-0104

Richters, K. (2012). *The Post-Soviet Russian Orthodox Church: Politics, Culture and Greater Russia*. Routledge. doi:10.4324/9780203115732

Rodriguez, L. (2019). 7 Obstacles to Girls' Education & How to Overcome them. Global Citizen. Available at: https://www.globalcitizen.org/en/content/barriers-togirls-education-around-the-world/

Rohingya Response, Bangladesh (2023). Joint Multi-Sectoral Needs Assessment (J-MSNA): Camp-level Findings. https://rohingyaresponse.org/wp-content/uploads/2024/04/BGD_CXB_J-MSNA-2023_Factsheets_Camps_Feb-2024.pdf

Rosenbaum. (2022). *From A Distance: A "Disciplined Democracy" Comes Undone in Myanmar*. Academic Press.

Rosenbaum, S. A. (2016). A New Day: Prime Time to Advance Afghan Clinical Education. *Journal of Legal Education*, *3*(1), 1–21. doi:10.1177/2322005815607131

RRRC. (2023). Report: At a glance. Office of The Refugee Relief and Repatriation Commissioner, Cox's Bazar. Retrieved from: https://rrrc.portal.gov.bd

Rubin, B., & Rudeforth, C. (2016). Enhancing access to education: Challenges and opportunities in Afghanistan. Center in International Cooperation, New York University.

Ruhani, A., Keshavarzi, S., Kızık, B., & Çakal, H. (2023). Formation of hatred emotions toward Afghan refugees in Iran: A grounded theory study. *Peace and Conflict*, *29*(4), 355–364. doi:10.1037/pac0000685

Ruttig, T. (2017, February 17). *Afghan Exodus: Afghan asylum seekers in Europe (3) – case study Germany*. reliefweb. https://reliefweb.int/report/germany/afghan-exodus-afghan-asylum-seekers-europe-3-case-study-germany

Sabieh, C. (2023). Teacher Education and Professional Development: Prepping Educators for Emergency Education Through Project-Based Learning and Technology. In D. Tafazoli & M. Picard (Eds.), *Handbook of CALL Teacher Education and Professional Development*. Springer. doi:10.1007/978-981-99-0514-0_24

Sadia, S. (2021). Experiences and challenges of Non-Governmental Organizations (NGO)s professionals in ensuring the right to education for the Rohingya refugee children in Cox's Bazar. *Instituto Universitário de Lisboa*. http://hdl.handle.net/10071/24014

Sadiq, A. (2022). Leading Literate Lives: Afghan Refugee Children in a First-Asylum Country. *Journal of Literacy Research*, *54*(1), 28–50. doi:10.1177/1086296X221076431

Salih, S.K. & Taniwall, N.J. (2020). Issues and challenges of E-learning system Adoption in a public university of Afghanistan: a case study of shaikh zayed university. *J. Human. Soc. Sci. (IOSR-JHSS),* 25(1), 63-69.

Save the Children. (2019a). *Ensuring Quality Education in Conflict-Affected Contexts: A Comprehensive Approach*. Save the Children.

Save the Children. (2019b). *Humanitarian Response Index: Education in Conflict and Crises*. Save the Children.

Schuetze, C. F. (2023, September). A Small German City Staves Off a Threat From the Far Right. *The New York Times*. https://www.nytimes.com/2023/09/24/world/europe/germany-afd-mayor-election.html

Schuster, L., & Shinwari, R. M. K. (2020). Migration in Afghan women's poetry. *Soundings*, *76*(76), 111–123. doi:10.3898/SOUN.76.08.2020

Seddighi, H., Naseh, M., Rafieifar, M., & Ilea, P. (2022). Education of Afghan refugee children in Iran: A structured review of policies. Academic Press.

Seyler, M. (2022). Single suicide bomber killed US troops and Afghans in ISIS-K attack at Kabul airport, Pentagon finds. ABC News. https://abcnews.go.com/Politics/single-suicide-bomber-killed-us-troops-afghans-isis/story?id=82676604

Shah, Q. (2023). Development Disrupted: The Case of Afghanistan Taliban 2.0 and Lessons Learned About Foreign Aid Management. Academic Press.

Shakir, L. (2021). 26 students, professor arrive in Sulaimani from Kabul. Rudaw. https://www.rudaw.net/english/kurdistan/300820211

Shakya, Y. B., Guruge, S., Hynie, M., Akbari, A., Malik, M., Htoo, S., Khogali, A., Mona, S. A., Murtaza, R., & Alley, S. (2010). Aspirations for higher education among newcomer refugee youth in Toronto: Expectations, challenges, and strategies. *Refuge: Canada's Periodical on Refugees*, *27*(2), 65–78. doi:10.25071/1920-7336.34723

Shams, F. (2021). The Sensorium of Exile: The Case of Elyas Alavi and Gloria Anzaldúa. *Losing Our Minds, Coming to Our Senses: Sensory Readings of Persian Literature and Culture*, 195-214.

Sharma, Y. (2024). A Relocated University Restores Hope for Female Students. University World News. https://www.universityworldnews.com/post.php?story=20240116075152546

Shohel, M. M. C. (2022). Education in emergencies: Challenges of providing education for Rohingya children living in refugee camps in Bangladesh. *Education Inquiry*, *13*(1), 104–126. doi:10.1080/20004508.2020.1823121

Shohel, M. M. C. (2023). Lives of the Rohingya children in limbo: Childhood, education, and children's rights in refugee camps in Bangladesh. *Prospects*, *53*(1–2), 131–149. doi:10.1007/s11125-022-09631-8 PMID:37197308

Shohel, M. M. C., Ashrafuzzaman, M., Babu, R., Akter, T., Tasnim, N., & Bayezid, A. (2023a). Access to Higher Education for Rohingya Refugees: Challenges, Opportunities and Future Directions. In S. Saeed & M. Zhang (Eds.), *Global Perspectives on the Difficulties and Opportunities Forced by Migrant and Refugee Students in Higher Education*. IGI Global. doi:10.4018/978-1-6684-7781-6.ch005

Shohel, M. M. C., Ashrafuzzaman, M., & Shabnam, S. (2024a). The Rohingya Crisis and the Global impact Refugee Influx. In P. Pietrzak (Ed.), *Analyzing Global Responses to Contemporary Regional Conflicts*. IGI Global.

Shohel, M. M. C., Babu, R., Ashrafuzzaman, M., Azim, F., & Bayezid, A. (2023b). Learning Competency Framework and Approach for the Displaced Rohingya Children Living in Bangladesh: A Critical Review. *Continuity in Education*, *4*(1), 50–66. doi:10.5334/cie.57

Shohel, M. M. C., Rashedujjaman, S. M., & Akter, T. (2024b). Relocation of the Rohingya Refugees to Bhasan Char: Human Rights, Government Policy and Refugee Crisis in Bangladesh. In P. Pietrzak (Ed.), *Dealing with Regional Conflicts of Global Importance*. IGI Global. doi:10.4018/978-1-6684-9467-7.ch014

Shohel, M. M. C., Roy, G., Ahsan, M. S., & Acharya, D. (2021). Access to Higher Education for Refugees in South Asia. In M. A. Prodip, G. Roy, & D. Nandy (Eds.), *Refugee Education in South Asia: Policies and Perspectives* (pp. 187–222). Nova Science Publishing House.

Shohel, M. M., Ashrafuzzaman, M., Azim, F., Akter, T., & Tanny, S. F. (2022). Displacement and Trauma: Exploring the Lost Childhood of Rohingya Children in the Refugee Camps in Bangladesh. In R. Throne (Ed.), *Social Justice Research Methods for Doctoral Research* (pp. 244–272). IGI Global. doi:10.4018/978-1-7998-8479-8.ch012

Sinclair, M. (2002). *Planning education in and after emergencies. UNESCO International Institute for Educational Planning*. UNESCO.

Skinner, B. (2009). The Western front of the Eastern church: Uniate and Orthodox conflict in 18th-century Poland, Ukraine, Belarus, and Russia. Northern Illinois Univ. Press.

Slattery, P. (2012). *Curriculum development in the postmodern era: Teaching and learning in an age of accountability*. Routledge.

Smith, A. B. (2020). Transformative Education in Conflict Zones: Strategies for Sustainable Change. *Comparative Education*, *56*(1), 89–108.

Smith, J. A. (2018). Conflict Zones and Their Impact on Education: A Comprehensive Analysis. *Journal of Conflict Studies*, *38*(2), 45–68.

Smith, L. H. (2017). Post-Conflict Reconstruction and Education: Opportunities and Challenges. *Comparative Education Review*, *61*(4), 625–648.

Smith, M. R. (2020). Fostering Teacher Efficacy and Well-being in Conflict-Affected Schools. *Journal of Educational Psychology*, *112*(4), 683–698.

Smith, V. J. (2009). Ethical and effective ethnographic research methods: A case study with Afghan refugees in California. *Journal of Empirical Research on Human Research Ethics; JERHRE*, *4*(3), 59–72. doi:10.1525/jer.2009.4.3.59 PMID:19754236

Southwick, S. M., Bonanno, G. A., Masten, A. S., Panter-Brick, C., & Yehuda, R. (2014). Resilience definitions, theory, and challenges: Interdisciplinary perspectives. *European Journal of Psychotraumatology*, *5*(1).https://doi.org/10.3402/ejpt.v5.25338

Special Inspector General for Afghan Reconstruction (SIGAR). (2023). *Status of Education in Afghanistan: Taliban policies Have Resulted in Restricted Access to Education and a Decline in Quality* (SIGAR 24-01 Evaluation Report).

Spink, J. (2005). Education and politics in Afghanistan: The importance of an education system in peacebuilding and reconstruction. *Journal of Peace Education*, *2*(2), 195–207. doi:10.1080/17400200500185794

Steffen, B., & Patt, A. (2022). A historical turning point? Early evidence on how the Russia-Ukraine war changes public support for clean energy policies. *Energy Research & Social Science*, *91*, 102758. doi:10.1016/j.erss.2022.102758

Stevens, K. E., Siraj, I., & Kong, K. (2023). A critical review of the research evidence on early childhood education and care in refugee contexts in low- and middle-income countries. *International Journal of Child Care and Education Policy*, *17*(1), 7. Advance online publication. doi:10.1186/s40723-023-00109-4

Sultana, N., Sultana, S., Saha, R., & Alam, M. M. (2023). The challenges and coping of Rohingya refugees: a comparative study of registered and non-registered Rohingya refugees in Bangladesh. Southeast Asia: A Multidisciplinary Journal, 23(2), 95-120. doi:10.1108/SEAMJ-04-2023-0033

Suma, J. T. (2022). Rohingya Refugees and Classroom Children: Cultivating a Lost Generation. In I. A. Hussain (Ed.), *Rohingya Camp Narratives*. Global Political Transitions. doi:10.1007/978-981-19-1197-2_11

Suslov, M. (2014). The Utopia of 'Holy Russia' in Today's Geopolitical Imagination of the Russian Orthodox Church: A Case Study of Patriarch Kirill. *Plural. History, Culture. Society*, (1-2), 81–97.

Swenson, G., & Sugerman, E. (2011). Building the Rule of Law in Afghanistan: The Importance of Legal Education. Academic Press.

Szporluk, R. (2020). *Russia, Ukraine, and the breakup of the Soviet Union.* Hoover Press.

Taheryar, H. (2017). Perceptions of quality in higher education in Afghanistan: A case study of Shaheed Rabbani Education University. Academic Press.

Talant, B. (2021). Afghan Students Make Daring Escape to Kyrgyzstan. Eurasianet. https://eurasianet.org/afghan-students-make-daring-escape-to-kyrgyzstan

*Taliban Orders Purge of Private University Libraries in Afghanistan.* (2023). https://kabulnow.com/2023/12/taliban-orders-purge-of-private-university-libraries-in-afghanistan/

Tayeb, R. (2021). Passport Department Resumes Services to Afghans. Tolo News. https://tolonews.com/afghanistan-174914

Teach For All. (2019). Supporting Teachers in Emergencies: A Resource Kit. Retrieved from https://teachforall.org/news/education-emergencies-connecting-across-borders

The American University of Afghanistan. (2024). About AUAF. https://auaf.edu.af/

The Diplomat. (2020). *Afghanistan's Higher Education System Has a Stark Geographic Divide.* https://thediplomat.com/2020/07/afghanistans-higher-education-system-has-a-stark-geographic-divide/

The Qur'an. 39:9. https://quran.com/39?startingVerse=9

The Qur'an. 96:1-2. https://quran.com/96

Thomas, N., Smith, M. B., Agerhem, S., Cadesky, J., Fadel, B., & Paz, B. V. (2018). Theme Paper 1: Everyday humanitarianism. Academic Press.

Thomas, M., & Bigatti, S. (2020). Perfectionism, impostor phenomenon, and mental health in medicine: A literature review. *International Journal of Medical Education*, *11*, 201–213. doi:10.5116/ijme.5f54.c8f8 PMID:32996466

Tosun, O. K., & Eshraghi, A. (2022). Corporate decisions in times of war: Evidence from the Russia-Ukraine conflict. *Finance Research Letters*, *48*, 102920. doi:10.1016/j.frl.2022.102920

Tovar-Gálvez, J. C. (2022). Acculturation in the Integration Course for Immigrants in Germany: A Study from the Cultural Bridge Approach. *Journal of Ethnic and Cultural Studies*, *9*(3), 72–91. https://www.jstor.org/stable/48710351. doi:10.29333/ejecs/1061

Ullah, A. K. M. A. (2016). Rohingya Crisis in Myanmar: Seeking Justice for the "Stateless". *Journal of Contemporary Criminal Justice*, *32*(3), 285–301. doi:10.1177/1043986216660811

UNESCO. (2018). Education in emergencies: Policy guidelines for inclusion and equity. Retrieved from https://unesdoc.unesco.org/ark:/48223/pf0000128214

UNESCO. (2018). Improving teacher support and participation in Local Education Groups. Retrieved from.https://www.globalpartnership.org/sites/default/files/2018-07-gpe-improving-teacher-support-in-legs.pdf

*UNHCR Projected Global Settlement Needs 2023*. (2023). UNHCR. https://www.unhcr.org/sites/default/files/legacy-pdf/62b18e714.pdf

UNHCR. (2021). *Refugee Education Report- Staying The Course-The Challenges Facing Refugee Education*. https://reliefweb.int/sites/reliefweb.int/files/resources/612f85d64.pdf

UNHCR. (2023). *Desperate Journeys: Rohingya Refugees in Search of Protection*. UNHCR Regional Bureau for Asia and Pacific (RBAP).

UNHCR. (2023). *The Right to Asylum between Islamic Shari'ah and International Refugee Law*. https://www.unhcr.org/sites/default/files/legacy-pdf/4a9645646.pdf

UNHCR. (2024). *Joint Government of Bangladesh - UNHCR Population Factsheet - Block Level as of March 2024*. https://data.unhcr.org/en/documents/details/107776.

UNICEF. (2018). Evaluation Office. Bangladesh - Rohingya Refugee Response: Education in Emergencies Sector Evaluation. Retrieved from https://reliefweb.int/report/bangladesh/evaluation-unicefs-response-rohingya-refugee-crisis-bangladesh

UNICEF. (2022). Girls' Education: Gender Equality in Education Benefits Every Child.

United Nations Children's Fund (UNICEF). (2020). *Education in Emergencies: A Handbook for the Field*. UNICEF.

United Nations Development Programme (UNDP). (2020). *Education for Sustainable Development in Post-Conflict and Post-Disaster Contexts: A Handbook*. UNDP.

United Nations Educational, Scientific and Cultural Organization (UNESCO). (2016). *Education for Peace, Reconciliation, and Conflict Prevention: Recommendations for a Holistic Approach*. UNESCO.

United Nations Educational, Scientific and Cultural Organization (UNESCO). (2021). *Education in Emergencies and Crises: Recommendations for Action*. UNESCO.

United Nations High Commissioner for Refugees (UNHCR). (2019). *Education for Displaced and Conflict-Affected Populations: Policy Recommendations*. UNHCR.

VadeanF. P. (2007). *Skills and remittances: The case of Afghan, Egyptian, and Serbian immigrants in Germany*. Social Science Research Network. https://dx.doi.org/ doi:10.2139/ssrn.1317135

Venet, A. S. (2023). *Equity-centered trauma-informed education* (1st ed.). Routledge. doi:10.4324/9781032677989

Viczko, M., & Matsumoto, R. (2022). Problematizing access to higher education for refugee and globally displaced students: What's the problem represented to be in Canadian university responses to Syrian, Afghan and Ukrainian crises? *Journal of Contemporary Issues in Education*, *17*(1), 40–56. doi:10.20355/jcie29504

Wallerstein, N., & Duran, B. (2010). Community-based participatory research contributions to intervention research: The intersection of science and practice to improve health equity. *American Journal of Public Health*, *100*(S1), S40–S46. doi:10.2105/AJPH.2009.184036 PMID:20147663

Wani, N. H., Dhami, J. K., & Sidana, N. (2022). Impact of education on poverty alleviation in Afghanistan: An empirical evaluation. *Kardan Journal of Social Sciences and Humanities*, *5*(1), 1–14. doi:10.31841/KJSSH.2022.47

Wardak, M. O. (2021). *Religion, Politics and Education in Afghanistan, 1950s-1980s: The Case of Habibia High School* (Doctoral dissertation).

Webb, M. E., & Paretti, M. C. (2023, June). Influences on Displaced Engineering Student Professional Identity Development: A Scoping Literature Review Across Forced Migration Contexts. *2023 ASEE Annual Conference & Exposition*. 10.18260/1-2--43696

Welch, A., & Wahidyar, A. (2020). Quality assurance in Afghan higher education: Achievements and challenges. *Asian Education and Development Studies*, *9*(4), 479–493. doi:10.1108/AEDS-09-2018-0146

White, S., & Feklyunina, V. (2014). Identities and foreign policies in Russia, Ukraine and Belarus. *New York. Palgrave Macmillan*, *10*, 9781137453112.

Williams, L. M. (2020). Social and Economic Forces in Conflict Zones: Effects on Education Access and Quality. *International Journal of Educational Development*, *82*, 102146.

Willner-Reid, M. (2016). IDPs in Afghanistan: A Confused National Glossary. *Refugee Survey Quarterly*, *35*(4), 79–96. doi:10.1093/rsq/hdw013

Wilson, E. K. (2019). Teacher Resilience in Conflict Zones: A Qualitative Study. *Teachers and Teaching*, *25*(5), 567–584.

World Bank. (2017). *Learning in the Face of Adversity: The Role of Education for Conflict-Affected Communities*. World Bank.

World Food Programme (WFP). (2017). *Investing in the Future: A Comprehensive Approach to Education in Emergencies*. WFP.

Yarar, B., & Karakaşoğlu, Y. (2022). "Scholars at Risk" in Germany: Forced Migration and Agency in Forced Migration Decision-Making. *Journal of Refugee Studies*, *35*(4), 1616–1637. doi:10.1093/jrs/feac044

Yosso, T. J. (2002). Toward a Critical Race Curriculum. *Equity & Excellence in Education*, *35*(2), 93–107. https://doi.org/10.1080/713845283

Yousofi, Z. (2016). The Millennium Development Goals: Women's Empowerment and Gender Equality in Afghanistan. Academic Press.

Zaman, S., Sammonds, P., Ahmed, B., & Rahman, T. (2020). Disaster risk reduction in conflict contexts: Lessons learned from the lived experiences of Rohingya refugees in Cox's Bazar, Bangladesh. *International Journal of Disaster Risk Reduction*, *50*, 101694. doi:10.1016/j.ijdrr.2020.101694 PMID:32518743

Zee, M., & Koomen, H. M. (2016). Teacher self-efficacy and its effects on classroom processes, student academic adjustment, and teacher well-being: A synthesis of 40 years of research. *Review of Educational Research*, *86*(4), 981–1015. doi:10.3102/0034654315626801

Zhan, R., Liu, X., Wong, D. F., & Chao, L. S. (2021, May). Meta-curriculum learning for domain adaptation in neural machine translation. *Proceedings of the AAAI Conference on Artificial Intelligence*, *35*(16), 14310–14318. doi:10.1609/aaai.v35i16.17683

Ziaulhaq Rashidi, D. G. U. G. (2019). Nation Building Process in Afghanistan. Academic Press.

Zreik, M. (2019). Geopolitical Situation in Northeast Asia: Development of Sino-Russian Relations. *RUDN Journal of Political Science*, *21*(3), 450–459. doi:10.22363/2313-1438-2019-21-3-450-459

Zreik, M. (2022). US-China Relations in The Era of Multi-Polarism: Trade War Between the Two Economic Giants. *Eurasian Research Journal*, *4*(2), 43–65. doi:10.53277/2519-2442-2022.2-03

Zreik, M. (2023a). Navigating the Dragon: China's Ascent as a Global Power Through Public Diplomacy. In S. Kavoğlu & E. Köksoy (Eds.), *Global Perspectives on the Emerging Trends in Public Diplomacy* (pp. 50–74). IGI Global. doi:10.4018/978-1-6684-9161-4.ch003

Zreik, M. (2023b). Analytical study on Foreign Direct Investment Divestment Inflows and Outflows in Developing Economies: Evidence of China. *Chinese Economy*, *56*(6), 1–16. doi:10.1080/10971475.2023.2193118

# About the Contributors

**Enakshi Sengupta** is an independent researcher and scholar. She is the CEO & Founder of Borders Without Barbed Wires, a not-for-profit organization that believes in the accessibility of education for the vulnerable and the marginalized. Dr. Sengupta brings 25 years of experience working both in the corporate world and in the academic environment. She is a PhD holder from University of Nottingham and has completed her MBA from the same university. She worked with American University of Afghanistan as the Chair of Business Department and Director of the MBA programme. Enakshi also has a master's degree in English Literature from Calcutta University and two undergraduate degrees in English literature and Education. She has several professional diplomas in marketing and public relations from the UK. Enakshi also works as the Associate Editor of the book series Innovations in Higher Education Teaching and Learning, Emerald Publishing and has till date worked as the lead editor and lead author of 38 books under this series. She is also the Managing Editor of Journal of Applied Research in Higher Education from Emerald Publishing. She has contributed several journals and book chapters for Routledge, Springer, Palgrave and Cambridge Publishing. Enakshi has been lauded with several international awards.

* * *

**Tahmina Akter** is a Commonwealth Scholar and Humanitarian & Developmental professional having more than 7 years of experiences in the field of Education, Education in Emergency (EiE), Disability and Inclusive Education, Gender Inclusion, Child Protection and Research. Skilled in project management & implementation, training & capacity building, budget phasing & monitoring, concept notes & project proposal development, documentation & donor reporting etc. Possess a remarkable record of working in different donors funded programs with several international and national NGOs i.e., Save the Children, World Vision, JICA, Room to Read, etc.

**Sara Aldous** is a dedicated disaster management specialist, whose passion lies at the intersection of enhancing student safety and well-being and optimizing school facilities for better learning outcomes. With a strong academic foundation in disaster management, Sara has devoted her career to ensuring that educational environments are not only safe havens but also conducive to learning and personal growth. Her work, particularly focused on the UAE but with implications globally, addresses the critical need for robust disaster preparedness and response strategies within schools. Sara's interest in student well-being and the physical and psychological aspects of learning environments has led her to pioneer innovative solutions that enhance educational spaces. These improvements aim to not only safeguard against emergencies but also to create atmospheres that inspire and support students' educational journeys. Through her contributions, Sara has become a beacon for change in how schools approach disaster management and student welfare, advocating for a holistic view that prioritizes safety without compromising on the quality of learning experiences. Her work exemplifies a commitment to creating resilient, supportive, and enriching educational environments for students around the world.

**Habib Al Badawi** has been a scholar of Japanese studies since 2005. His M.A. thesis and Ph.D. dissertation covered the modern history of Japan (Meiji-tenno, Taisho-tenno, and Showa-tenno). Dr. Badawi is currently a professor at Lebanese University and a freelance researcher in international relations. In 2016, he became the coordinator of the "United States of America - History and Civilization" course at the History Department for the university's five branches. Dr. Badawi earned the "Academic Excellence Award in the Arab World" in the category of "Academic Personality of the Year 2018" for his role in introducing "Japanese studies" to the Arab world as a stand-alone academic field of "Human Sciences." .

**Setha Fayaz** is a student at the American University of Afghanistan. She has studied Bachelor of Arts in Law (LLB). She has worked in law firms and has experience in research.

**Davida Finger** is the René August and Mary Jane Pastorek Distinguished Clinical Professor of Law and Director of the Gillis Long Poverty Law Center at Loyola New Orleans College of Law. She has taught the Community Justice section of the Law Clinic, the externship course, and the Law & Poverty course. Davida and her clinic students represent clients in cases on landlord-tenant, post-disaster housing, housing discrimination, and other civil rights matters. In addition to litigating cases, Davida strives to provide support and collaboration on community advocacy for anti-poverty and justice initiatives.

**Ghazi Hashimi** was born in Nangarhar Province, Afghanistan in 1980. Prior to his departure from Afghanistan, he held the position of Associate Professor at Kabul University Faculty of Law and Political Sciences since 2009. Currently, he is a Clinical Fellow with the Bluhm Legal Clinic's Center for International Human Rights at Northwestern University Pritzker School of Law and Visiting Scholar with the University's Buffett Institute for Global Affairs. In 2013, Ghazi earned his LLM from the University of Washington School of Law. He has collaborated on various projects with the National Center for Policy Research since 2006, showcasing his commitment to academic research and policy development. His published works, such as "Helping Afghanistan's Informal Dispute Resolution Systems Follow Afghan Law in Criminal Matters" and "Defending the Principle of Legality in Afghanistan," reflect his dedication to advancing legal discourse. In addition to his academic pursuits, Ghazi has authored numerous articles and textbooks in Afghanistan's official languages (Dari and Pashto), emphasizing the pivotal role of legal education in fostering development and strengthening the rule of law in Afghanistan. As a legal scholar and advocate for justice, his journey continues to impact legal education and contribute to the global understanding of Afghan law.

**Shukria Hashimi** is a Women Rights Activist and Member, Research Society Kardan University, Kabul, Afghanistan.

**Ghaith Ismail** is a disaster management graduate with a strong commitment to student safety. His passion for this area stems from a belief that every child deserves to learn in a secure and protected environment. Ghaith doesn't just hold a degree in disaster management; he actively utilizes his knowledge and expertise in his career. This might involve conducting risk assessments for school facilities, implementing emergency preparedness plans, or developing safety protocols for various situations. He stays up-to-date on best practices in school safety and actively seeks opportunities to improve school infrastructure and response procedures. Ghaith's dedication goes beyond technical expertise. He understands that a safe learning environment isn't just about physical security; it's also about fostering a culture of safety awareness within the school community. This might involve developing educational programs for students and staff, or creating a collaborative environment where everyone feels empowered to identify and address potential safety concerns. Ultimately, Ghaith's work ensures a secure learning environment for students. By proactively mitigating risks and fostering a culture of safety, he allows students to focus on what matters most: thriving in their academic journey.

**Mustafa Kayyali**, Ph.D., is an ardent advocate for excellence in higher education, driven by a relentless pursuit of quality, recognition, and innovation. As the

Manager of HE Higher Education Ranking, he leads initiatives that play a crucial role in shaping the landscape of academia.

**Negina Khalili** is a Visiting Professor at Loyola University New Orleans. Before her departure from Afghanistan, she was Chief Prosecutor for the Attorney General's Office and before that was the first female prosecutor in her native Ghor province. Negina is a renowned lawyer and expert on the law of the elimination of violence against women, with over 6 years of experience in the field of prosecution, legal affairs and international law practices. She has received numerous awards and recognition for her work. Negina was also a law professor at Rana and Fanoos Universities in Kabul. She earned her degree from Herat University Faculty of Shari'a and her LLM (Rule of Law and Good Governance) from Ohio Northern University. Her professional areas of interest also include the promotion of good governance and women, gender equality, peace and security, and Afghan women in leadership.

**Farha Khan** is an Assistant Professor, Anil Surendra Modi School of Commerce SVKM's Narsee Monjee Institute of Management Studies, Mumbai.

**Sadaf Khan** has extensive teaching experience of more than ten years. A core academician and a prolific researcher, she has done incredible work in the domain of English Literature, Cultural Studies, and Afghanistan Studies. Having an international teaching experience, she has been felicitated multiple times as 'Best Teacher' by various academic institutions in which she had worked. Besides, she has also served as the Joint Editor of Women's College Annual magazine of Aligarh Muslim University for the session 2007-2008.

**Saad Ullah Khan** has more than a decade long extensive experience at university level teaching and research. Currently he is working at Amity School of Communication, Amity University Rajasthan (India). He has authored various research papers in Journals of national as well as international repute. He was awarded 'Rajat Jayanti Vigyan Sancharak Fellowship-2009' by DST, Ministry of Science and Technology(Government of India) for his contribution to the field of communication and its social implications. Further he was bestowed with MANF Fellowship 2011 by UGC to carry out his research work. A PhD from prestigious Aligarh Muslim University, Dr. Saad Ullah Khan has presented 90 Research Papers in various National and International Conferences. He was awarded 'Best Researcher Award' twice (2020 & 2023) and 'Best Journalism Teacher Award' in 2021.

**Ahmad Suhaib Majeed** is a senior student of Business Administration at the American University of Afghanistan with a concentration in Finance. At the same time Remote Ambassador of Open Society University Network and TEDx Darulaman and TEDx Kabul Delegate.

**S. M. Rashedujjaman**, a humanitarian professional with over six years of experience, is dedicated to enhancing educational opportunities, particularly in emergency contexts like the Rohingya Crisis. His expertise includes Early Childhood Development (ECD), Education in Emergency (EiE), Psychosocial Support (PSS), and Social and Emotional Learning (SEL). Rashed has worked with organizations like International Rescue Committee (IRC), Save the Children International (SCI) and Relief International (RI), and has achieved notable achievements like integrating SEL into the Accelerated Learning Program and leading field-testing of the INEE PSS-SEL Framework Project in the Rohingya emergency setting. Rashed holds a Knowledge Fellowship from ECDAN as well. Research, evaluation, learning and capacity building are of Rashed's interest to professional growth. He holds a Master of Education (M.Ed) in Educational Psychology and Guidance (EPG) and a Bachelor of Education (B.Ed) Hons. in Science, Mathematics, and Technology Education (SMTE) from the University of Dhaka.

**Stephen A. Rosenbaum**, MPP, JD, is Frank C. Newman Lecturer at UC Berkeley Law. He is also a Visiting Researcher Scholar at the University's Othering & Belonging Institute and Research Affiliate at the Berkeley Interdisciplinary Migration Initiative. As a University of Washington Visiting Senior Lecturer in Law (2012-14), Stephen was a member of the State Department-funded LESPA (Legal Education Support Program--Afghanistan) faculty team and conducted workshops for Afghan law faculty and senior students in Herat, Mazr-i-Sherif and Kabul. He has served as a legal education consultant to NGOs funded by USAID, European Union and British Council and has published articles on clinical legal education, and law and development in Afghanistan, the Middle East, West Africa, Southeast Asia and Europe. Stephen is currently teaching on-line courses, under the auspices of Classrooms without Walls in partnership with Monash University, to girls and women still living in Afghanistan.

**Fahim Sadat** is an Afghan professional with nearly a decade long experience at national and international organisations and also served at strategic technical positions in erstwhile Islamic Republic of Afghanistan. He was also associated with Afghanistan's largest development project, the National Solidarity Program (NSP). Later he joined academia and joined Kardan University, as assistant professor and head of international relations Department. He also remained managing editor for Kardan Journal of Social Sciences and Humanities during his stay in Kardan University until he was evacuated and left Afghanistan. He is frequently invited by international media organisations like Aljazeera, CGTN America, TRT, and publishes Op-eds on different outlets such as South Asia Monitor and etc as an expert/panelist on Afghanistan's security and policy related issues. Currently he is engaged in providing consultancy to various stakeholders and is looking for a challenging opportunity to contribute for peace, development and positive change in his native nation.

**M. Mahruf C. Shohel** is currently working at the University of Roehampton. He is an academic researcher with special interests in education, childhood studies, international development, technology-enhanced learning and social science research methods. He has written extensively on development issues in the Global South and conducted research on disadvantaged children including socioeconomically deprived children, street children, sex worker's children and displaced refugee children. Currently, Dr Shohel is engaged in the fields of education in emergencies, education for sustainable development and global citizenship, emerging technologies in education, students' learning journeys and their engagement, and teaching and learning in higher education.

**Mahbuba Rahman Sofin** is a Research Assistant in the College of Business Administration Department at the International University of Business Agriculture and Technology, who has experience as an author of research publications. Her research interests include Artificial intelligence, Education, Higher Education, M-learning, Teachers' professional development, Human Resource Management, Entrepreneurship and Business, etc.

**Mohamad Zreik**, a Postdoctoral Fellow at Sun Yat-sen University, is a recognized scholar in International Relations, specializing in China's Arab-region foreign policy. His recent work in soft power diplomacy compares China's methods in the Middle East and East Asia. His extensive knowledge spans Middle Eastern Studies, China-Arab relations, East Asian and Asian Affairs, Eurasian geopolitics, and Political Economy, providing him a unique viewpoint in his field. Dr. Zreik is a proud recipient of a PhD from Central China Normal University (Wuhan). He's written numerous acclaimed papers, many focusing on China's Belt and Road Initiative and its Arab-region impact. His groundbreaking research has established him as a leading expert in his field. Presently, he furthers his research on China's soft power diplomacy tactics at Sun Yat-sen University. His significant contributions make him a crucial figure in understanding contemporary international relations.

# Index

## E



## F



## G



## H



## I



## L



## M

## N



## O



## P



## Q



## R



## S

# T



# U



# W